SPRINGER
LAB MANUAL

Springer-Verlag Berlin Heidelberg GmbH

Dietmar Tietz (Ed.)

Nucleic Acid Electrophoresis

With 62 Figures and 10 Tables

 Springer

DIETMAR TIETZ

8167 Shoal Creek Drive
Laurel, MD 20724-2950
USA

ISBN 978-3-642-47763-8

Library of Congress Cataloging-in-Publication Data
Nucleic acid electrophoresis / Dietmar Tietz (ed.). – Berlin ;
Heidelberg ; New York : Springer, 1998
  (Springer lab manual)
  ISBN 978-3-642-47763-8   ISBN 978-3-642-58924-9 (eBook)
  DOI 10.1007/978-3-642-58924-9

© Springer-Verlag Berlin Heidelberg 1998
Originally published by Springer-Verlag Berlin Heidelberg New York in 1998

Production: B. Reichenthaler, A. Schimmel-Sevim, Springer-Verlag, D-69121 Heidelberg
Cover design: design & production GmbH, D-69121 Heidelberg
Typesetting: Mitterweger Werksatz GmbH, D-68723 Plankstadt
SPIN 10519580   39/3133   5 4 3 2 1 0 – Printed on acid free paper

Dedicated to the memory of John S. Fawcett († February 2, 1998),
an outstanding human being, a highly creative scientist,
a good friend and a very helpful colleague

Preface

Modern publications are generally written in a very concise fashion with the main focus on results rather than experimental procedures. Therefore, when Springer-Verlag approached me to edit a book on nucleic acid electrophoresis for their lab manual series, I welcomed this as an opportunity to publish all those usually hidden technical details so important for the success of many projects in molecular biology, medicine, analytical biochemistry, agriculture, forensic sciences and other important disciplines.

The book's concept is to focus on laboratory electrophoretic procedures and to assist a wide readership by describing experimental techniques. It addresses the experienced scientist and technician in search of a new electrophoretic method, as well as the student who wants to become familiar with this originally "blue fingers" discipline that is rapidly changing into a high-tech enterprise.

This lab manual does not follow the path of a traditional textbook and makes no attempt to provide an exhaustive coverage of as many aspects as possible. Instead, we took the approach of inviting experts in the most important electrophoretic techniques and asked them to write a methodological chapter of their own choice. The rationale was to obtain the best contributions by letting authors write about their favorite subject. We also seek to encourage use of the new electronic media by providing e-mail addresses and data about helpful Internet newsgroups and Web sites. In addition, a Web site will be created to accompany this book and provide more information and comments from our readers: http://www.his.com/~djt/labmanual.html.

At this point we wish to thank all the contributors for providing their articles preformatted for electronic publication and for giving their time and effort without payment in order to produce a more accessible book. This lab manual project was started in the Department of Biometry and Population Genetics at Justus-Liebig University in Gießen, Germany, and was continued using the scientific resources of the National Institutes of

Health (NIH) in Bethesda, Maryland, USA. I wish to gratefully acknowledge the support of these two institutions and their staff. In particular, I would like to mention Prof. Dr. Wolfgang Köhler (Gießen), Dr. Andreas Chrambach (NIH) and Dianeth Dorris (NIH). Special thanks are also due to Dr. Jutta Lindenborn and other members of Springer-Verlag for their very helpful support of this venture and for their patience. We hope that the contributions will benefit a number of scientific projects.

Bethesda, December 1997   Dietmar Tietz

Corresponding address:
DR. DIETMAR TIETZ
8167 Shoal Creek Drive
Laurel, Maryland 2072-4-2950
USA
e-mail: djt@his.com

Table of Contents

Authors' Addresses

NANCY C. STELLWAGEN (*Chapter 1*)
Nancy C. Stellwagen, Department of Biochemistry, University of Iowa, 4403 Bowen Science Building, Iowa City, Iowa 52242, USA (*phone* +1-319-335-7932; *fax* +1-319-335-9570; *e-mail* stellwag@blue.weeg.uiowa.edu)

ALEXANDER KRAEV (*Chapter 2*)
Alexander Kraev, Laboratory of Biochemistry III, Swiss Federal Institute of Technology (ETH), Universitätsstrasse 16, 8092 Zürich, Switzerland (*phone* +41-1-632-31-47; *fax* +41-1-632-12-13; *e-mail* kraev@bc.biol.ethz.ch; *web site* http://www.bc.biol.ethz.ch/BiochemistryIII/Sasha/kraev.html)

R. J. SNOWDON and A. LANGSDORF (*Chapter 3*)
R.J. Snowdon, A. Langsdorf, Department of Biometry and Population Genetics, Justus-Liebig Universität, Ludwigstrasse 37, 35390 Giessen, Germany)
Correspondence to R.J. Snowdon: phone +49-641-99-37542; fax +49-641-9937549; *e-mail* Rod.Snowdon@agrar.uni-giessen.de)

MICHAEL B. COULTHART, WENDY M. JOHNSON, FRASER E. ASHTON (*Chapter 4*)
Correspondence to Michael B. Coulthart, Laboratory Centre for Disease Control, Room 240, HPB Building #7, PL 0700F, Ottawa, Ontario, Canada K1A 0L2 (*phone* +1-613-952-7312; *fax* +1-613-941-2408; *e-mail* mike-coulthart@inet.hwc.ca)

CARL R. MERRIL, KAREN M. WASHART, ROBERT C. ALLEN *(Chapter 5)*
Carl R. Merril, Karen M. Washart, Laboratory of Biochemical Genetics, National Institute of Mental Health, NIH, Building 10, Room 2D54, Bethesda Maryland 20892, USA
Robert C. Allen, Department of Pathology and Laboratory Medicine, Medical School of the University of South Carolina, 171 Ashley Ave. Charleston, South Carolina, 29425, USA
Correspondence to Carl R. Merril, 2 Winder Court, Rockville, MD 20850, USA
(*phone* +1-301-435-3583; *fax* +1-301-480-9862;
e-mail merrilc@helix.nih.gov)

SHIRIN S. JOSEPH *(Chapter 6)*
Shirin S. Joseph, The Sanger Centre, Hinxton Hall, Hinxton, Cambridge-shire, CB10 1SA, UK
(*phone* +44-1223-494843; *fax*: +44-1223-494919;
e-mail shirin@sanger.ac.uk)

JOHN M. BUTLER and DENNIS J. REEDER *(Chapter 7)*
John M. Butler, Dennis J. Reeder, DNA Technologies Group, Biotechnology Division, Chemical Science and Biotechnology Laboratory, Bldg. 222, National Institutes of Standards and Technology, Gaithersburg, Maryland 20899, USA
Correspondence to John M. Butler, GeneTrace Systems, Inc., 333 Ravenswood Avenue, PN088, Menlo Park, California 94025, USA
(*phone* +1-650-859-3051; *fax*: +1-650-859-2654;
e-mail butler@genetrace.com)

SAMEER A. SAKALLAH, ROBERT W. LANNING, DAVID L. COOPER
(Chapter 8)
Correspondence to Sameer A. Sakallah, Division of Molecular Diagnostics, Department of Pathology, 728 Scaife Hall, University of Pittsburgh Medical Center, Pittsburgh, Pennsylvania 15261, USA
(*phone* +1-412-648-7549; *fax* +1-412-383-9594;
e-mail ssa@med.pitt.edu)
Robert W. Lanning, Molecular Genetics Laboratory, Carolina Medical Center, Charlotte, North Carolina, 28232, USA
David L. Cooper, Quest Diagnostics, Inc., 33608 Ortega Highway, San Juan Capistrano, California 92690, USA

HILDEGARD HAAS-ROCHHOLZ *(Chapter 9)*
Hildegard Haas-Rochholz, Institute of Legal Medicine, Frankfurter Strasse 58, 35392 Giessen, Germany
(*phone* +49-641-99-414-27 or 26; *fax* +49-641-99-414-19; *e-mail* Hildegard.Haas-Rochholz@forens.med.uni-giessen.de)

MICHAEL G. FRIED and MARK M. GARNER *(Chapter 10)*
Correspondence to Michael G. Fried, Dept. of Biological Chemistry, Pennsylvania State University Medical College, P.O. Box 850, Hershey, Pennsylvania 17033, USA
(*phone* +1-717-531-8585; *fax* +1-717-531-7072; *e-mail* mfried@bcmic.hmc.psu.edu)
Mark M. Garner, FMC BioProducts, 191 Thomaston St., Rockland Maine 04841, USA

JUN XIAN and MICHAEL G. HARRINGTON *(Chapter 11)*
Michael G. Harrington, Biology 139/74, California Institute of Technology, Pasadena, California 91125, USA
Jun Xian, Cereon Genomics, L. L. C., One Kendall Square, Building 200, Cambridge, MA 02139, USA
Correspondence to Michael G. Harrington, 4548 Leland Place, La Canada, California 91011-2129, USA
(*phone* +1-818-952-2959; *fax* +1-818 957 2641; *e-mail* mike-97@pacbell.net)

ANDREAS KYAS, WINFRIED MÄUELER, JOERG T. EPPLEN *(Chapter 12)*
Andreas Kyas, Winfried Mäueler, Joerg T. Epplen, Molecular Human Genetics, Ruhr-University, 44780 Bochum, Germany
Correspondence to Joerg T. Epplen: *phone* +49-234-700-3839; *fax* +49-234-709-4196; *e-mail* epplejbz@rz.ruhr-uni-bochum.de

DAVID WHEELER *(Chapter 13)*
David Wheeler, Laboratory of Molecular and Cellular Biology, NIDDK, The National Institutes of Health, Bethesda, Maryland 20892, USA
Correspondence to David Wheeler, National Center for Biotechnology Information, Bldg 38 A, 8N800, The National Institutes of Health, Bethseda, MD 20892, USA
(*phone* +1-301-594-3193; *fax* +1-301-594-3193; *e-mail* wheeler@ncbi.nlm.nih.gov)

Chapter 1

DNA Gel Electrophoresis

NANCY C. STELLWAGEN

Introduction

The development of gel electrophoresis as a method of separating and analyzing DNA has been one of the forces driving the revolution in molecular biology for the last 20 years. In principle, DNA gel electrophoresis is conceptually easy to understand and technically easy to execute. In practice, there are a lot of small details that affect the accuracy and reproducibility of the results. This chapter presents a detailed description of the experimental methods used for DNA gel electrophoresis, designed as a guide for the investigator with little or no experience with this technique. The methods described here are those used every day in the author's laboratory; additional protocols and ancillary techniques may be found in Sambrook et al. (1987). All discussions refer to the separation of double-stranded DNA molecules in slab gels, using unidirectional electric fields and fluorescent detection methods. The pulsed field gel electrophoresis (PFGE) of large DNA molecules (Birren and Lai 1993) and DNA capillary electrophoresis (Righetti and Gelfi 1996) are described in detail elsewhere.

The organization of this chapter is as follows. First, the theory of gel electrophoresis will be described briefly. Next, the preparation and use of agarose and polyacrylamide gels, the two types of gel matrices most commonly used for DNA gel electrophoresis, will be described. The advantages and disadvantages of each type of gel matrix will be outlined, as well as additional factors which need to be considered when using electrophoretic mobilities to construct Ferguson plots. Finally, the last section is devoted to troubleshooting: technical difficulties, possible causes and suggested remedies.

Nancy C. Stellwagen, Department of Biochemistry, University of Iowa, 4403 Bowen Science Building, Iowa City, Iowa 52242, USA
(*phone* +1-319-335-7932; *fax* +1-319-335-9570; *e-mail* stellwag@blue.weeg.uiowa.edu)

Theory

The electrophoretic mobility observed for a macromolecule in solution is determined by the ratio between the force exerted on the macromolecule by the electric field and the resistance to its motion caused by friction with the solvent. When these two forces are in balance, the macromolecule migrates with an equilibrium velocity, v, equal to:

$$v = q \, E/f \tag{1}$$

where q is the total charge of the macromolecule, f is its frictional coefficient and E is the electric field strength. The electrophoretic mobility, μ, is defined as the velocity per unit field strength, according to Eq. (2):

$$\mu = v/E = d/E \, t \tag{2}$$

where d is the distance migrated in cm, t is the time in seconds, and E is the electric field strength in V/cm. Hence, electrophoretic mobilities are usually expressed in units of $cm^2V^{-1}s^{-1}$, rather than SI units. For large DNA molecules, both q and f increase approximately linearly with DNA molecular weight. Hence, the velocity in the electric field (Eq. 1) and the electrophoretic mobility in solution (Eq. 2) are expected to be independent of molecular weight, as observed experimentally (Olivera et al. 1964). For this reason, DNA mixtures cannot be fractionated by electrophoretic methods in free solution.

Fortunately, the situation is markedly improved when electrophoresis takes place in a supporting gel matrix. In gels, the electrophoretic mobility of a macromolecule is thought to be determined primarily by the volume fraction of pores within the gel that the macromolecule can enter (Ogston 1958; Rodbard and Chrambach 1970). Since small DNA molecules can fit into more pores than large DNA molecules (Fig. 1a), small DNAs will migrate through the matrix faster than large DNAs. Hence, a mixture of DNA molecules of different sizes will separate into discrete bands during electrophoresis, as shown schematically in Fig. 1b.

The electrophoretic mobility observed for a macromolecule migrating through a gel matrix is given approximately by Eq. (3):

$$\mu = \mu_o \exp \, (-K_R \, C) \tag{3}$$

where μ is the mobility observed in the gel, μ_o is the free solution mobility, C is the gel concentration, and K_R is a constant, specific for each macromolecule, called the retardation coefficient. K_R is proportional to $(R+r)^2$, where R is the radius of the macromolecule (assumed to be spherical) and r is the radius of the gel fibers (Rodbard and Chrambach

Fig. 1.a Cross section of a gel, with DNA molecules of various sizes approaching the pores in the gel matrix. **b** One lane in the same gel after several hours of electrophoresis, illustrating the separation of DNA molecules of different molecular weights

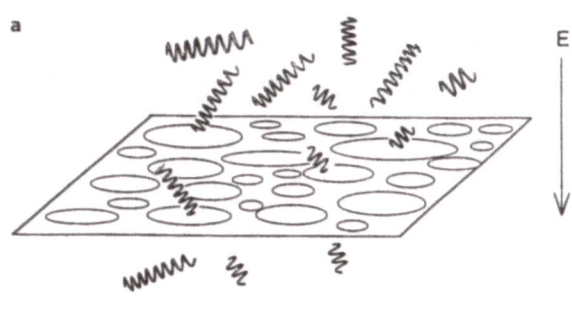

1970; Tietz 1988). For DNA molecules, which have a constant free solution mobility, Eq. (3) predicts that semilogarithmic plots of the mobility vs gel concentration should be linear and should extrapolate to a common intercept, as shown schematically in Fig. 2. Semilogarithmic plots of mobility vs gel concentration are usually called Ferguson plots, after the investigator who first noted this correlation (Ferguson 1964). The applicability of Eq. (3) to the electrophoresis of linear DNA molecules in agarose and polyacrylamide gels and the interpretation of the K_R values in terms of molecular structure are matters of debate in the current literature (see "Ferguson Plots"). Several other theoretical treatments of DNA gel electrophoresis have also been proposed (Duke et al. 1994; Kozulic 1995; Slater and Guo 1996).

Large DNA molecules tend to become oriented in the electric field during electrophoresis and migrate end-on through the gel matrix (Fritsch and Lerman 1982; Lumpkin and Zimm 1982). Since end-on

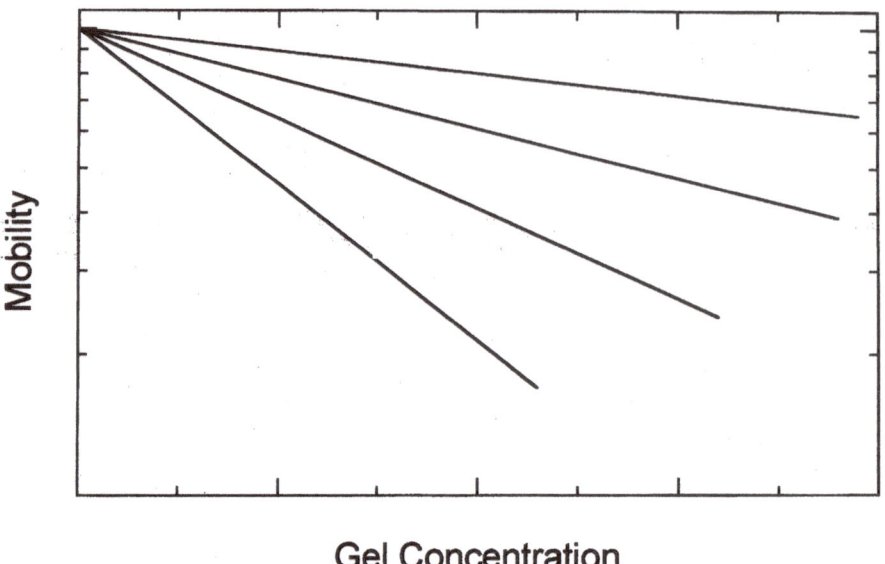

Gel Concentration

Fig. 2. An idealized Ferguson plot for DNA molecules of different molecular weights

migration is similar to the motion of a snake crawling through grass, it is called reptation. The electrophoretic mobility of DNA in the reptation regime can be described approximately by Eq. (4):

$$\mu = \frac{q}{3\,f}\left[\frac{1}{N} + \frac{E'}{3}\right] \tag{4}$$

where q is the total charge on the DNA molecule, f is the frictional coefficient, N is proportional to DNA length and E' is proportional to the electric field strength (Duke et al. 1994). Since q and f are both proportional to DNA molecular weight, the molecular weight dependence of electrophoretic mobility in the reptation regime is determined by the relative importance of the two terms in brackets. When the electric field strength is relatively small, the 1/N term is dominant and the electrophoretic mobility decreases linearly with increasing DNA molecular weight, in contrast to the behavior predicted by Eqs. (1) and (2). When the DNA molecular weight and/or electric field strength is large, the second term in brackets in Eq. (4) becomes large with respect to the first, causing the electrophoretic mobility to reach a constant value that is independent of DNA molecular weight. Molecular weight separation in this size range can only be achieved by using pulsed electric fields (Birren and Lai 1993).

1.1
Agarose Gel Electrophoresis

Agarose, a high molecular weight polysaccharide extracted from the walls of certain marine red algae, is an alternating copolymer of 1,3-linked β-D-galactose and 1,4-linked 3,6-anhydro-α-L-galactose, as shown in an idealized representation (Fig. 3). The galactose residues are occasionally substituted with negatively charged groups such as sulfate and pyruvate, giving the agarose fibers a fixed negative charge. Agarose is insoluble in cold water but dissolves readily in boiling water (see "Agarose Gel Prepation"). Upon cooling, the agarose chains form side-by-side aggregates which condense into a three-dimensional, interlocking network held together by noncovalent hydrogen bonds. Because each individual gel fiber contains many agarose chains, agarose gels are relatively strong; typical gels contain 1 % agarose or less.

Two major types of agarose are available commercially: unmodified agaroses and hydroxyethylated agaroses. Unmodified agaroses with various trade names have the same basic structure shown in Fig. 3 but differ in molecular weight and/or the density of negatively charged groups in the backbone chain, two factors which determine the strength of the gel and its gelation and melting temperatures. Hydroxyethylated agaroses form gels with lower gel strength and significantly lower melting and gelation temperatures than gels cast from unmodified agarose. Although the hydroxyethylated (low melting point) agaroses are relatively expensive, they are useful for isolating and purifying specific DNA fragments. The DNA can be recovered from the gel by simply excising the appropriate band and melting the agarose; the gel melts at a lower temperature than required to denature the DNA. A partial list of several types of agarose available from one commercial supplier is given in Table 1, along with some characteristics of each type of gel and typical applications for which it is used.

Agarose gels have become the "work horses" of the molecular biology laboratory because they are easy to cast, quick to run, and unused lanes

Fig. 3. Structure of agarose, an alternating copolymer of 1,3-linked β-D-galactose and 1,4-linked 3,6-anhydro-α-L-galactose

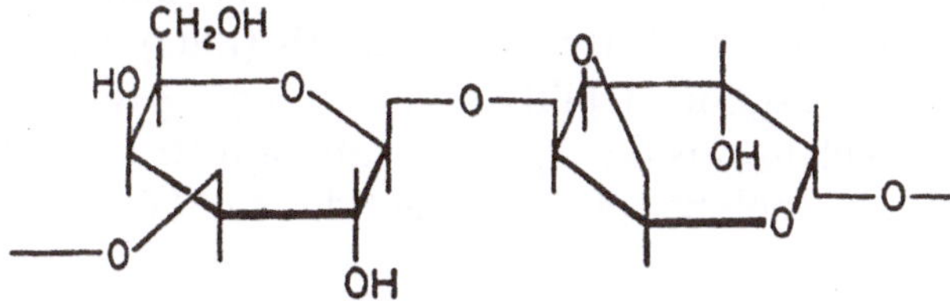

Table 1. Partial listing of various types of agarose used for electrophoresis[a]

Agarose type	Seakem Gold	Seakem LE	NuSieve GTG	SeaPlaque	Metaphor
Gelling temp.[b]	36 °C (1.5 %)	36 °C (1.5 %)	≤36 °C (4 %)	26–30 °C (1.5 %)	≤35 °C (3 %)
Melting temp.[b]	–	–	≤65 °C (4 %)	≤65 °C (1.5 %)	≤75 °C (3 %)
Sulfate	≤0.10 %	≤0.15 %	≤0.15 %	≤0.10 %	
EEO	≤0.05 %	0.09–0.13	≤0.15 %	≤0.10 %	≤0.05 %
Gel strength[b]	≥1800 (1 %)	≥1200 (1 %)	≥500(4 %)	≥200 (1 %)	≥300 (3 %)
Typical applications	Low conc. gels, large DNAs	Analytical gels	Small DNAs	Preparative gels	PCR products, small DNAs

[a] FMC BioProducts Catalog 1997.
[b] Concentration-dependent properties.

Table 2. Advantages and disadvantages of agarose gel electrophoresis

Advantages	Disadvantages
Nontoxic gel medium	Electroendosmosis
Gels are quick and easy to cast	High cost of agarose
Unused lanes can be used in subsequent experiments	Fuzzy bands
Mobilities are not sequence-dependent	Poor separation of low
Good for separating large DNA molecules	molecular weight samples
Can recover samples by melting the gel, digesting with the enzyme agarase (FMC), or treating with chaotropic salts	

can be loaded and run in later experiments. These and other advantages and disadvantages of using agarose gels for DNA electrophoresis are summarized in Table 2. One of the more serious problems associated with agarose gels is due to the fact that the gel fibers are negatively charged. The positive counterions required for electroneutrality tend to migrate toward the cathode (negative electrode) during electrophoresis, carrying along buffer and solvent. This bulk flow of solvent toward the cathode is called electroendosmosis (EEO). Because of the EEO flow, the electrophoretic mobility observed for a given DNA molecule in an agarose gel, μ_{obs}, is the algebraic sum of its true electrophoretic mobility, μ, and the mobility of the solvent due to EEO, μ_{EEO}, according to Eq. (5):

$$\mu_{obs} = \mu + \mu_{EEO} \tag{5}$$

The effect of EEO can be neglected for routine DNA separations in agarose gels. However, for quantitative work, and especially for the construction of Ferguson plots, the EEO flow must be measured and the true DNA electrophoretic mobilities calculated from Eq. (5).

Materials

Horizontal gel forms suitable for submarine gel electrophoresis can be **Gel Forms** purchased from many commercial suppliers. Some of the commercially available gel forms have provisions for leveling the gel bed, cooling the gel and/or recirculating the buffer from one electrode compartment to another, which is useful for preventing pH changes in the buffer compartments during electrophoresis. Miniaturized gel forms are also available for the high-speed analysis of DNA restriction digests and other routine separations. A partial list of suppliers offering a large selection of gel forms is given at the end of this chapter. However, in the author's laboratory we use simple, home-made gel forms constructed primarily from Plexiglas (Perspex in Europe), based on the general design of McDonell et al. (1977). These gel forms have the advantage of being versatile and rugged. They are also inexpensive, can be constructed in a variety of sizes and require relatively little running buffer.

Caution: Electrical hazard! The gel forms and buffer tanks described here are not enclosed in a protective shield and do not have an electrical interrupt circuit, so care must be taken not to touch the gels or buffer tanks when the gels are running.

The gel form used in the author's laboratory is based on a 6"×10.5"×1/8" glass plate (15.2×26.7×0.3 cm) mounted on a Plexiglas support, as shown schematically in Fig. 4. The legs of the Plexiglas support contain L-shaped channels 1/4" (0.6 cm) wide, which hold agarose wicks connecting the gel bed to the buffer chambers.

Plexiglas components needed to construct gel form

- 2 pieces 11.5"×0.75"×1/4" (29.2×1.9×0.6 cm)
- 2 pieces 6"×1.75"×1/4" (15.2×1.9×0.6 cm)
- 2 pieces 6.5"×0.75"×1/4" (16.5×1.9×0.6 cm)
- 2 pieces 1.25"×0.75"×1/4" (3.2×1.92×0.6 cm)
- 2 pieces 6"×1.25"×1/4" (15.2×3.2×0.6 cm)
- Also needed: glass plate 6"×10.5"×1/8" (15.2×26.7×0.3 cm)

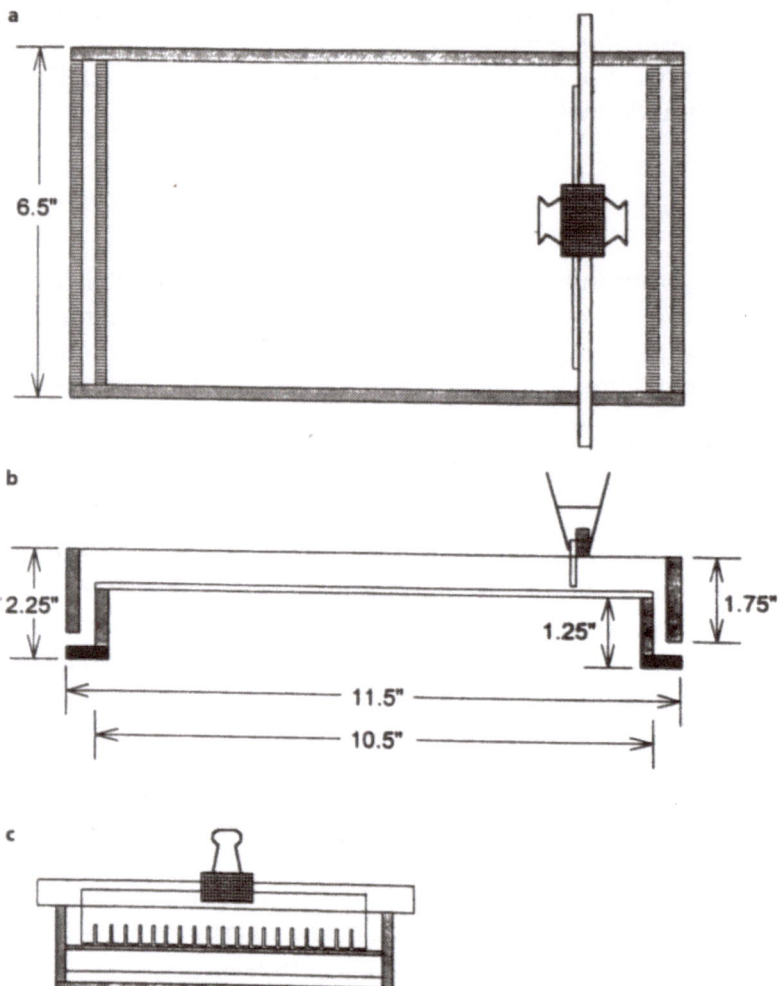

Fig. 4a–c. The agarose gel apparatus described in the text, with dimensions indicated: **a** top view of the agarose gel form, with the mounting bar and comb in place; **b** side view of the gel form, detailing the L-shaped legs holding the wicks; **c** end view, showing the assembly of the mounting bar and comb

The following method is used to construct the gel form:

1. Assemble the legs from the appropriate Plexiglas pieces and glue together with dichloroethane, airplane glue or, preferably, a thick solution of Plexiglas dissolved in chloroform.

Note: All assembly work must be carried out in a fume hood.

2. Attach the legs to the side rails.

3. After the glue has set, mount the glass plate on the upper inside edge of the legs and seal in place with silicone rubber glue (e.g., Dow Corning General Purpose Sealant).

The buffer tanks are constructed from 1/4" (0.6 cm) thick Plexiglas, **Buffer Tanks** except for one of the end plates which is 3/4" (1.9 cm) thick. The outer dimensions of the buffer tanks are 7.75"×2.75"×2.25" (19.7×7.0×5.7 cm), as shown schematically in Fig. 5a. The thick end plates are drilled or

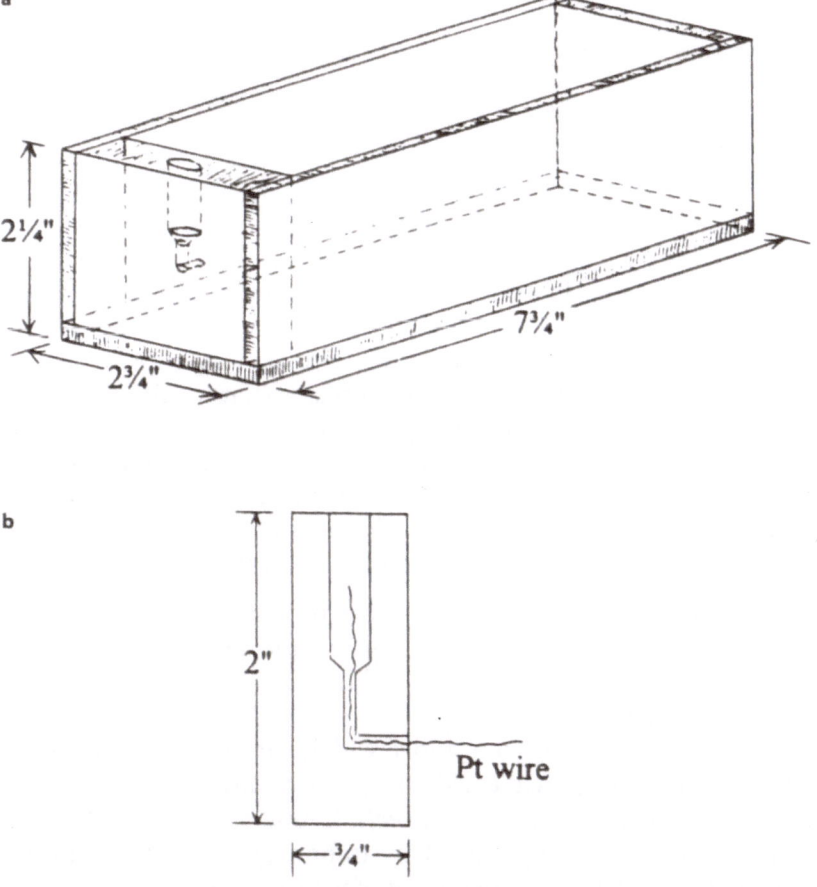

Fig. 5a,b. The agarose buffer tanks: **a** assembled buffer tank, with dimensions indicated; **b** side view of the thick end plate, with a hole drilled out to hold a banana jack. The banana jack is not shown, but the placement of the platinum wire attached to the bottom of the banana jack is illustrated

reamed out to hold a banana jack and connecting wire, as shown in Fig. 5b. After assembling the buffer tanks, platinum wires, long enough to stretch across the bottom of each buffer tank, are soldered to the bottoms of the banana jacks and threaded through the openings in the thick end plates. After making sure that everything fits, the banana jacks are screwed and/or glued in place with an all-purpose glue. The platinum wires are then stretched along the sides and bottoms of the buffer tanks and held in place with a few dabs of silicon rubber glue. The openings where the Pt wires enter the buffer tanks are also sealed with silicone rubber glue.

Plexiglas components for buffer tanks

- 6 pieces 7.75"×2.25"×1/4" (19.7×5.7×0.6 cm)
- 2 pieces 2.25"×2.25"×1/4" (5.7×5.7×0.6 cm)
- 2 pieces 2.25"×2.25"×3/4" (5.7×5.7×1.9 cm)
- Also needed: 2 banana jacks, platinum wire

Combs   An assortment of combs is useful, so that the number and size of the wells can be varied for different experiments. We usually use combs constructed from flat sheets of Plexiglas, attached by spring clips to rectangular Plexiglas mounting bar(s) resting across the top rails of the gel form, as shown in Fig. 4c. The advantage of this arrangement is that the combs can be readily exchanged and the teeth can be raised or lowered, depending on the depth of the gel bed. The same combs can also be used for agarose and polyacrylamide gels, which is an important consideration since combs are expensive to buy and tedious to make.

Note: When mounting the comb, care must be taken that the comb is held straight on the Plexiglas mounting bar and the bar is placed parallel to the end of the gel form. Marks on the side rails will aid in the proper placement of the comb(s).

The dimensions of two of our most widely used combs are given below. Plexiglas dividers, which divide the gel bed into long, narrow subsections, are also useful. Several of the teeth in one of the 28-well combs can be removed to allow the divider bars to be inserted in the gel form before casting the gel. The modified comb has three groups of eight teeth separated by spaces 1 cm wide.

Plexiglas comb assemblies and divider bars

- Mounting bars: 7"×1/2"×1/4" (1.3×17.8×0.6 cm)
- Divider bars: 11"×1/2"×1/4" (1.3×27.9×0.6 cm)
- Combs: 1"×5.5"×1/16" (2.5×14.0×0.15 cm)
  - 18 teeth: 3/16" wide, 3/8" deep (0.5 cm wide×1.0 cm deep)
  - 28 teeth: 1/8" wide, 1/2" deep (0.3 cm wide×1.3 cm deep)
- Also needed: spring clips

Power supplies with continuously adjustable output voltages are versatile **Power Supplies** and convenient to use for a variety of electrophoresis studies. In the author's laboratory we use regulated high voltage power supplies (model 2717 A) from Heath Co. (Bentonsport, Michigan), with output voltages continuously adjustable from ~20–250 V DC. Unfortunately, these units are no longer commercially available. However, all of the commercial companies that sell gel forms also sell equivalent power supplies; a partial list of such suppliers is given at the end of this chapter. Insulated cables with banana jack plugs are used to connect the power supplies to the electrodes in the buffer chambers. They can be home-made or purchased with the power supply.

Agarose Types

A partial listing of the various types of agarose used for DNA electrophoresis is given in Table 2.

Running Buffers

Two running buffers are commonly used for DNA gel electrophoresis, Tris-acetate-EDTA (TAE) and Tris-borate-EDTA (TBE). Most investigators use similar recipes for TAE, but two different recipes, which differ in concentration by approximately a factor of 2, are used for TBE. Unfortunately, both are called TBE, so care must be taken when reading the literature if you want to duplicate the buffer conditions. The recipes given below are those used in the author's laboratory. Other versions are given in Sambrook et al. (1987).

Note: The recipe for TBE in Sambrook et al. (1987) contains 89 mM Tris-borate.

It is convenient to prepare the running buffers as 10× stock solutions and dilute as necessary immediately before the electrophoresis experiment. The 10× stock solutions can be stored for long periods of time at room temperature in screw capped glass bottles.

10 × TAE (1 l)

48.44 g Tris base (0.40 M)
3.72 g Na$_2$EDTA (10 mM)

Adjust to pH 8.0 with glacial acetic acid (~13 ml); add distilled water to 1 l. **Caution:** Glacial acetic acid must be handled in the fume hood.

10× TBE (1 l)

20 g boric acid (0.32 M)
60.5 g Tris base (0.50 M)
3.72 g Na$_2$EDTA (10 mM)

Add distilled water to 1 l. The final pH should be ~8.6.

Procedure

Agarose Gel Preparation

Dissolving Agarose

Agarose can be dissolved in boiling water (or buffer) by a variety of methods: autoclaving, stirring on a hot plate, placing in a boiling water bath, or microwaving. For most electrophoresis studies, the method of dissolving the agarose is not important. However, three factors must be kept in mind: (1) Dissolution is a relatively slow process. (2) Typical agarose samples contain particles that dissolve at different rates; therefore, care must be taken to see that the sample is completely dissolved. (3) Heating the agarose solution will cause some of the solvent to evaporate; therefore, to know the concentration of the gel accurately, it is necessary to weigh the solution before and after heating and add enough distilled water to bring the solution back to its original weight.

The following microwave-based procedure can be used to prepare 200 ml of a 1% agarose solution, sufficient to cast a gel in the above-described gel form (the basic recipe). One-half the basic recipe is required for the wicks. Since solvent exchange within a gel matrix is very slow, **the agarose must be dissolved in the running buffer to be used for electrophoresis.** Step 9 is optional, and is included for those who wish to incorporate an intercalating dye, such as ethidium bromide, into the gel matrix. An intercalating dye is useful for routine measurements (e.g., analyzing DNA minipreps or restriction enzyme digests), because the DNA can be visualized without staining the gel in a separate step. In addition, the progress of the electrophoretic separation can be followed during the course of the experiment, by illuminating the gel briefly with UV light (see "Detection of DNA"). However, DNA electrophoretic mobilities are somewhat altered ($\sim$10 %) in the presence of ethidium bromide, so the dye should not be incorporated into the gel matrix for quantitative work.

1. Dilute 20 ml 10$\times$ TAE or 10$\times$ TBE to 200 ml in a 500 ml Erlenmeyer flask.

Basic Recipe for 1.0 % Agarose Gel

Note: The flask must be two to three times larger than the volume of solution in order to prevent boiling over in the microwave oven.

2. Weigh out 2.0 g agarose powder and add to the buffer solution.

Note: Although all agarose samples contain residual water, gel concentrations are usually calculated on the basis of the gross weight of the agarose powder.

3. Weigh the agarose/buffer solution in the flask; tare or record the weight.

4. Heat in a microwave oven at 100 % power (with a 750 Watt microwave oven) for 3 min in 1 min increments, swirling the solution gently between between heating cycles to release trapped air and resuspend any agarose particles caught on the side of the flask. **Caution:** Swirl the flask carefully when the solution begins to get hot and protect your hands from boil-overs by wearing heavy rubber gloves.

5. After the third 1 min heating period, observe the solution as it comes to rest after swirling the flask. Undissolved agarose particles usually appear as translucent dots moving through the liquid.

6. Continue heating and swirling in 15 s increments until the translucent particles can no longer be seen.

7. Weigh the flask, add distilled water to bring the solution to its original weight and swirl.

8. Let the agarose solution cool on the bench top until it reaches a temperature of $\sim$50–55 °C (slightly hot to the hand, or use a thermo-

meter). Swirl the flask occasionally, to keep the contents at a uniform temperature and prevent the agarose from gelling at the bottom of the flask. Swirl gently, to prevent the formation of bubbles.

9. **Optional**: If desired, add 50 μl of a 10 mg/ml ethidium bromide solution to the warm agarose after it has cooled to ~50 °C. Swirl the flask gently to disperse the dye.

---
10 mg/ml ethidium bromide (20 ml)
---
200 mg ethidium bromide (Sigma)
20 ml distilled water

Stir slowly with a magnetic stirrer until dissolved. Store in a brown bottle at 4 °C.

Caution: Ethidium bromide is a potent mutagen and suspected carcinogen.

Always wear latex gloves and a face mask when weighing out the dry powder. Always wear latex gloves when handling ethidium bromide solutions and stained gels. Dispose of ethidium bromide solutions and stained gels in an environmentally safe manner (see "Detection of DNA").

Casting Wicks

1. Carefully tape the outside edges of the L-shaped openings in the empty gel form, using a very sticky tape, e.g., strapping tape or electrical tape (**not masking tape**). Place the gel form on a tray, to catch spilled agarose if necessary.

2. Prepare 100 ml of a 1 % agarose solution in the running buffer to be used for the experiment (1/2 the basic recipe), omitting the ethidium bromide in optional step 9. Cool the agarose solution to ~45–50 °C with occasional swirling.

Note: For structural stability, the wicks should contain at least 0.8–1 % agarose.

3. Pour the warm agarose solution into the two vertical openings beside the glass plate in the gel form. Watch for leaks at the taped edges of the openings in the legs. Use finger pressure or more tape to stop the leaks, if necessary. Alternatively, place the gel form in a cold box at 4 °C to force the agarose to gel quickly.

4. After the agarose has gelled, scrape any excess agarose from the gel bed and remove the tape from the legs of the gel form.

5. The wicks may be reused for subsequent gels by scraping the old gel off the top of the gel form and pouring a new gel.

Note: The wicks must always be replaced if the new gel is cast in a different running buffer.

1. Fill the buffer tanks with fresh running buffer (~500 ml buffer in each buffer tank) and place the legs of the gel form (with wicks in place) in the buffer tanks.

2. Place a small leveling device on the bed of the gel form to see whether the bench top is level. If not, the gel form can be leveled by inserting two to three small pipette tips between the bottom of the gel form and the top of one of the buffer tanks.

Note: Remove the pipette tips after the gel is poured and hardens.

3. Place the comb(s) at the desired position(s) on the gel form. The teeth should not touch the gel bed; check by pushing a folded paper towel beneath the teeth. Two or even three combs may be placed parallel to each other for routine analyses (e.g., DNA minipreps or restriction enzyme digests). Use only one comb for quantitative measurements or when separating fragments that differ widely in molecular weight.

4. Dissolve the agarose and cool to 50–55 °C as described above. Vary the amount of agarose in the solution to vary the concentration of the gel.

5. Pour the warm agarose solution slowly into the gel form, to prevent bubble formation, being careful not to dislodge the comb(s). If bubbles form, they can often be removed by poking them with the pointed end of a spatula before the gel has set.

Note: The gel spreads across the gel form more evenly if the agarose solution is at the warm end of the above temperature range. Alternatively, the gel form can be heated to ~50 °C before pouring the agarose solution, but this is often inconvenient and the described method works just as well.

6. Pour concentrated agarose gels at a higher temperature of ~60–65 °C to minimize bubble formation. Alternatively, use NuSieve agarose or NuSieve/LE agarose mixtures to cast concentrated gels. The lower average molecular weight of NuSieve agarose decreases the viscosity of concentrated agarose solutions, trapping fewer bubbles. To decrease cost and increase gel strength, a 3:1 ratio of NuSieve/LE agarose is often recommended.

7. After the gel has set for 30–60 min, flood the top of the gel with buffer and gently remove the comb(s) by wiggling back and forth gently and then lifting up. If the gel is not to be used immediately, cover with a

Casting the Gel

piece of Saran wrap (or other flexible plastic wrap) or with a flat glass plate to prevent solvent evaporation.

Casting Sea Plaque Gels Low melting point, hydroxyethylated agarose gels are useful for separating and purifying individual DNA fragments. For economy and ease of handing, the low melting point (e.g., SeaPlaque) gel is often poured within a larger frame of LE agarose, as shown schematically in Fig. 6.

1. Pour an LE agarose gel as described above.

2. After the gel has set, cut out a rectangular portion of the LE gel, extending from the comb to about half-way toward the bottom of the gel.

Note: The rectangular cut-out portion must be long enough to achieve complete separation of the fragment of interest in SeaPlaque agarose.

3. Dissolve a suitable quantity of SeaPlaque agarose in the desired running buffer, cool and pour into the cut-out area of the LE gel.

4. Wait at least 1 h for the SeaPlaque agarose to harden.

5. Flood the gel with running buffer, remove the comb and load the sample of interest in one or more of the SeaPlaque lanes.

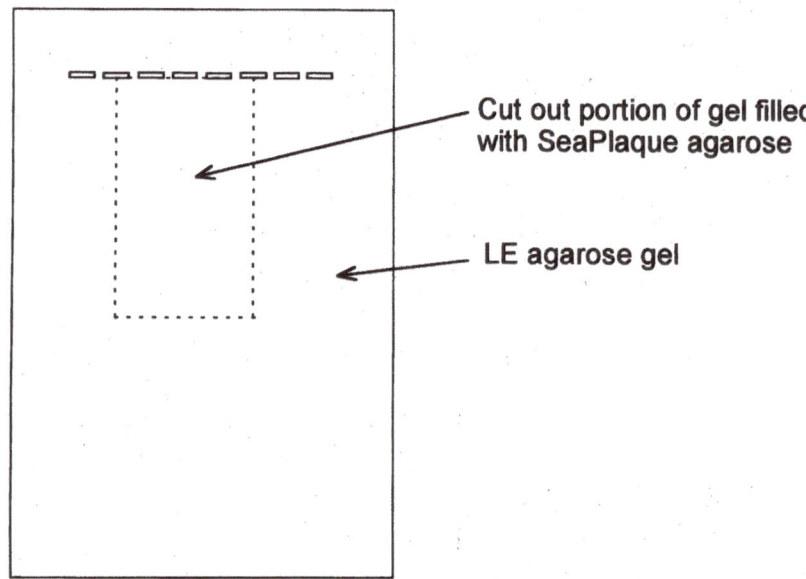

Fig. 6. A two-part SeaPlaque/LE agarose gel

6. Electrophorese as usual.

7. After staining the gel and visualizing the bands with long wavelength UV light (see "Detection of DNA"), cut out the band(s) of interest with a scalpel or sharp razor blade, taking care to minimize the amount of extra agarose around each band.

8. Place the excised band(s) in a suitable number of microcentrifuge tubes and incubate in a water bath at 60–70 °C to melt the agarose.

Note: Ligations and restriction enzyme digestions can be carried out directly in the melted agarose/DNA solution(s), after cooling to 37 °C.

DNA Sample Preparation

1. Place 50–500 ng DNA into each sample tube, the exact amount depending on gel thickness, the cross-sectional area of the wells, the DNA molecular weight, the number of DNA fragments in each sample, and the sensitivity of the fluorescent dye to be used for detection.   **DNA Samples**

Note: A good starting concentration is 50 ng DNA per band, using the gel form described above and ethidium bromide detection. After a trial run the DNA concentration should be adjusted to optimize the intensity of the signal and the sharpness of the bands.

2. Dilute the DNA sample to 9 µl with T 1/10 E buffer (see below), if necessary.

3. Add 1 µl glycerol dye mix to the solution.

Note: If the total volume of the DNA solution is larger or smaller than 9 µl, add 1/10 volume dye mix.

4. Instead of staining the gel with ethidium bromide, either before or after electrophoresis, the dyes ethidium homodimer, propidium iodide, TOTO, YOYO or SYBR green (all from Molecular Probes) may be added to the DNA samples. These dyes bind very tightly to the DNA and are not displaced during electrophoresis, making it possible to detect the DNA bands without staining the gel. The cyanine dyes TOTO, YOYO and SBYR green also offer much more sensitive detection limits than ethidium-related dyes, as shown in Table 3. A reasonable dye/DNA ratio to start with would be 0.25 (i.e., 1 dye molecule per 4 base DNA base pairs). Adjust the ratio as necessary to optimize sensitivity.

Table 3. Properties of fluorescent dyes used for DNA detection[a]

Dye	Optimal dye: DNA ratio	Detection limit per band[b,c,d]	$\lambda_{extinction}$[e]	$\lambda_{emission}$[e]
Ethidium bromide	1:1	500 pg	510 nm	595 nm
Ethidium homodimer	1:4	25	528	617
Propidium iodide	1:4	–	536	617
TOTO	1:4	4	509	533
YOYO	1:4	4	491	509
SYBR green I	–	20	494	521

[a] Haugland (1992).
[b] Glazer et al. (1990).
[c] Rye et al. (1992).
[d] Skeidsvoll and Ueland (1995).
[e] Bound to DNA.

Note: DNA-dye complexes migrate with reduced electrophoretic mobilities that vary, depending on the dye/DNA ratio; for quantitative studies of DNA electrophoretic mobilities, no DNA-binding dyes should be added to the samples.

Sample Buffer

T 1/10 E Buffer (100 ml)

1 ml 1 M Tris-HCl buffer
200 µl 0.5 M EDTA
99 ml distilled water

1 M Tris-HCl buffer (100 ml)

12.2 g Tris base
1 N HCl to pH 8.0 (~40 ml)

Dilute to 100 ml with distilled water.
0.5 M EDTA (10 ml)

1.86 g disodium EDTA
distilled water to 10 ml

Add solid NaOH to pH 8 with stirring or heat with stirring until dissolved.

Glycerol dye mix (10 ml)
_____

6 ml glycerol
20 mg bromophenol blue (and/or xylene cyanol FF)
1 ml 10× TAE or 10× TBE
3 ml distilled water
_____

Mix thoroughly.

Note: The buffer may be replaced by additional distilled water.

Glycerol and other density modifiers are added to increase the density of the DNA solutions, so that the samples will stay at the bottom of the wells after loading.

Note: If edge-tailing of the DNA bands is a problem (i.e., the individual bands are U-shaped and trail backwards at each edge), switch to a Ficoll-based dye mix.

Ficoll dye mix (5 ml)
_____

2.5 g Ficoll (Sigma, type 400L, MW ~400,000)
7–8 mg bromophenol blue (and/or xylene cyanol FF)
distilled water to 5 ml
5–10 µl 1 N NaOH
_____

Add Ficoll very slowly to ~2 ml distilled water with constant stirring. Add distilled water to 5 ml and stir. Add bromophenol blue and stir. Add NaOH with stirring to pH 8.

Note: 1× TAE or 1× TBE may be used instead of distilled water and NaOH.

The marker dyes bromophenol blue and/or xylene cyanol FF are added to indicate how far the DNA fragments have migrated into the gel during the electrophoresis experiment. The dyes do not usually interfere with the measurements unless they happen to co-electrophorese with a DNA fragment of interest. In agarose gels, bromophenol blue co-electrophoreses with ~200–400 bp DNA; xylene cyanol FF co-electrophoreses with ~2–4 kb DNA. In polyacrylamide gels containing 7–8 %T, 3 %C, bromophenol blue co-electrophoreses with ~40–50 bp DNA; xylene cyanol FF co-electrophoreses with ~150–200 bp DNA. The exact co-electrophoresis values depend on the type of gel matrix and the gel concentration, electric field strength and temperature.

Electrophoresis

Loading the Gel

1. Remove the gel form from the buffer tanks and pour off excess buffer from the top of the gel, leaving enough buffer to fill the wells and provide a very thin layer on top of the gel.

Note: This step may be omitted.

2. Fill the buffer tanks with fresh buffer (~500 ml in each buffer tank) if necessary.

Note: If the gel was previously used, check the pH in the buffer tanks with pH paper. If the pH is equal in both buffer tanks, buffer replacement is not necessary. However, without buffer recirculation, buffer replacement is always necessary after several hours of electrophoresis.

3. If ethidium bromide was incorporated into the gel matrix (optional step 9 in "Agarose Gel Preparation"), and the duration of electrophoresis will be several hours, ethidium bromide (2.5 μg/ml) should be added to the anodic buffer tank (positive electrode) to prevent depletion of the dye from the gel during electrophoresis.

4. Replace the gel form in the buffer tanks.

5. With a Pipetman, carefully layer each DNA sample under the buffer in the required number of wells. Expel the samples from the pipette with slow and steady pressure to minimize air bubble formation. Except for extremely accurate work, it is not necessary to replace the pipette tip between samples. The wells of gels cast from the basic recipe typically have a volume of 25–35 μl and easily contain a 10–15 μl DNA sample.

Note: Place a piece of dark paper beneath the gel form to make the wells more visible. Also, be careful not to perforate the bottom of the wells while loading the samples.

Starting the Run

1. When all the DNA samples have been loaded, turn on the power supply voltage. With the gel form described above, it is convenient to apply a voltage of 100 V (~2.5 V/cm). At this voltage, resistive heating of the gel by the electric field is approximately balanced by evaporative cooling, so that the temperature of the gel remains essentially equal to room temperature (~23 °C) during electrophoresis.

Note: For checking restriction enzyme digests and other routine work, it is often convenient to use a higher voltage (200–250 V), to see the results more quickly.

2. After the DNA samples have entered the gel (as indicated by the movement of the marker dye into the gel matrix), flood the gel with buffer to a depth of 3–5 mm so that the gel will not dry out during the electrophoresis experiment.

3. For quantitative work, the exact temperature in the gel should be measured with, e.g., an Omega digital thermocouple, model 450 AKT. The exact voltage can be measured using, e.g., a Fluke digital multi-meter, model 75. Other manufacturers make comparable instruments for measuring the temperature and voltage.

Caution: The gel and the buffer in the buffer tanks are electrically live. Always use extreme care whenever working around a running gel. Pour extra buffer onto the gel surface from a suitable beaker or flask held several centimeters above the surface of the gel. Never place your fingers near the gel surface or the buffer in the buffer tanks. Be sure that only the probes of the thermocouple and multimeter (and not your fingers) touch the gel. Never bail extra buffer from the buffer tanks while the gel is running. If an overflow or bad leak occurs, turn off the power supply, wait for the voltage to drop to zero and then clean up.

Important Cautions

Measurement of Electroendosmosis

For accurate measurement of DNA mobilities, and especially for the construction of Ferguson plots in agarose gels (see Sect. 1.3), the electro-endosmotic mobility of the solvent must be measured and the observed DNA electrophoretic mobilities corrected to their true values according to Eq. (5). It is convenient to use a highly colored, electrically neutral marker such as vitamin B12 to measure the EEO flow.

1. Cast an agarose gel with the wells located about 1/3 of the way down the gel.

Measuring EEO

Note: To prevent excessive diffusion of the vitamin B12 marker band during electrophoresis, choose a gel concentration between 1 and 2 %.

2. Add 1 µl vitamin B12 stock solution to the DNA sample (see "DNA Sample Preparation"). Also prepare a control DNA sample by adding 1 µl distilled water instead of the vitamin B12 solution, to be sure that the marker does not perturb the electrophoretic mobility of the DNA.

3. Electrophorese as usual.

4. Turn off the electric field and wait for the voltage to drop to zero.

5. Measure the displacement of the red vitamin B12 band toward the cathode (with a ruler) before staining and photographing the gel (see "Detection of DNA").

6. Calculate the electroendosmotic mobility of the solvent from the distance migrated by the vitamin B12 band and Eq. (2).

7. Measure the distance migrated by the DNA from the gel or photograph and calculate the true electrophoretic mobility of the DNA using Eqs. (2) and (5).

Vitamin B12 stock solution (2 ml)

---

25 mg vitamin B12
2 ml distilled water

Stir until dissolved. Store in a brown bottle at 4 °C.

DNA Detection

After the electrophoresis experiment is finished, which can require from 1 to more than 24 h depending on the type of separation, temperature, applied electric field strength and DNA molecular weight, turn off the power supply, wait until the voltage drops to zero, and lift the gel form out of the buffer tanks. If ethidium bromide was not included in the gel matrix (optional step 9 in "Agarose Gel Preparation"), or the optional dyes described in step 4 of "DNA Sample Preparation" were not added to the DNA samples, the gel should be stained with ethidium bromide.

Note: Ethidium bromide and other aromatic fluorescent dyes are potent mutagens and suspected carcinogens. Always wear latex gloves when handling solutions and gels containing these dyes. Dispose of ethidium bromide solutions and stained gels as hazardous waste.

Staining the Gel

1. Pour off the running buffer from the top of the gel.

2. Add 100 ml of a solution containing 2.5 µg/ml ethidium bromide in distilled water to the top of the gel.

Note: If the gel is to be reused, the ethidium bromide should be diluted with running buffer.

3. Let stand 15–30 min.

4. Decant the dilute ethidium bromide solution into a suitable beaker or container.

5. To destain, add 100 ml distilled water to the top of the gel.

Note: Destaining is usually not necessary, unless the DNA concentration is very low or the agarose concentration is very high.

6. Let stand another 15–30 min.

7. Pour off the water.

All solutions containing ethidium bromide or other aromatic fluorescent dyes must be decontaminated before being discarded. The procedure used in the author's laboratory is based on decontamination with activated charcoal.

1. Add 100 mg activated charcoal/ml dye solution.

2. Let stand at least 1 h, with occasional stirring.

3. Filter through Whatman No. 1 filter paper in a large-mouth funnel sitting on top of a 1 l Erlenmeyer flask. Pour the clear filtrate into the sink. Place the filter paper containing the dye/charcoal mixture into the biohazard waste.

4. Put all gels stained with ethidium bromide into the biohazard waste.

When the aromatic dyes listed in Table 3 are intercalated into double-stranded DNA and illuminated with ultraviolet (UV) light, the dyes become highly fluorescent, marking the positions of the DNA bands in the gel. The gels are usually photographed to provide a permanent record of the electrophoresis experiment. Many suitable photographic systems are available from various commercial suppliers. In the author's laboratory, we use hand-held UV illuminators and a platform-mounted Polaroid camera. This system has the advantage that the UV light is always directed downward, away from the operator's face. A schematic diagram of the photographic set-up is shown in Fig. 7.

1. Place the gel form with the stained (and destained, if necessary) gel on a piece of black Plexiglas 12"×13" (30.5×33 cm) in size.

2. Illuminate **analytical** gels with short wavelength UV light, using, e.g., 254 nm light generated by Mineral-light model R-52 (Ultraviolet Products Inc.). The wand is held above and to one the side of the gel, moving and tilting it if necessary to achieve even illumination.

Disposal of Ethidium Bromide

Photographic Detection

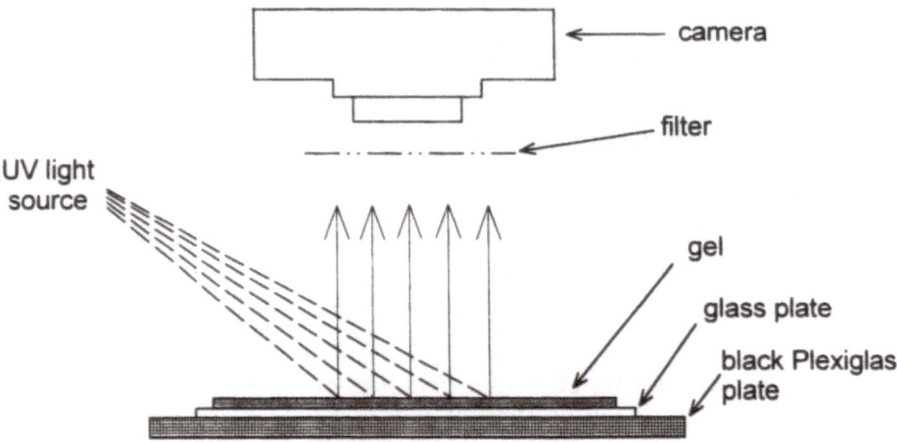

Fig. 7. The camera set-up, with side illumination from a hand-held UV lamp

Note: Ultraviolet light is very damaging to the eyes. Safety goggles or glasses must be worn at all times. If a transilluminator is used to visualize the DNA bands, a complete face shield should be worn.

3. Illuminate **preparative** gels with long wavelength UV light, using, e.g., 365 nm light generated by Mineral-light model 56 (UVP Inc.), to prevent nicking of the DNA/dye complexes.

Note: The positions of the DNA bands of interest in preparative gels can also be identified by staining narrow strips cut from the edge of the gel. The stained strips are returned to their original place(s) in the gel matrix and the fragment(s) of interest cut out of the unstained portion of the gel with a sharp scalpel or razor blade.

4. Place a ruler with fluorescent markings on the gel to provide the magnification factor.

5. Photograph the gel with, e.g., a Polaroid MP-4 Land camera, using a No. 23a Wrattan filter (red-orange) to eliminate scattered light (ethidium bromide fluorescence has a maximum at 595 nm) and Polaroid type 57 high speed film to obtain high resolution images. Typical exposure times are 10–15 s, depending on the DNA concentration and the gel concentration and thickness.

Note: A yellow filter is required for the cyanine dyes YOYO, TOTO and SYBR green.

6. Measure the migration distances of the various bands directly from the photographs, and calculate the electrophoretic mobilities from

Eq. (2). The magnification factor is determined from the ruler photographed simultaneously with the gel.

7. To analyze the intensities of the DNA band(s), and/or to measure the migration distances more quantitatively, remove the gel from the gel form and place it directly on the imaging plate of, e.g., a BioRad GS-670 Imaging Densitometer. The migration distances and fluorescence intensities can then be quantified as described by the manufacturer. Equivalent densitometers are available from other suppliers.

A typical plot of the electrophoretic mobilities observed for DNA fragments of different molecular weights in a 1 % agarose gel, cast and run in TAE buffer, is illustrated in Fig. 8. The logarithm of the DNA molecular weight, in bp, is plotted as a function of the distance migrated in the gel. Semilogarithmic plots of this type are convenient for displaying DNA mobilities over a wide range of molecular weights. Note that the mobilities of all the DNA fragments fall on a single smooth curve (in contrast to polyacrylamide gels, illustrated in Fig. 12). Also note that the curve is approximately linear for fragments between ~300 and 3000 bp, making it relatively easy to estimate the molecular weights of unknown DNA

Fig. 8. Typical electrophoretic separation of DNA restriction fragments in a 1 % agarose gel, cast and run in TAE buffer. The logarithm of DNA molecular weight, in base pairs (*BP*), is plotted as a function of the distance migrated in the gel, *D*. The different symbols correspond to different restriction enzymes that were used to digest the plasmid DNA

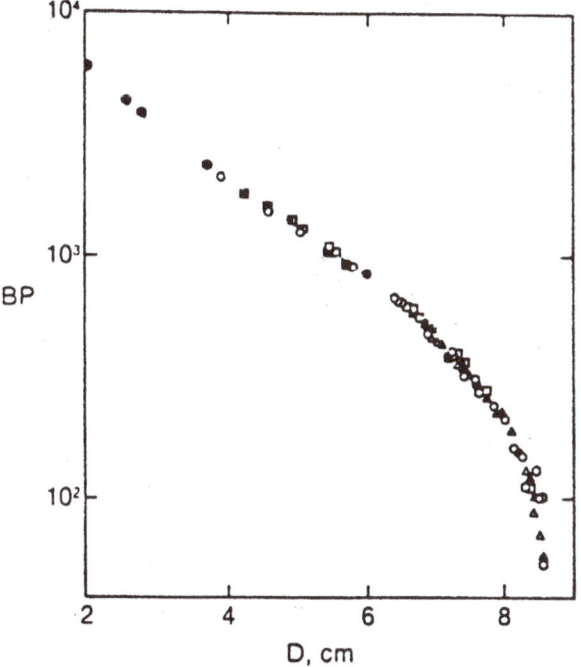

samples in this molecular weight range. Fragments smaller than ~200 bp migrate with similar mobilities in 1 % agarose gels; separation can only be achieved in more concentrated gels.

1.2
Polyacrylamide Gel Electrophoresis

Polyacrylamide gels are chemically cross-linked gels formed by the polymerization of acrylamide with a cross-linking agent, usually N,N'-methylenebisacrylamide (Bis), as shown schematically in Fig. 9. The reaction is a free radical polymerization, usually carried out with ammonium persulfate as the initiator and N,N,N',N'-tetramethylethylenediamine (TEMED) as the catalyst. The gel composition is usually described

Fig. 9a–c. Components in the polyacrylamide gel reaction: **a** acrylamide; **b** N,N'-methylenebisacrylamide (Bis); **c** a cross-linked polyacrylamide gel

a $CH_2=CH-\underset{\underset{O}{\|}}{C}-NH_2$

b $CH_2=CH-\underset{\underset{O}{\|}}{C}-NH-CH_2-NH-\underset{\underset{O}{\|}}{C}-CH=CH_2$

c

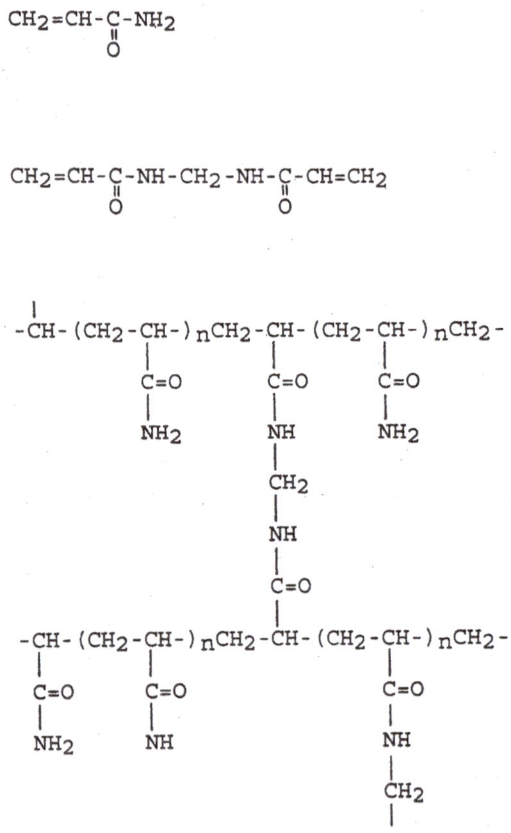

Table 4. Advantages and disadvantages of polyacrylamide gel electrophoresis

Advantages	Disadvantages
Stable chemically cross-linked gel	Toxic monomers
Sharp bands	Gels are tedious to prepare and often leak
No EEO	Must pre-electrophorese
Sensitive to DNA curvature	Sequence-dependent DNA mobilities
Good separation of low molecular weight fragments	Need new gel for each experiment

in terms of %T, the total w/v concentration of acrylamide and Bis in the gel mixture, and %C, the w/w percentage of Bis in %T. Some of the advantages and disadvantages of using polyacrylamide gels for DNA electrophoresis are summarized in Table 4.

Polyacrylamide gels must be pre-electrophoresed before use, in order to eliminate charged impurities (assumed to be ammonium persulfate derivatives) from the gel matrix. If this is not done, the charged impurities will form complexes with the DNA fragments, changing their electrophoretic mobilities in a non-reproducible manner. If highly purified acrylamide and Bis are used to cast the gels, the gel fibers will be uncharged and electroendosmosis will not occur. Hence, DNA electrophoretic mobilities measured in polyacrylamide gels do not have to be corrected for electroendosmosis.

Materials

Polyacrylamide gels are usually cast between glass plates to prevent oxygen from inhibiting the polymerization reaction, and are run in a vertical format. Many commercial companies sell suitable gel cabinets and the required glass plates for casting the gels. Several companies that offer a variety of gel cabinets of different sizes are listed at the end of this chapter. However, in the author's laboratory we use home-made gel cabinets constructed from Plexiglas (Perspex in Europe), based on the original design by Studier (1973). The gel cabinet illustrated in Fig. 10 has been found to be convenient for running a variety of analytical and preparative polyacrylamide gels.

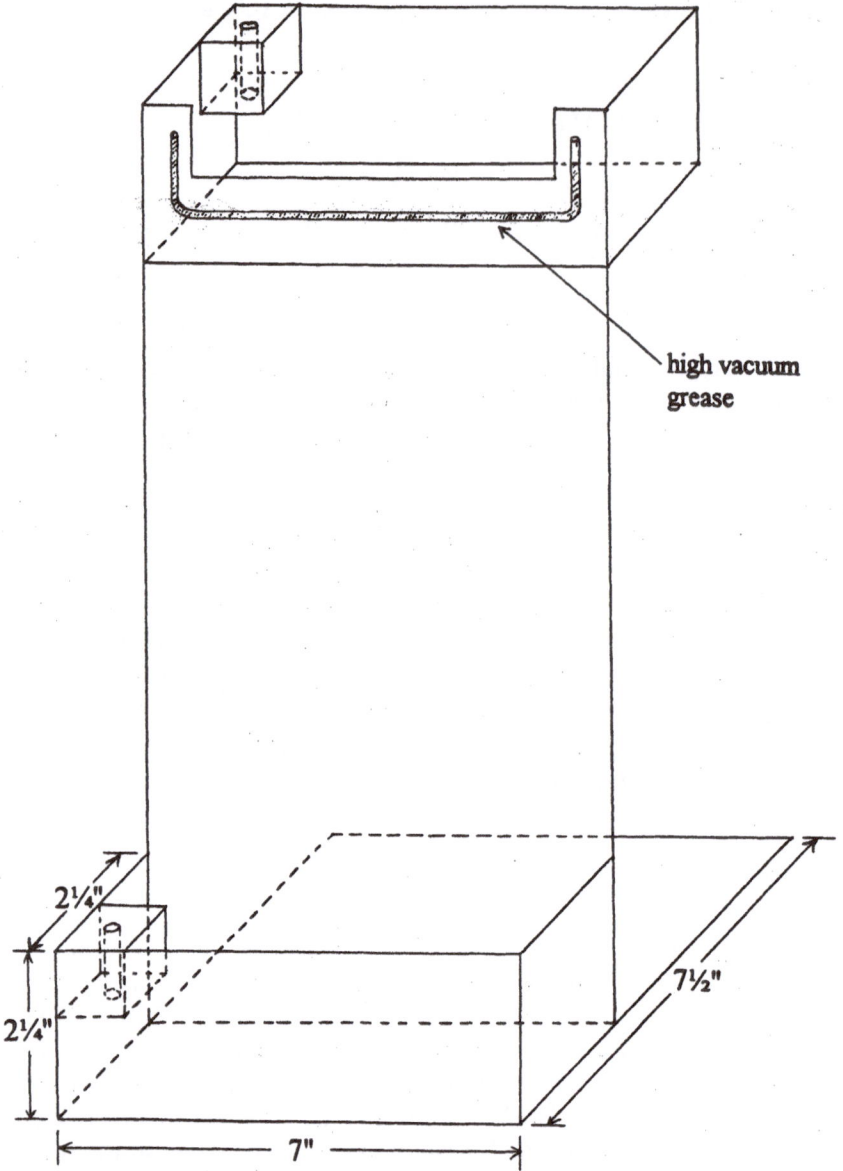

Fig. 10. The polyacrylamide gel cabinet described in the text, with dimensions indicated. The *shaded, U-shaped line* represents the thick band of grease that forms the seal between the gel form and the gel cabinet

Plexiglas pieces required to assemble gel cabinet

- 4 pieces 6.75"×2.25"×1/8" (17.1×5.7×0.3 cm)
- 4 pieces 2.25"×2.25"×1/8" (5.7×5.7×0.3 cm)
- 1 piece 7"×7.5"×1/4" (17.8×19×0.6 cm)
- 1 piece 7"×13"×1/4" (17.8×33×0.6 cm) with a 1" (2.5 cm) deep notch 5.5" (14 cm) long, cut out of the center of one of the short sides
- 2 cubes 1"×1"×1" (2.5×2.5×2.5 cm)
- Also needed: 2 banana jacks, platinum wire

1. Glue the sides, back and bottom of the upper buffer tank together, using dichloroethane, airplane glue or, preferably, a thick solution of Plexiglas dissolved in chloroform.

Note: All assembly work must be carried out in a fume hood.

2. Attach the upper buffer tank to the notched edge of the long Plexiglas plate.

3. Glue the sides and front of the lower buffer tank to the front of the base plate.

4. Check to make sure that the back of the lower buffer tank will fit squarely against the lower side of the notched plate. Note that the upper and lower buffer tanks are on opposite sides of the notched plate.

5. Glue the notched plate to the base plate and the lower buffer tank.

6. Cut a hole through the center of each cube and ream the upper portion of each hole so that it will hold a banana jack.

7. Solder a piece of platinum wire long enough to stretch across the side and bottom of each buffer tank to the base of each banana jack. Thread the platinum wire through the holes in the Plexiglas cubes and screw the banana jacks in place. Attach the banana jacks with all-purpose glue, if desired.

8. Glue the Plexiglas cubes into the buffer tanks as shown in Fig. 10. Stretch the platinum wire along the side and bottom of each buffer tank and attach with dabs of silicone rubber glue (Dow Corning). Also use silicone rubber glue to seal the openings where the platinum wires enter the buffer tanks.

Procedure

Polyacrylamide Gel Forms

The polyacrylamide gels used with the above gel cabinet are cast between two rectangular glass plates of different lengths, separated by Plexiglas spacers, as shown schematically in Fig. 11. Plexiglas "ears" are used to form the seal between shorter glass plate and the gel cabinet, avoiding the need to make a rectangular notch in one of the glass plates. Notched glass plates tend to get broken easily.

Note: The thickness of the spacers determines whether the gel is an analytical gel or a preparative gel. For analytical gels, the spacers and combs are usually 1/16" (0.15 cm) thick. For preparative gels, spacers (and combs) ranging up to 1/2" thick (1.8 cm) can be used, depending on the amount of DNA to be loaded onto the gel.

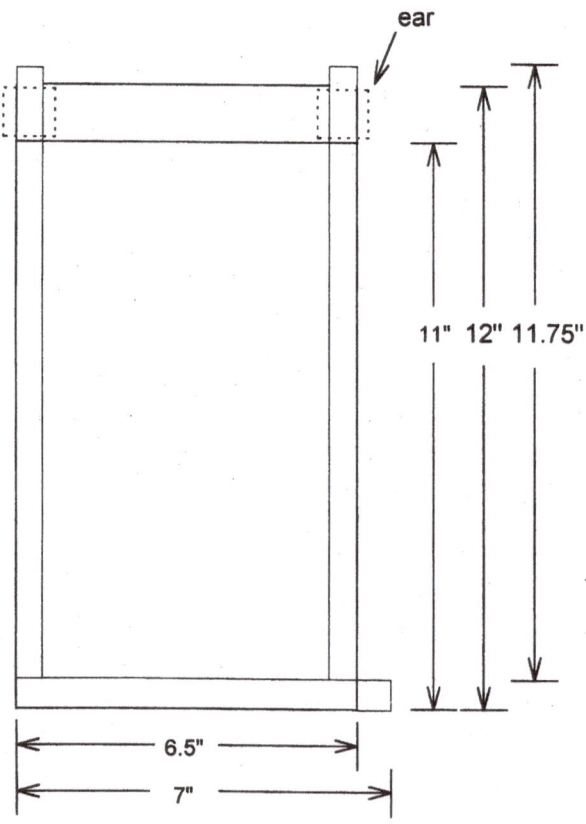

Fig. 11. The assembled gel form, with the dimensions of the spacers and glass plates indicated. The *dashed squares* correspond to the "ears" mounted on the gel form

Note: A well with a $1\,\mathrm{cm}^2$ cross-sectional area (perpendicular to the direction of electrophoresis) can be loaded with $\sim 5\,\mu g$ DNA.

Equipment needed to cast an analytical polyacrylamide gel

Analytical Gel Form

- 1 glass plate: 6.5"×12"×1/8" (16.5×30.5×0.3 cm)
- 1 glass plate: 6.5"×11"×1/8" (16.5×28×0.3 cm)
- 2 Plexiglas spacers: 11.75"×0.5"×1/16" (30×1.3×0.15 cm)
- 1 Plexiglas spacer: 7"×0.5"×1/16" (18×1.3×0.15 cm)
- 2 Plexiglas ears: 1"×1"×1/8" (2.5×2.5×0.3 cm)
- 1 Plexiglas comb: 5.12"×1"×1/16" (13×2.5×0.15 cm) with teeth 1/4" (0.6 cm) wide×3/8" (1 cm) deep, separated by 1/16" (0.15 cm), as shown in Fig. 4c.
- Also needed: nine large binder clips and one small binder clip
- Optional: high vacuum grease (Dow Corning), red fingernail polish

Note 1: The edges of the glass plates should be smoothed to prevent cuts.

Note 2: Because glass is often fluorescent, inspect the glass plates with UV light before assembling the gel form. One side of a glass plate is usually much more fluorescent than the other side. Mark the fluorescent side with red fingernail polish, to remind you to keep the fluorescent side of the glass on the outside of the assembled gel form.

Note 3: Larger glass plates can be used to cast larger-format gels, but the glass should be 3/16" (0.5 cm) thick to prevent breakage.

Note 4: If spacers thicker than $\sim 3/8$" (1 cm) are used, the combs must have air holes drilled in the teeth to prevent the formation of a partial vacuum when the combs are removed from the gel. The air holes should be filled with distilled water before the gel is poured.

1. Clean the glass plates and Plexiglas spacers thoroughly with soapy water. Rinse with distilled water and dry. Spray with ethanol to remove all traces of grease and wipe dry. Place the glass plates on the bench top with their nonfluorescent sides up.

Assembly of Gel Form

2. Place spacers along the sides and bottom of the larger glass plate (on the nonfluorescent side). Check the junction between the spacers. You should be able to run your finger over the spacers and feel little or no difference in height at the junctions. Since Plexiglas varies somewhat in thickness, you may have to turn the spacers end-for-end to achieve the best fit. Exact matching of spacer thickness is important for preventing leaks.

Note: The thickness of the comb must also closely match the spacer thickness.

3. If desired, place a drop of high vacuum grease at the junctions between the side and bottom spacers. Use only a small amount of high vacuum grease because the grease inhibits polymerization.

4. Place the shorter glass plate on top of the spacers (nonfluorescent side down). Clamp the bottom spacer in place with three large spring clamps.

5. Push the side spacers firmly against the bottom spacer. Clamp the side spacers in place with three large spring clamps on each side.

6. Check the fit of the comb in the top of the gel form. It should slide in and out easily.

7. Incline the assembled gel form at ~20–30° to the horizontal. It is convenient to set the gel form in a shallow glass dish (like a glass cake pan) to contain spills.

8. Spacers thicker than 1/16" (0.15 cm) must be sealed in place with high vacuum grease. Fill a 5 ml syringe (without a needle) with high vacuum grease and slowly streak a narrow line of grease close to the sides and bottom of the bottom gel plate. Set the spacers on the lines of grease and squeeze gently to be sure there is a continuous film of grease between the spacers and the bottom gel plate. Streak another narrow line of high vacuum grease on the spacers and set the upper gel plate on top of the spacers. Press down gently on the upper gel plate to spread the grease evenly. Check to make sure that the lines of grease are continuous. Clamp the glass plates together as described above.

The polyacrylamide gels used for the electrophoresis of double-stranded DNA fragments usually contain ~3%C, because the mobilities are reasonably rapid in gels with this cross-linker ratio and the DNA bands are sharp and well spread out. The following basic recipe will make enough gel mixture to cast an analytical 6.9%T, 3%C gel in the gel form described above, using 1/16" (0.15 cm) thick spacers.

Basic Recipe for Analytical Polyacrylamide Gels

6.9%T, 3%C Polyacrylamide gel (75 ml)

5.02 g acrylamide
150 mg Bis
7.5 ml 10× TBE or 10× TAE
75 mg $(NH_4)_2S_2O_8$
75 µl TEMED
distilled water to 75 ml

Note: Acrylamide is a potent neurotoxin and suspected carcinogen. Always wear latex gloves when handling acrylamide solutions and gels. Wear a face mask when weighing out the dry powder. Dispose of acrylamide solutions and gels as hazardous waste.

1. Dissolve the acrylamide and Bis in ~60 ml distilled water in a graduated cylinder and stir magnetically for at least 1 h.

Casting the Gel

Note: Bis dissolves more easily if it is first pulverized with a metal spatula.

2. Filter the solution through a 0.4 μm filter (Millipore) to remove undissolved particles.

3. Add the required volume of 10× TBE or 10× TAE buffer (see "Running Buffers") and enough distilled water to bring the volume to 73 ml. Mix gently by inversion (no air bubbles) three to four times.

Note: Degassing the solution to remove dissolved oxygen is usually not necessary.

4. Add the ammonium persulfate, freshly dissolved in 2 ml distilled water. Mix gently by inversion three to four times (no air bubbles).

5. Add the TEMED and again mix gently by inversion three to four times (no air bubbles).

6. Pour the gel mixture slowly into the previously prepared gel form, until the liquid level reaches the top of the upper (shorter) glass plate.

Note: If air bubbles become trapped in the flowing liquid, they can often be dislodged by tapping gently on the upper glass plate or by raising the gel form to a vertical position.

7. Insert the comb between the two glass plates and clamp in place with a small spring clip.

8. Dribble a little extra gel mixture over the junction between the comb and the glass plates, using a Pasteur pipette, to compensate for volume shrinkage upon gelation and/or small leaks. Repeat as necessary.

Note: Gels cast from the basic recipe should polymerize within 10–15 min. If polymerization occurs significantly more slowly, increase the amount of ammonium persulfate and/or TEMED in the gel mixture. Polymerization occurs with the evolution of heat; the glass plates will become noticeably warm to the touch.

9. Let the gel stand 1–2 h to solidify.

Note: Gels with low %C should be allowed to polymerize undisturbed for 3–4 h.

10. Pour any leftover polyacrylamide gel mixture into a disposable test tube and allow to solidify. Place a Pasteur pipette into the test tube to judge the progress of polymerization; when the gel solidifies, the pipette can no longer be lifted from the test tube. Dispose of the excess polyacrylamide as hazardous waste.

Note: Never pour the left-over polyacrylamide gel mixture down the drain. Besides being environmentally unsafe, the gel mixture will solidify and clog the pipes.

11. If a bad leak occurs, and most of the gel mixture leaks out of the gel form (the bane of polyacrylamide gel electrophoresis), you must take the gel form apart and start over.

Mounting the Gel Form

After the gel has solidified for 1–2 h, the comb can be removed and the gel mounted in the gel cabinet.

Note: The polymerization reaction requires ~24 h to reach completion. Therefore, for accurate, quantitative determination of DNA electrophoretic mobilities, polyacrylamide gels should be aged 24 h.

1. Scrape excess polyacrylamide (if any) from the top of the gel form. Remove the comb by gently wriggling it and lifting out. Be careful not to break the dividers between the wells when removing the comb.

2. Wash unpolymerized acrylamide out of the wells by squirting with distilled water. Tip the gel form upside down to drain or remove the water from the wells with a Pasteur pipette.

Note: The washing step is particularly important for gels with low %C.

3. Remove the spring clips and bottom spacer from the gel form.

4. Attach the "ears" to the gel form. Place narrow lines of high vacuum grease ~1 cm from each edge of the long glass plate, between the top of the short glass plate and the top of the long glass plate. Also place a narrow line of grease along the bottom edge of each ear. Place the ears on the greased strips on the long glass plate, with the greased edge of each ear touching the top edge of the short glass plate, as shown schematically in Fig. 11. Press the ears firmly into place, effectively creating a notched gel plate.

5. Place a thick band of high vacuum grease (Dow Corning) parallel to the U-shaped notch in the gel cabinet (see Fig. 10).

Note: It is convenient to dispense the high vacuum grease from a 5 ml syringe (no needle), moving the syringe slowly to create a wide band of grease.

6. Place the gel form against the thick band of grease, with the top edge of the shorter glass plate even with the notch in the top of the gel cabinet. Clamp the gel form to the gel cabinet with two large spring clamps.

Note: It is often convenient to rest the gel form on a support temporarily placed in the bottom buffer tank while positioning it.

7. Fill the lower buffer tank with the desired buffer (e.g., 1× TBE or 1× TAE). The running buffer must be the same as the buffer in which the gel was cast. To prevent the formation of air bubbles in the space where the bottom spacer was originally located, tilt the gel cabinet and pour the buffer slowly into the lower buffer tank. Gradually decrease the angle of tilt as the buffer tank gets full.

Note: Use a bent wire to perforate and remove air bubbles, if necessary.

8. Fill the upper buffer tank and remove any air bubbles from the wells of the gel by squirting with running buffer.

9. Check for leaks (visible beads of buffer running down the gel from the upper buffer tank). If necessary, add more grease and/or additional spring clamps to improve the seal between the gel form and the gel cabinet.

Note: Slow leaks can be detected by marking the levels of the buffer in the upper and lower buffer tanks and checking to see if the levels change as the gel stands in the gel cabinet.

10. Allow the gel to equilibrate with the buffer in the gel cabinet for at least 1 h. If the gel is to be run at a temperature other than room temperature, place the gel cabinet in the desired location and allow to equilibrate for several hours (or overnight).

Pre-electrophoresis

All polyacrylamide gels must be pre-electrophoresed to remove polar impurities.

1. Connect the buffer tanks to the electrophoresis power supply, using insulated cables with banana jack plugs. The power supplies used in

the author's laboratory are Heath regulated high voltage power supplies, described in "Equipment." Equivalent power supplies may be purchased from many suppliers.

2. Pre-electrophorese all polyacrylamide gels for 2 h, at a voltage of ~3 V/cm. (We use an applied voltage of 100 V, corresponding to an electric field strength of 3.3 V/cm with the gel cabinet described above.) During pre-electrophoresis, the current will decrease and then become essentially constant.

Important Cautions

Caution: The gel, buffer chambers and any leaking buffer are electrically live. Use great care not to touch the gel or the buffer tanks while the gel is running. **Never** bail liquid from one buffer tank to another while the gel is running. If a bad leak occurs, turn off the power supply, wait until the voltage drops to zero, and then clean up. Leaks or cracks in the buffer chambers are particularly serious and must be repaired immediately. **If too much buffer leaks out of the buffer tanks or the buffer completely evaporates, the gel cabinet may melt and catch on fire.** Careful inspection is required, especially before leaving the system unattended.

Electrophoresis

1. After the pre-electrophoresis is finished, turn off the high voltage, wait until the voltage decays to zero and load the DNA samples in the wells. The preparation of the DNA samples has been described in "DNA Sample Preparation." Because polyacrylamide gels are usually very thin, tapered pipette tips or Pasteur pipettes must be used to layer the DNA samples under the buffer in the wells.

Note: To get sharp bands, the DNA samples should be 1 mm or less in height after loading.

2. Set the high voltage to the desired value and start the electrophoresis run. With the gel form and gel cabinet described here, Joule heating is negligible when the applied voltage is 3.3 V/cm or less (corresponding to a setting of 100 V on the power supply). Under these conditions, the DNA samples migrate with a constant mobility across the entire width of the gel.

Note: When Joule heating becomes important, the temperature near the center of the gel is higher than the temperature near the edges. The resulting difference in the viscosity of the solvent causes the DNA bands to migrate faster near the center of the gel, giving a "smiling" effect when the same sample is loaded in all the wells.

Note: The electric field must not be set so high that the buffer completely evaporates from the buffer tanks or the gel cabinet may catch on fire. Check the buffer levels periodically when the gel is running.

3. Measure the temperature of the gel and the electric field strength, if desired.

Note: Because acrylamide gels are cast between glass plates, it is somewhat difficult to measure the exact voltage and temperature during electrophoresis. The temperature of the gel can be monitored with a digital liquid crystal thermometer strip (Cole-Parmer) attached to the outer glass plate. Alternatively, the temperature can be measured by thrusting a thermister (e.g., an Omega digital thermister) deep into the gel immediately after the high voltage is turned off. The approximate voltage drop across the gel can be estimated from the applied voltage and the distance between the electrodes in the gel cabinet (not the length of the gel). The exact voltage should be measured in separate control experiments, using supplementary electrodes embedded in the gel. Care must be taken that the placement of the supplementary electrodes does not affect the measured electric field strength.

DNA Detection

1. After the electrophoresis experiment is finished, turn off the power supply, wait until the voltage has decayed to zero and remove the cables connecting the buffer tanks to the power supply.

2. Pour off the buffer in the buffer tanks.

Note: If the buffer contained ethidium bromide, the buffer must be saved and decontaminated as described in "Disposal of Ethidium Bromide."

3. Remove the side binder clamps and pry the gel from the gel cabinet by placing a metal spatula between the back glass plate and the gel cabinet and twisting gently.

4. Place the gel form flat on the bench top and remove one of the side spacers.

5. Separate the top (short) glass plate from the gel by introducing air between the glass plate and the gel, using a razor blade or thin spatula to carefully raise the top glass plate. The gel usually remains attached to the bottom glass plate, making it possible to slowly lift top glass plate and rotate it away from the gel.

Note: The gel occasionally sticks to the top glass plate, making it necessary to turn the gel form over and remove the bottom (long) glass plate instead.

6. If the optional dyes described in step 4 of "DNA Sample Preparation" were not added to the DNA samples, place the glass plate holding the polyacrylamide gel in 500 ml of a solution containing 1.5 µg ethidium bromide/500 ml and soak for 15 min. The preparation of the ethidium bromide stock solution is described in "Agarose Gel Preparation." Destain the gel 15 min in distilled water if necessary, and photograph the gel as described in "Photographic Detection," simultaneously photographing a ruler to determine the magnification factor. Quantitative analysis of band intensities can be made using, e.g., a BioRad GS-670 Imaging Densitometer.

Note: Ethidium bromide is a potent mutagen and suspected carcinogen. Always wear latex gloves when handling ethidium bromide solutions and stained gels. Always decontaminate all ethidium bromide solutions when finished and dispose of stained gels in an environmentally safe manner (see „Disposal of Ethidium Bromide").

A plot of the electrophoretic mobilities typically observed for DNA restriction fragments in a 6.9 %T, 3 %C polyacrylamide gel, cast and run in TBE buffer, is illustrated in Fig. 12. Note that some of the fragments in the 300–500 bp range migrate anomalously slowly, and exhibit mobilities that fall below the dotted (straight) line that describes the mobility of the majority of the DNA fragments in this molecular weight range. The anomalously slowly migrating DNA fragments are thought to be bent or curved, making it relatively difficult for them to migrate through the polyacrylamide gel matrix during electrophoresis. The same fragments exhibit normal electrophoretic mobilities in agarose gels, as shown in Fig. 8, presumably because of the larger average pore size of agarose gels. Therefore, agarose gels should be used to estimate the molecular weights of unknown DNA samples, to avoid the sequence-dependent mobilities observed in polyacrylamide gels (Fig. 12).

Also note that polyacrylamide gels are much more useful than agarose gels for separating small DNA fragments, because the mobilities of small fragments are well spread out in polyacrylamide gels (compare Figs. 8, 12). In fact, DNA mobilities are well spread out over the entire molecular weight range from 50 to 1000 bp in polyacrylamide gels containing ∼7 %T and 3 %C, as shown in Fig. 12. The separation of fragments in other molecular weight ranges can be optimized by changing %T and/or %C.

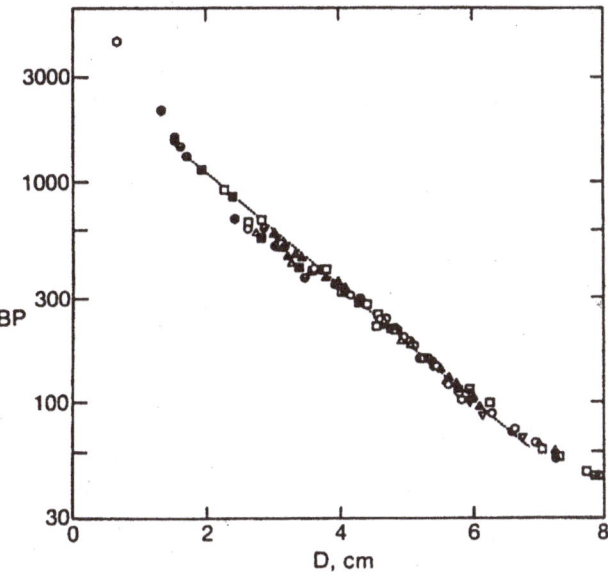

Fig. 12. Typical electrophoretic separation of DNA restriction fragments in a 6.9 %T, 3 %C polyacrylamide gel, cast and run in TBE buffer. The logarithm of DNA molecular weight, in base pairs (*BP*), is plotted as a function of the distance migrated in the gel, *D*. The various symbols correspond to different restriction enzymes used to digest the plasmid DNA; the *dotted line* corresponds to a straight line drawn through many of the data points in the central part of the gel

1.3
Ferguson Plots

Most DNA electrophoresis experiments are designed to give "yes-or-no" answers to simple questions such as: Did the restriction enzyme cut the DNA? Did the ligation work? Is there any DNA in the miniprep? How pure is the DNA sample? What is the size of the DNA? Does a particular protein or antibody bind to the DNA?

To answer such questions, electrophoresis only needs to be carried out at a single gel concentration, in the type of gel matrix best suited to separate the DNA molecules of interest. The experimental conditions are not very important, as long as sufficient separation of the desired fragments is achieved. However, it is sometimes useful to construct Ferguson plots (logµ vs gel concentration) to analyze the gel matrix and/or the macromolecules undergoing electrophoresis (Stellwagen 1987; Tietz 1988). Since, by definition, Ferguson plots require the comparison of electrophoretic mobilities in several different gels with different gel concentrations, the accuracy and reproducibility of the electrophoretic measurements becomes critical.

A certain variability will always be observed in the electrophoretic mobilities measured for DNA molecules in agarose and polyacrylamide gels, because of the inherent randomness of the structure of the gel matrix. Therefore, it is important to eliminate other sources of variability in the electrophoresis experiment before constructing Ferguson plots.

Procedure

Ferguson Plot Measurements

1. Temperature: Measure the exact temperature in the gel during electrophoresis using, e.g., an Omega digital thermometer. For agarose gels, correct all mobilities to a common temperature using Eq. (6):

$$\frac{\mu_{T1}}{\mu_2} = \frac{T_1 + 30}{T_2 + 30} \tag{6}$$

where μ_{T1} is the calculated mobility at the common (standard) temperature T_1 and μ_{T2} is the mobility observed at the temperature of measurement T_2 (West 1987). Equation (6) is valid over the temperature range 15–45 °C. Outside this temperature range, or when using polyacrylamide gels, correct the mobilities to a common temperature using Eq. (7):

$$\frac{\mu_{T1}}{\mu_2} = \frac{\eta_{T2}}{\eta_{T1}} \tag{7}$$

where μ_{T1} and μ_{T2} have been defined above and η_{T2} and η_{T1} are the viscosity of water at the temperature of the measurements and the standard temperature, respectively.

2. Electric field strength: The electric field strength in the gel must be known exactly in order to calculate accurate electrophoretic mobilities from Eq. (2). Measure the electric field in the gel using, e.g., a Fluke digital multimeter, either during the electrophoresis experiment (agarose gels) or in separate control gels run under identical conditions (polyacrylamide gels). The electrophoretic mobilities should also be measured at several different electric field strengths at each gel concentration, correcting the mobilities to a common temperature, if necessary, using Eq. (6) or Eq. (7). If the corrected mobilities are independent of electric field strength, average the mobilities.

However, if the electrophoretic mobilities vary with the electric field strength, the mobilities observed for each DNA fragment at each gel concentration should be extrapolated to zero electric field strength and the extrapolated values used to construct the Ferguson plots.

3. DNA concentration: Representative gels should be run at different DNA concentrations, to be sure that there is no dependence of the electrophoretic mobilities on DNA concentration.

4. Gel concentration: For reasonably accurate extrapolation of the Ferguson plots to zero gel concentration, at least five gel concentrations should be measured; six to seven gel concentrations are better. Ferguson plots in polyacrylamide gels must be constructed from gels with constant %C.

5. Gel aging: Chemically cross-linked polyacrylamide gels require approximately 24h for the polymerization reaction to reach completion. Hence, for accurate work, all polyacrylamide gels should be aged 24h before use. Aging of the gel matrix is not necessary for agarose gels.

6. Electroendosmosis: Because agarose gels have negatively charged gel fibers, it is very important to measure the electroendosmotic mobility of the solvent ("Measurement of Electroendosmosis") and calculate the true DNA mobility at each agarose gel concentration from Eq. (5). Although electroendosmosis is relatively unimportant for small DNA fragments in dilute agarose gels, the electroendosmotic mobility becomes dominant at high gel concentrations where the DNA mobility becomes very small (Stellwagen 1992). Since the electroendosmotic mobility is also buffer dependent, it must be taken into account before interpreting the apparent DNA mobilities observed in different running buffers. Corrections for the electroendosmotic mobility are not usually necessary for polyacrylamide gels.

Typical Ferguson plots measured for DNA fragments in agarose and polyacrylamide gels, cast and run in TAE buffer, are illustrated in Fig. 13. It can be seen that the Ferguson plots are linear only over a limited range of gel concentrations and do not extrapolate to a common intercept at zero gel concentration, as expected from Eq. (3). The intercepts of the Ferguson plots in polyacrylamide gels are also highly dependent on the cross-linker ratio (Holmes and Stellwagen 1991a, b). As a result, the experimental Ferguson plots look very different from the idealized version illustrated in Fig. 2. The explanation of these effects, and the applicability of Eq. (3) to the electrophoresis of DNA in agarose and polyacrylamide gels, are matters of debate in the current literature.

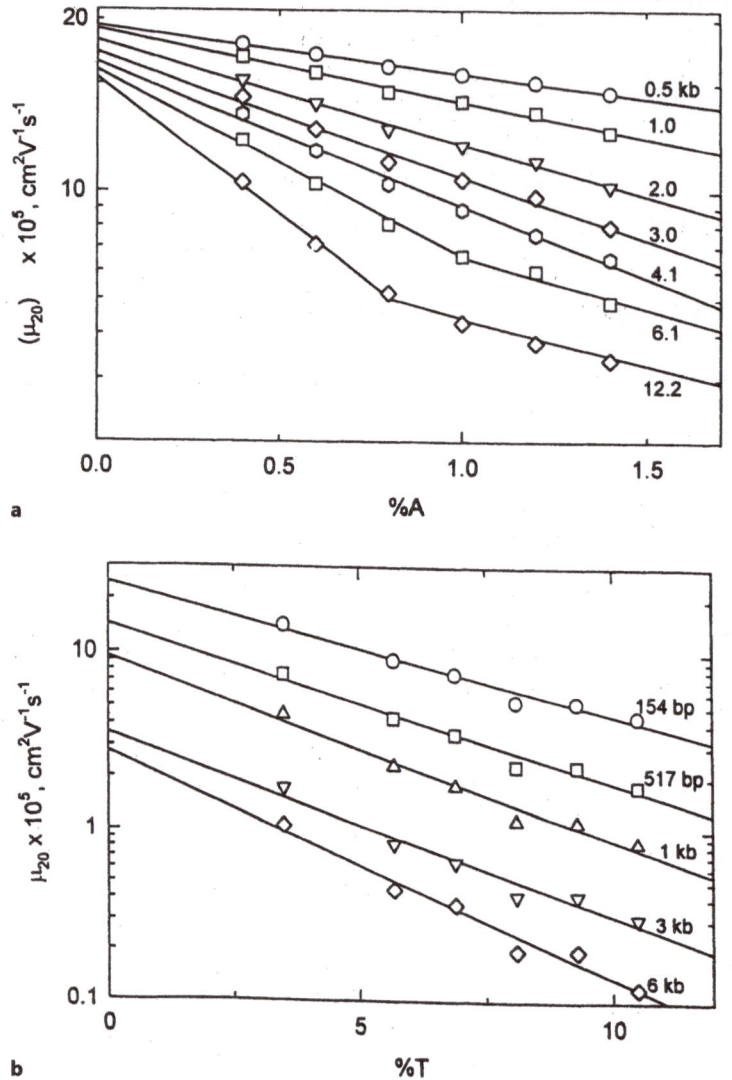

Fig. 13a,b. Ferguson plots of DNA in agarose and polyacrylamide gels. The logarithm of the absolute mobility, logμ, is plotted vs: **a** agarose gel concentration (%A) or **b** polyacrylamide gel concentration (%T). The polyacrylamide gels contained 3 % C. The sizes of the DNA molecules are given beside each curve

▪ Troubleshooting

- The sample foams out of the wells after loading.
  Problems with
Agarose Gels
  - The DNA sample contains too much ethanol, so that the viscosity of the solution is too low.
    - Heat the DNA sample at 65 °C for 15 min to evaporate the ethanol, then add the dye mix.
  - Too much running buffer on top of the gel.
    - Pour off excess buffer.
  - Too little glycerol or Ficoll in the dye mix.
    - Recheck the recipe.

- The sample forms large streaks back toward the wells.
  - Too much running buffer evaporated during the run and the gel ran dry.
    - Always be sure that the gel is covered by at least 3–5 mm running buffer or use a submarine gel apparatus.

- Cannot find the expected number of restriction fragment bands.
  - The enzyme cuts very inefficiently at one or more locations.
    - Redigest, using more enzyme or a longer digestion time or more optimal buffer conditions. Alternatively, choose a different restriction enzyme.
  - Two bands with complementary "sticky ends" have formed a complex.
    - Heat at 65 °C for 15 min to denature the noncovalent hydrogen bonds. Certain λ DNA fragments are particularly prone to form complexes with each other.
  - The DNA fragments are large (≥ 5 kb) and migrate with virtually identical mobilities.
    - Decrease the gel concentration and the applied voltage to enhance the mobility differences between large DNA molecules.
  - The DNA fragments are small (≤ 150 bp) and two or more fragments comigrate.
    - Increase the gel concentration by using NuSieve agarose or a 3:1 NuSieve:LE agarose gel mixture. Alternatively, switch to polyacrylamide gels.

- The DNA bands exhibit edge tailing (i.e., each band is somewhat U-shaped). This effect is sometimes called smiling, which creates confusion because smiling is more commonly used to describe the linearity of migration across the width of a gel.

- The gel is overloaded.
  - Decrease the sample concentration.
- The sample dye mix contained glycerol instead of Ficoll.
  - Switch density modifying agents.

- The DNA bands are "smeary," not sharp.
  - The wells have penetrated all the way through to the glass plate at the bottom of the gel.
    - Do not place the comb so close to the glass plate when pouring the gel.
  - The comb was removed before the agarose had sufficiently hardened, making the wells irregular in shape.
    - Let the gel solidify longer.
  - The sample solution contained too much ethanol and/or the dye mix contained too little glycerol or Ficoll.
    - See foaming of the sample, above.
  - The gel form was bumped before the DNA had penetrated into the gel.
    - Be careful not to bump the gel form.
  - The gel is too thick, and/or the electric field strength is too high, causing a temperature difference between the top and bottom of the gel because of the opposing effects of Joule heating and evaporative cooling. Such temperature differences can lead to band tilting and loss of sharpness.
    - Pour a thinner gel and/or decrease the electric field strength.

with
mide
Gels

- The DNA samples do not stay in the wells after loading.
  - The DNA solution contains too much ethanol, so that the sample does not stay layered under the buffer after loading.
    - Heat the sample at 65 °C for 15 min to evaporate the ethanol, then add the dye mix and load the samples.
  - The wells contain polyacrylamide "strings" and/or excess acrylamide polymerized at the bottom of each well.
    - Carefully wash all unpolymerized acrylamide out of the wells after removing the comb by squirting the wells with distilled water or running buffer. If the gel contains < 3 %C, let the gel polymerize at least 4 h before removing the comb and washing out the wells.

- The gel contains trapped air bubbles.
  - The gel mixture was poured into the gel form too rapidly.

- There are several solutions: Pour the gel mixture more slowly. Place the gel form at an angle of ~15° to the horizontal, rather than ~30°. Stop pouring and tap the top glass plate gently after bubble formation, or raise the gel form toward the vertical position, to make the bubble(s) rise to the top of the liquid and burst.
- The gel solidified too rapidly.
  - Decrease the amount of TEMED and/or ammonium persulfate to decrease the rate of polymerization.
- The glass plates were not sufficiently clean.
  - Meticulously clean all traces of grease from the glass plates, spacers and your hands. Repour the gel.
- The gel mixture contained bubbles after adding the TEMED and/or persulfate.
  - Invert the graduated cylinder containing the gel mixture very slowly after each addition and allow any air bubbles to dissipate before pouring the gel.

- The gel does not polymerize or polymerizes very slowly.
  - The gel mixture is incorrect.
    - Check the amounts of acrylamide, Bis, TEMED and persulfate in the basic recipe. Be sure that ammonium sulfate was not accidentally substituted for ammonium persulfate.
  - The ingredients are correct but the mixture still does not polymerize.
    - Remake all stock solutions, especially the persulfate. Open a new bottle of persulfate if the old one has been open for a long time. If all else fails, increase the TEMED and/or the persulfate concentration in the basic recipe until the mixture polymerizes in 10–15 min.
  - A lot of air was introduced into the gel mixture by stirring in an open beaker.
    - Degas the gel mixture for 10 min.
  - Some of the high vacuum grease came into contact with the gel mixture, inhibiting polymerization. This problem is especially serious if the grease is near the comb.
    - Place all high vacuum grease near the outer edges of the glass plates and spacers. Use the minimum amount of grease necessary to prevent leaks.

- The gel form leaks. This problem is the bane of polyacrylamide gel electrophoresis.
  - The spacers at the bottom of the gel form do not have exactly the same thickness.

- – There are several possible solutions: Match the thickness of the spacers more carefully. Place a drop of high vacuum grease at the junction of the side and bottom spacers. Put more clamps near the junction of the side and bottom spacers. If all else fails, place a continuous streak of grease along each side of each spacer. This works, but makes the gel form more difficult to clean up.
- – The spacers are thicker than 0.15 cm.
  - – Place a continuous streak of grease along each side of each spacer and clamp the glass plates together at frequent intervals.
- – The gel mixture polymerizes very slowly.
  - – Increase the TEMED or persulfate concentration so that the gel mixture polymerizes in ~15 min.
- – The comb is thicker than the spacers, causing the gel mixture to leak out of the gel form near the junction of the long and short glass plates.
  - – Carefully match the thickness of the comb and spacers.
- – The glass plates have beveled edges because they have been smoothed, and the gel mixture leaks out of the top of the gel form through the beveled edges.
  - – Possible solution: Put more clamps near the top of the gel form. If necessary, place a small amount of high vacuum grease on the very outside edges of the glass plates near the junction of the long and short glass plates, to fill the beveled edges.

Note: Be careful not to get the high vacuum grease on the flat surfaces of the glass plates, because the grease may interfere with polymerization of the gel near the wells.

- • The comb is very difficult to remove.
  - – The comb is slightly thicker than the side spacers.
    - – There are several possible solutions: Choose a comb better matched to the thickness of the spacers. Only clamp the glass plates half way up each side, so that there is a little more flexibility at the top of the gel form. Test the ease of extraction of the comb before pouring the gel.
  - – Thick spacers and combs (>0.6 cm) are used, so that a partial vacuum is created by trying to remove the comb.
    - – Use combs with air holes drilled in the middle of each tooth. Fill the air holes with distilled water before pouring the gel. Before removing the comb, remove the water from the air holes with a Pasteur pipette.

- The gel form leaks at the junction of the "ears" and the glass plate.
  - The ears are not firmly pushed down on the short glass plate.
    - Push the ears down firmly; clamp in place if necessary.

Note: Be sure that the clamp does not sit in the buffer solution. Otherwise, the clamp will rust and Fe^{3+} ions will be added to the buffer.

  - The ears are not the correct thickness.
    - Replace with ears having the same thickness as the glass plates.
  - The band of grease between the ears and the gel support, and/or between the glass plate and the gel support, is not continuous, allowing buffer from the upper buffer chamber to leak and run down the gel cabinet.
    - Apply more grease and/or use more clamps.

- The DNA bands are smeary and/or uneven.
  - The gel mixture was not thoroughly mixed, so that polymerization did not take place evenly.
    - Stir the acrylamide/Bis mixture longer using a magnetic stir bar. Mix thoroughly at every addition step.
  - The bottoms of the wells were not flat.
    - Carefully remove all unpolymerized acrylamide from the wells after removing the comb.
  - The gel was bumped after loading the DNA samples, or the gel location was changed before running the samples into the gel.
    - Be careful not to bump the gel cabinet after loading.
  - The acrylamide concentration in the gel mixture was not high enough to give sharp bands.
    - Increase %T. For very small DNA molecules, also increase %C.
  - The DNA sample volume was too large, so that the depth of the sample was more than 1 mm in the bottom of the wells.
    - Use a smaller volume of a more concentrated DNA solution, or cast a gel with wider wells.

- The dye front and/or the band pattern is not even across the width of the gel.
  - The most common cause of this band pattern is that the temperature in the center of the gel is higher than at the edges, because Joule heating effects are more easily dissipated at the edges of the gel. Since electrophoretic mobilities vary inversely with the viscosity of the solvent, DNA samples in the center of the gel will migrate faster than the same samples electrophoresed in wells near the edge.

- Decrease the applied voltage, so that Joule heating is balanced by air cooling. Alternatively, use a gel cabinet with thermostated temperature control.
- An air bubble was inadvertently trapped between the glass plates at the bottom of the gel when the lower buffer chamber was filled.
  - Inspect the bottom of the gel form for air bubbles before loading the DNA samples. Remove any bubbles by perforating them with a bent wire and forcing them to the edge of the gel.

- The DNA bands are U-shaped and trail backward at each edge.
  - The gel is overloaded.
    - Decrease the DNA concentration.
  - Glycerol was used in the dye mix.
    - Use Ficoll in the dye mix.

- Some DNA fragments migrate with the "wrong" mobility.
  - Some of the DNA molecules are bent or curved, causing them to migrate anomalously slowly.
    - Identify anomalously slowly migrating fragments by comparing the observed mobilities with standard curves generated from DNA molecules known to migrate with normal electrophoretic mobilities. If necessary, use other methods to identify the anomalous fragments: digest with additional restriction enzymes or blot with proteins or antibodies known to bind to the DNA in question. Alternatively, switch to agarose gels.
  - Complementary "sticky ends" of two fragments have annealed together, causing two fragments to migrate as one larger fragment.
    - Heat the DNA sample at 65 °C for 15 min to dissociate the complex before loading the gel.
  - The DNA sequence is incorrect, or the wrong restriction enzyme has been used to digest the sample.
    - Do more restriction digests with other enzymes and compare the results.

lems
- The DNA bands are very faint or no bands are seen (detection problems).
  - The DNA concentration is too low.
    - Use a higher DNA concentration or a more sensitive dye (e.g., TOTO or SYBR green instead of ethidium bromide). Alternatively, switch to a low wavelength UV lamp (254 nm) for ethidium bromide detection.
  - The dye concentration is too low.

- Restain with a more concentrated dye solution or stain for a longer time.
- The gel and/or the glass plate scatters too much light to see the fluorescent DNA bands clearly.
  - Destain the gel in distilled water. If the problem is the glass plate, transfer the gel to a nonfluorescent glass plate, if possible. Later, mark the fluorescent sides of all glass plates and cast all gels on the nonfluorescent sides.
- The ethidium bromide stock solution is too old.
- Remake the stock solution.

- No DNA bands are seen (nondetection problems).
  - The electrodes were connected backwards, so that the DNA molecules migrated toward the short end of the gel.
    - The presence of the marker dye in the upper buffer chamber is diagnostic of this problem. Be sure the positive terminal of the power supply is connected to the bottom buffer chamber.
  - Forgot to add the DNA.
  - Remake the samples.

- The DNA bands appear as closely spaced doublets when the dyes TOTO or YOYO are added to the sample before electrophoresis.
  - These dyes equilibrate very slowly on the DNA, leading to different DNA/dye binding ratios in the same sample (Carlsson et al. 1995).
    - Heat the DNA solution for 2 h to equilibrate the dye.

Note: Heating for such a long time will probably cause some nicking or degradation of the DNA.

Acknowledgments. Useful discussions with Kurt Strutz and assistance in preparing the figures are gratefully acknowledged. Financial support from grant GM29690 from the National Institute of General Medical Sciences is also acknowledged.

References

Birren B, Lai E (1993) Pulsed field gel electrophoresis: a practical guide. Academic, Harcourt Brace Jovanovich, San Diego, California
Carlsson D, Jonsson M, Akerman B (1995) Double bands in DNA gel electrophoresis caused by bis-intercalating dyes. Nucleic Acids Res 23:2413–2420
Duke T, Viovy J-L, Semenov AN (1994) Electrophoretic mobility of DNA in gels. I. New biased reptation theory including fluctuations. Biopolymers 34:239–247
Chrambach A (1985) The Practice of Quantitative Gel Electrophoresis, VCH, Weinheim Germany

Ferguson KA (1964) Starch-gel electrophoresis – application to the classification of pituitary proteins and polypeptides. Metabolism 13:985–1002

Glazer AN, Peck K, Mathies RA (1990) A stable double-stranded DNA-ethidium homodimer complex: application to picogram fluorescence detection of DNA in agarose gels. Proc Natl Acad Sci USA 87:3851–3855

Haugland RP (1992) Handbook of fluorescent probes and research chemicals, 5th ed, Molecular Probes, Eugene, Oregon, pp 221–229

Holmes DL, Stellwagen NC (1991a) Estimation of polyacrylamide gel pore size from Ferguson plots of normal and anomalously slowly migrating DNA fragments. I. Gels containing 3 % N,N'-methylenebisacrylamide. Electrophoresis 12:253–263

Holmes DL, Stellwagen NC (1991b) Estimation of polyacrylamide gel pore size from Ferguson plots of linear DNA fragments. II. Comparison of gels with different crosslinker concentrations, added agarose and added linear polyacrylamide. Electrophoresis 12:612–619

Kozulic B (1995) Models of gel electrophoresis. Anal Biochem 231:1–12

Lerman LA, Frisch HL (1982) Why does the electrophoretic mobility of DNA in gels vary with the length of the molecule? Biopolymers 21:995–997

Lumpkin OJ, Zimm BH (1982) Mobility of DNA in gel electrophoresis. Biopolymers 21:2315–2316

McDonell MW, Simon MN, Studier FW (1977) Analysis of restriction fragments of T7 DNA and determination of molecular weights by electrophoresis in neutral and alkaline gels. J Mol Biol 110:119–146

Ogston AG (1958) The spaces in a uniform random suspension of fibers. Trans Faraday Soc 54:1754–1757

Olivera BM, Baine P, Davidson N (1964) Electrophoresis of the nucleic acids. Biopolymers 2:245–257

Righetti PG, Gelfi C (1996) Capillary electrophoresis of DNA. In: Righetti PG (ed) Capillary electrophoresis in analytical biotechnology, CRC, Boca Raton, pp 431–476

Rodbard D, Chrambach A (1970) Unified theory of gel electrophoresis and filtration. Proc Natl Acad Sci USA 4: 970–977

Rye HS, Yue S, Wemmer DE, Quesada MA, Haugland RP, Mathies RA, Glazer AN (1992) Stable fluorescent complexes of double-stranded DNA with bis-intercalating asymmetric cyanine dyes: properties and applications. Nucleic Acids Res 20:2803–2812

Sambrook J, Fritsch EF, Maniatis T (1987) Molecular cloning: a laboratory manual. 2nd ed, Cold Spring Harbor Laboratory, New York, New York

Skeidsvoll J, Ueland, PM (1995) Analysis of double-stranded DNA by capillary electrophoresis with laser-induced fluorescence detection using the monomeric dye SYBR Green I. Anal Biochem 231:359–365

Slater GW, Guo HL (1996) An exactly solvable Ogston model of gel electrophoresis: I. The role of the symmetry and randomness of the gel structure. Electrophoresis 17:977–988

Stellwagen NC (1987) Electrophoresis of DNA in agarose and polyacrylamide gels. Adv Electrophoresis 1:177–228

Stellwagen NC (1992) Agarose gel pore radii are not dependent on the casting buffer. Electrophoresis 13:601–603

Studier FW (1973) Analysis of bacteriophage T7 early RNAs and proteins on slab gels. J Mol Biol 79:237–248

Tietz D (1988) Evaluation of mobility data obtained from gel electrophoresis: Strate-
    gies in the computation of particle and gel properties on the basis of the extended
    Og ston model. Adv Electrophoresis 2:109–169
Tietz, D, Chrambach, A (1992) Concave Ferguson plots of DNA fragments and con-
    vex Ferguson plots of bacteriophages: evaluation of molecular and fiber proper-
    ties, using desktop computers. Electrophoresis 13:286–294
West R (1987) The electrophoretic mobility of DNA in agarose gel as a function of
    temperature. Biopolymers 26:607–608

Suppliers

Partial List of Suppliers of General Electrophoretic Equipment

BioRad Laboratories
2000 Alfred Nobel Drive
Hercules, California 94547, USA
Phone: 1-800-424-6723
Fax: 1-800-879-2289
Web site: http://www.bio-rad.com

Life Technologies
8400 Helgerman Court
P.O. Box 6009
Gaithersburg, Maryland 20898-9980, USA
Phone: 1-800-828-6686
Fax: 1-800-331-2286
e-mail: info@lifetech.com
Web site: http://www.lifetech.com

Pharmacia Biotech Inc.
800 Centennial Avenue
Piscataway, New Jersey 08855-1327, USA
Phone: 1-800-526-3593
Fax: 1-800-FAX-3593
e-mail: ts@ep.pharmaciabiotech.com
Web site: http://www.biotech.pharmacia.se

Other Suppliers Mentioned in the Text

Cole-Parmer Instrument Company
625 E. Bunker Court
Vernon Hills, Illinois 60061-1844, USA
Phone: 1-800-323-4340
Fax: 1-847-549-7676
e-mail: info@coleparmer.com
Web site: http://www.coleparmer.com

Dow Corning Corp.
2200 W. Salzburg Road
Auburn, Michigan 48611, USA
Phone: 1-800-248-2481
Fax: 1-517-496-8026

John Fluke Mfg. Co. Inc.
P. O. Box 9090
Everett, Washington 98206-9090, USA
Phone: 1-800-443-5853
Web site: http://www.fluke.com

FMC BioProducts
191 Thomaston Street
Rockland, Maine 04841, USA
Phone: 1-800-341-1574
Fax: 1-800-362-5552
e-mail: BIOTECHSERV@FMC.COM
Web site: http://www.bioproducts.com

Millipore Corporation
80 Ashby Road
Bedford, Massachusetts 01730, USA
Phone: 1-800-645-5476
Fax: 1-617-533-8873
e-mail: order@millipore.com
Web site: http://www.millipore.com

Molecular Probes
4849 Pitchford Avenue
Eugene, Oregon 97402-9165, USA
Phone: 1-541-465-8300
Fax: 1-541-344-6504
e-mail: order@probes.com
Web site: http://www.probes.com

Omega Engineering Inc.
One Omega Drive
Stamford, Connecticut 06907-0047, USA
Phone: 1-800-826-6342
Fax: 1-203-359-7700
e-mail: info@omega.com
Web site: http://omega.com

Polaroid Corporation
549 Technology Square
Cambridge, Massachusetts 02139, USA
Phone: 1-800-225-1618
Fax: 1-617-386-5605
Web site: http://www.polaroid.com

Sigma
P. O. Box 14508
St. Louis, Missouri 63178, USA
Phone: 1-800-325-3010
Fax: 1-800-325-5052
e-mail: custserv@sial.com
Web site: http://www.sigma.sial.com

UVP, Inc.
5100 Walnut Grove Ave.
San Gabriel, California 91778, USA
Phone: 1-800-452-6788
Fax: 1-909-946-3597
e-mail: uvp@dial.pipex.com
Web site: http://www.uvp.com

Denaturating Electrophoresis in DNA Sequencing: A Brief History and Current Protocols

ALEXANDER KRAEV

Introduction

Two common DNA sequencing methods rely on the principle of measuring relative distances of individual nucleotides from a fixed point in a DNA chain (Sanger and Coulson 1975; Maxam and Gilbert 1977). Base-specific sequencing reactions produce nested sets of fragments, sharing one end and terminating at all positions of a given base, for example guanine, in a DNA chain. The ability to determine a sequence is thus dependent on a separation system which is capable of resolving two DNA fragments that differ in length by a single nucleotide. Until very recently, the only system with adequate resolution, suggested in the 1960s, was based on electrophoresis in a flat polyacrylamide gel with a Tris-borate buffer system (Peacock and Dingman 1967). Electrophoresis was conducted between two glass plates in a simple apparatus, based on the design of Studier (1973). After a certain time, controlled by the mobility of marker dyes, the electrophoresis was terminated and the gel was exposed to an X-ray film or to a phosphor-imaging screen to detect DNA bands labelled with ^{32}P, ^{35}S or ^{33}P (batch technique). The nucleotide sequence was then read from the image using band order and specificity of the sequencing reaction (track identity) (Fig. 1A). Improvements of this system over two decades were achieved using gels with extreme dimensions (Garoff and Ansorge 1981; Ansorge and Barker 1984) and electric field gradients (Biggin et al. 1983; Ansorge and Labeit 1984; Olsson et al. 1984; Sheen and Seed 1988). However, for a long time the basic gel system remained the same, due to the fact that the chemical sequencing method (Maxam and Gilbert 1977) most commonly used

Alexander Kraev, Laboratory of Biochemistry III, Swiss Federal Institute of Technology (ETH), Universitätsstrasse 16, 8092 Zürich, Switzerland (*phone* +41-1-632-31-47; *fax* +41-1-632-12-13; *e-mail* kraev@bc.biol.ethz.ch; *web site* http://www.bc.biol.ethz.ch/Carafoli/Sasha/kraev.html)

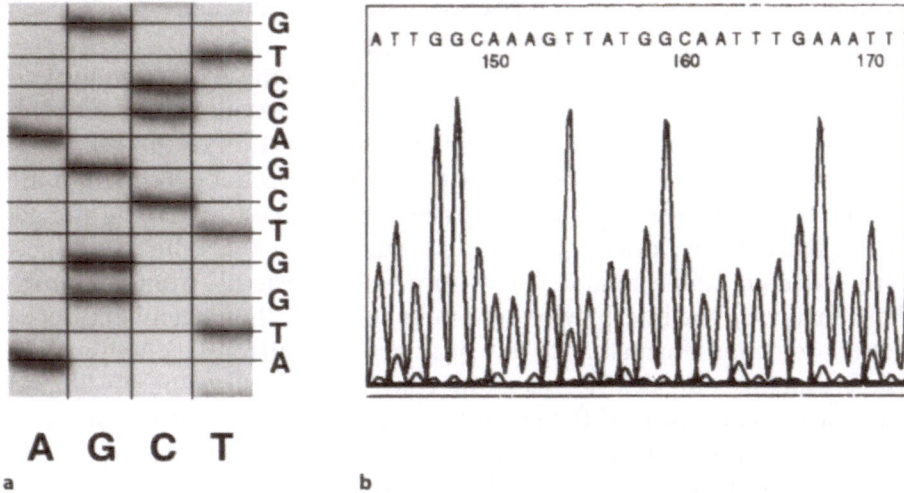

Fig. 1A, B. Two principal ways of data output from sequencing gel electrophoresis. **A** Autoradiogram image. The grid, superimposed on the X-ray film image, imitates the way in which the human eye or a software interprets the image. *Vertical lines* denote the borders of adjacent tracks, in which products of sequence specific reactions were fractionated into bands; *horizontal lines* show band locations. Intersection of the band locations with the track identity results in the DNA sequence, shown at the *right side*. **B** A software reading of the autoradiogram image (after scanning the film) as well as the band pattern on-line transforms bands into peak profiles, shown in four-colour code in the original output (in this case from the ABI Prism 310 CGE sequencer). The sequence read by the software is shown above the peaks

before the mid-1980s was tedious and time-consuming; hence productivity was not limited by the gel electrophoresis system itself. In the mid-1980s, when methods of oligonucleotide synthesis and single-stranded DNA production were improved, chemical sequencing practically vanished from routine use, while enzymatic sequencing became sufficiently efficient to consider automation of the process. The use of an optical detector to record fluorescent DNA bands into a computer during electrophoresis (Ansorge et al. 1986; Smith et al. 1986; Prober et al. 1987; Nishikawa and Kambara 1991; Middendorf et al. 1992; Nishikawa and Kambara 1992) suggested a way of automating not only data collection but also data interpretation. Although gel preparation and sample loading are done exactly as in the standard batch technique, these automated sequencers produce a chromatogram-like output, showing bands as peaks, from which the integrated software automatically "calls" bases (Fig. 1B). Some models also reconstruct an autoradiogram-like image to simplify user control and comparison with data obtained by the batch technique.

Relative to the batch technique, automated systems achieve equivalent DNA band separation using much shorter gels, because the bands are recorded by a detector positioned in the lower portion of the gel. However, for the same reason the gels have to be run much longer than in the batch technique. The gel dimensions are also limited by having the gel plates made from optical quality glass, which precludes the use of gels with extreme dimensions in most commercial systems. Complex problems involving the interplay between the DNA labelling chemistry, separation/detection format and the peak-calling software actually place these systems outside the realm of conventional electrophoresis, and extensive comparison of the advantages and drawbacks of specific systems is therefore outside the scope of this chapter. Furthermore, active commercialisation of this format limits detailed descriptions of technical developments in the standard scientific literature. For example, published studies on the limits of these systems, such as the one made at Hitachi Optical (Nishikawa and Kambara 1991), largely emphasise the problems associated with the gel format, while reducing discussion of the subtleties of the optical system and the software to mere statements.

A different approach to improving gel resolution was direct blotting electrophoresis (DBE). In this system, a small ultrathin gel is used to separate DNA fragments which are continuously adsorbed on the moving nylon membrane at the lower edge of the gel (Beck and Pohl 1984). It took a long time to commercially develop this idea, and a likely application of this separation technique, multiplex sequencing (Church and Kieffer-Higgins 1988), was not very popular due to its complexity. Current commercial DBE devices are marketed as a "poor man's automated sequencer" in conjunction with hapten-based nonradioactive labelling and detection of DNA (Richterich et al. 1989). With radioactive label this system makes efficient use of radioactivity, gel chemicals and X-ray film space, but the overall level of automation is low compared to on-line fluorescence detection. High throughput versions of DBE are currently being developed (Cherry et al. 1994).

Early studies on the limits of the conventional separation system showed that using thermostated gels with dimensions close to 80 cm long and 0.1 mm thick (Ansorge and Barker 1984) fragments 600 nucleotide long could still be analysed with one-base resolution. Such gels, however, are rarely employed in practice and commercial devices offering this technique with 55–60 cm gels (Pharmacia Macrophor and Genomix Corp. GenomixLR) are expensive and not widely used. Studies on the limits of noncommercial automated sequencers came to the conclusion that the most important factor limiting band separation was uneven heat dissipation within the gel sandwich (Nishikawa and Kambara 1991). Similar studies led to the introduction of (Drossman et al. 1990) and research into capillary gel electrophoresis (CGE). CGE allows the use of high field gradients, as well as complete automation when coupled to a fluorescence detector (Pentoney et al. 1992). However, further research on CGE revealed certain problems, such as limited gel capacity and matrix degradation (Swerdlow et al. 1992) which had already been noticed in early studies using ultrathin slab gels (Ansorge and Barker 1984).

A primitive electrophoresis device with a horizontal gel allowing efficient heat exchange was already suggested in 1979 (Kutateladze et al. 1979), but this publication was apparently well ahead of its time since a similar device was re-invented in 1991 (Brumley and Smith 1991) and converted to the on-line fluorescence detection format (Kostichka et al. 1992). An electronically advanced version of this device was commercialized in 1997 by Genesys Technologies, Inc. A similar device, based on a capillary array cassette, is to be produced in 1998 by Molecular Dynamics. Such devices typically analyse 96 samples in 90 min, with up to 600 bases read per sample. Though they approach the practical limit of current electrophoresis technology in terms of productivity for standard sequencing applications, because of their price they are limited to specialised commercial sequencing facilities.

Separation formats, capitalising on the speed of separation (and concomitant low capacity gels), obviously do not make much sense with radioactive label which must be detected in dry gels by lengthy autoradiography or storage-phosphor technology. This factor and the limited commercial availability of affordable systems with large gel dimensions explains why gel formats similar to the one suggested by Sanger in 1978 (Sanger and Coulson 1978) are still widely used, particularly in sequencing-related techniques such as transcript mapping, studies of DNA-protein interactions, and studies of RNA processing. These methods are neither easily convertible to nonradioactive labelling nor easily automatable. Similarly, stretching the limits of sequencing gel electrophoresis with pulsed fields (Brassard et al. 1992; Mouradian et al. 1994) will only be possible when commercial devices capable of handling submillisecond pulses of $>10\,kV$ become available, but such devices are unlikely to be commonly used. Although gel electrophoresis in any polymer matrix has a built-in limit of resolution (Slater and Drouin 1992), it remains to be determined whether a different gel matrix (Ruiz-Martinez et al. 1993; Tietz and Chrambach 1993) and/or buffer system may significantly push quantitative limit of resolution in DNA sequencing, which currently is ~1000 nucleotides.

The multitude of separation and signal detection formats for DNA sequencing electrophoresis that have evolved in the past two decades makes it impossible to include all protocols in a short article. Methods of high throughput standard sequencing are constantly improving (Hawkins et al. 1992; Voss et al. 1993; Krishnan and Chaplin 1994; Favello et al. 1995; Fulton et al. 1995). The current trend is already to delegate sequencing reactions of certain kinds to a central service facility. In addition, an automated sequencer comes with a manual describing essential protocols which have relatively little room for modification, due to the balance between the gel system, optical system and the base-calling software. DBE devices also work properly only within a limited range of operating conditions, which are included in the instrument's manual. Template preparation methods for sequencing can be found in several recent papers and reviews (Voss et al. 1990; Wilson et al. 1990; McCombie et al. 1992; Pogue et al. 1993; Wilson 1993; Hengen 1995). Consequently, this chapter focuses on a three-gel system for the batch technique (i.e. manual sequencing), compiled by the author as a result of 20 years of experience. An attempt was made to eliminate a "ritual" component of the procedures by describing various equivalent versions at different stages of the process. For the same reason, a general discussion of the factors affecting the quality of separation precedes the "Procedure".

The separation system described here is useful for all experiments requiring single-base resolution of oligonucleotides 35–1000 bases long. Characterisation of longer fragments is made by reconstruction from a series of overlapping shorter fragments. The way to break down the fragment of interest into a series of overlapping blocks analysed with single-base resolution can be random (Anderson 1981; Deininger 1983; Ahmed 1987; Bankier et al. 1988; Kieleczawa et al. 1992) or non-random (Barnes et al. 1983; Henikoff 1984), and it does not always require physical separation of the blocks (e.g. by subcloning). As the prices for custom oligonucleotide synthesis steadily go down, the approach of primer walking (Voss et al. 1993) may well become the method of choice for small sequencing projects. However, for analysis of large genomic fragments, primer walking, besides being expensive, may be hampered by the presence of repetitive sequences, so that multistep subcloning of the large target fragment may be required (Wilson et al. 1992; Sulston et al. 1992).

Factors Affecting the Quality of Separation

Since the first experiments with DNA sequencing started in the 1970s, **Reagent Quality** when commercial reagent quality was variable, successful recipes coming from different laboratories have openly contradicted each other – if compared from a logical standpoint. While researchers in the 1970s often had to purify key reagents, e.g. deionise urea and recrystallize gel monomers, now these reagents are commercially available as concentrated solutions prepared from ultrapure components, or even as ready-to-pour gel mixtures containing all components except polymerisation catalyst (e.g. Sequagel-XR from National Diagnostics). The latter formulations are generally targeted for use in specialised sequencing facilities which perform many identical experiments every day. Researchers there are more likely to pay an increased price for the advantage of elimination of certain preparatory steps and stable quality. Pre-made sequencing gels were marketed by Pharmacia for a short time, but were later discontinued due to unstable quality. Purchase of large quantities of pre-mixed liquid gel reagents for infrequent use is not recommended, as the "stabilised" acrylamide solution nevertheless deteriorates faster than the solid chemical. A word of caution should be said against buying ready-for-use gel solutions at discount prices if they are not going to be used up within a short time after purchase. A good price/quality compromise for low to medium use is to purchase a 40 % acrylamide-bis-acrylamide concentrate and prepare small quantities of a ready-to-pour formulation (e.g. for ten gels), which is stable for at least 3 months at 4 °C.

Special gel kits (Singel) are available from FMC and include chemicals and water in separate compartments of a sealed plastic package. The components are mixed by removing the compartment seals and the solution is squeezed out into the gel mold through an integral cigarette-type filter. Although these kits are expensive, they provide consistent quality for infrequent users and reduce exposure to toxic acrylamide solutions.

Traditionally, denaturing gels ran at an elevated temperature, produced **Run Parameters** by dissipation of the Joule heat within the gel. Inefficient heat dissipation creates the so-called smiling effect (higher mobility in the centre of the gel plate relative to the edges), which seriously affects the quality of the separation. A simple way to overcome this problem was to attach two metal plates at both sides of the gel, which effectively redistributed the heat. This method (also called passive thermostating) is still used frequently, since it is simple and cost-effective. Another way of passive thermostating is to use a contacting upper buffer chamber as the heat sink

(BioRad Sequigene apparatus). It was suggested (Garoff and Ansorge 1981) to cast ultrathin gels in direct contact with the flat chamber (thermostatic plate), which was connected to a circulating water bath with the desired temperature during the run. This approach was implemented in a commercial device (Pharmacia Macrophor). Actively thermostated devices such as this allow use of field gradients that are about twofold higher than those in passively thermostated devices. They also allow runs to be performed at temperatures up to 70 °C, although at this temperature gel degradation can become a problem. Since the speed of the sequencing gel run shows a strong dependence on the gel temperature of about 2 % per degree Celsius (Nishikawa and Kambara 1991), insufficient or directional heat dissipation would create a temperature gradient across the gel thickness and result in excessive band broadening or even smearing. For DNA fragments >400 nucleotides competition between band separation distance and band broadening becomes the most critical factor (Kambara 1992). For the resolution beyond a kilobase the gel should be run relatively cool (40 °C) and with a field gradient of <30 V/cm, largely to minimise effects of thermal band broadening. An ideal gel therefore should be as thin as practical and be sandwiched between the two symmetrical thermostating devices (Nishikawa and Kambara 1991). However, there is only one commercial device on the market with this construction (ABI Prism 377), while the most popular way of thermostating is still to use aluminium alloy plates clamped to the gel mold. Many commercial devices feature a thermostating metal plate only at one side. While this type of construction may be efficient with ultrathin gels (<0.25 mm), it will not allow efficient heat exchange with thicker gels. In addition, such devices should be protected from strong air currents.

The best way to run passively thermostated gels is to use a power supply with a temperature feedback circuit set at the temperature of 40–45 °C, otherwise it has to be run in the constant power mode. The exact power setting in the latter case, however, has to be determined by a test run of an actual gel, measuring the gel temperature with an external thermocouple.

Gel Length and Casting Techniques

Band separation distance is proportional to the square root of the migration distance, i.e. the gel length (Nishikawa and Kambara 1991; Slater and Drouin 1992). During band migration band broadening competes with band separation, reaching a limit at a certain molecular size at which resolution is not improved by increasing gel length. Automated sequencers are characterised by the well-to-read (WTR) distance, which is typically 25–35 cm. Such a gel approximates the combined resolution range of two 50–60 cm gels run for different times in the batch technique. There are only a few commercial automated sequencers with a larger WTR distance using thermostated ultrathin gels (ABI Prism 377 and LI-COR 4000L). These are expected to go over the separation range of the batch technique (Nishikawa and Kambara 1992; Voss et al. 1992), coming close to the limit of the separation matrix.

The traditional sequencing gel casting technique imitated those used in protein electrophoresis. To prepare very long and thin gels two novel techniques were suggested (Garoff and Ansorge 1981; Barnes et al. 1983) which made use of the capillary force of thin gels. Since perception of capillary force is not intuitive , these two techniques do look unusual to the novice; for the same reason, pouring gels with extreme dimensions (such as 0.1 mm thick and 1 m long) using these techniques is not a common practice. This fact, rather than any theoretical considerations, explains why most researchers (as well as device manufacturers) working with the batch technique use gels of compromised dimensions, such as 0.3–0.4 mm thick and 35×43 cm large, fitting onto a common X-ray film. However, only about two thirds of the film surface contain usable data after exposure in this case (in the upper part the bands are not resolved). Using a short gel for reading past 400 bases in manual sequencing is very inefficient, requires multiple loadings of the same sample and in the end only about half of the film area contains usable data. However, on the market there are also a few holders for longer gels (BioRad Sequigene, IBI Base Runner and others), as well as large gel dryers, X-ray films in roll format and oversized exposure cassettes.

Gel Capacity

Sequencing gel capacity was never, to the best of this author's knowledge, defined in quantitative terms, as every sequencing track resolves >1000 bands and the concentration of DNA in a band was always considered to be too low to be taken into consideration. The problem of gel capacity is serious in relatively exotic sequencing techniques, such as genomic sequencing (Church and Gilbert 1984) and to a lesser extent in multiplex sequencing (Church and Kieffer-Higgins 1988). However, the problem of capacity of ultrathin and capillary gels was always underestimated, particularly since studies of the limits of automated sequencers, promoted and supported by the onset of genome projects, largely dealt with specific sequencing chemistries producing an excess of labelled fragments and used ultrapure template DNA. Linear amplification sequencing (Murray 1989; for a recent discussion see also Fulton and Wilson 1994) greatly increased the area of usability of automated fluorescent equipment (which now includes a CGE sequencer ABI Prism 310), because it largely eliminated problems associated with the necessity to use large amounts of ultrapure template DNA in fluorescent sequencing chemistries. It did not, however, eliminate the problem of the quantitative balance between gel capacity and detection sensitivity. Thus users of CGE equipment are likely to encounter problems associated with gel capacity, such as overloading with template (Swerdlow et al. 1992). One can also observe the result of overloading with template in conventional gels in direct sequencing of very large DNA, such as bacterial chromosomal DNA.

Choice of Gel Composition and Concentration

Two papers describing DNA sequencing methods also suggested two gels systems derived from Peacock and Dingman (1967). As the chemical sequencing method required separation of fragments ideally from a dinucleotide, 20 % gels were suggested (Maxam and Gilbert 1977); for primer-based methods, 10 % gels were more appropriate (Sanger and Coulson 1975). These two systems were slavishly reproduced by many researches and have also evolved over time into many slightly different "religions". If one neglects oligonucleotides two to approximately 35 bases long, more data can be read from a single run on a 5–6 % gel with a field gradient (Biggin et al. 1983; Ansorge and Labeit 1984; Olsson et al. 1984; Sheen 1988). Such a gel resolves more bases with increasing length, with a compromise length of 50 cm resolving 300±20 bands in an approximately 3.5 h run. A modified field (electrolyte) gradient gel (Sheen and Seed 1988) is described as "standard" in the "Procedure" section, as it indeed can be used for most separations.

Resolution beyond 400 bases requires a substantial increase in separation length and time using gels with T<5 % (Ansorge and Barker 1984; Nishikawa and Kambara 1991; Nishikawa and Kambara 1992). These types of polyacrylamide gels are very fragile and difficult to handle in the batch technique, but they can be used in automated equipment and are standard for DBE. In the two latter techniques the gel is covalently bound to one or both glass plates during polymerisation. Introduction of a new gel matrix, Long Ranger (FMC), allows for an improved resolution beyond 400 nucleotides. From a number of geometries possible for such a gel, an 85 cm gel of 5 % Long Ranger, conveniently run overnight, typically resolves fragments 350–750 nucleotides long, with minimum overlap with the standard gel. Combined data from the two gels typically yield at least 700 bases of contiguous sequence from a single sample, while with an increased run time up to 1000 bases can be read (but the overlap with the standard gel is lost). With a 15 h run time only the lower part of this gel needs to be exposed to a standard film and the entire film surface contains usable data. It should be stressed here that the common practice of multiple loadings on a relatively short gel is wasteful, particularly when the separation times differ less than fourfold, because the band separation distance is dependent on the square root of the band migration distance and the mobility of any given fragment may be considered a linear function of time (Nishikawa and Kambara 1991).

Fragments with residual secondary structure, appearing as an abnormal distance between bands (band compression), often cannot be resolved on gels containing urea (such as the two gels described above). Although the use of formamide resolves most compressions, the DNA mobility is greatly reduced and the gels are difficult to cast and handle. Therefore the use of a formamide-urea composite gel as a complementary approach is suggested. For this gel using 8 % Long Ranger is recommended, as it provides for a relatively fast run compared to polyacrylamide gel. The band spacing in this gel is increased relative to the standard gel, allowing for easier identification of partially compressed regions. It is advisable to design special primers so as to place the compression region within 250 bases from the primer.

Buffer Choice

The buffer system used in the following procedures was first suggested by Anderson (1981). Compared to the original (Peacock and Dingman 1967) Tris-borate buffer, this one (HTBE, high-pH Tris-borate-EDTA) has a higher pH ($\sim$9.0 instead of 8.3) and molarity, which results in an apparently higher capacity and sharper bands in long runs.

Note: A certain amount of confusion has been created over time with respect to concentrated buffer solutions. For example, a 0.89 M Tris-borate buffer is often referred to as 10×, while the different gel recipes may use a concentration of 0.5–1.0×, and the tank buffer may have the same concentration as in the gel or lower. The buffer concentration in the gel affects the optimal power that has to be applied to the gel, so that mistakes in this area may result in gel overheating and breaking of the glass plates. In this chapter, all concentrated buffer solutions are referred to as 20×, while the gel usually contains a 2× dilution and the tank buffer a 1× or a 2× dilution.

HTBE is compatible with all three gels suggested here, and there is no need to use original TBE in short runs, where the two buffers deliver practically the same results. A twofold difference in the buffer concentration in the gel and the running buffer speeds up the short run and reduces costs, but for long runs a uniform buffer concentration is recommended. For sequencing kits containing high concentrations of glycerol in the sequencing reactions, an alternative buffer (TTE, Tris-taurine-

EDTA), is marketed as glycerol tolerant buffer by USB/Amersham. In the author's experience it is not compatible with either salt gradients or long runs. A discontinuous buffer system (Carninci et al. 1990) results in somewhat faster runs, however, experience shows that the gain in resolution is small compared to the use of a different gel matrix, i.e. Long Ranger. While the Long Ranger can also be used in short runs, it does not provide an increased resolution relative to acrylamide gel with a field gradient (the standard gel), but it is more expensive than an equivalent acrylamide concentrate.

Sample Well Geometry

Two kinds of sample well molds are in use. The rake comb should be inserted in the gel during polymerisation and removed before loading the sample. The second kind, the shark teeth comb, consists of two parts. The part with the flat surface is inserted in the gel during polymerisation and forms a single wide slot. The second part has sharp plastic teeth and is inserted into the flat gel surface shortly before loading the sample. The sample is then loaded in the space between adjacent teeth. The shark teeth comb is preferable, since it uses the maximum amount of gel space and results in straight bands with all known sequencing chemistries. The rake comb type is not recommended for samples with the T7 DNA polymerase (Sequenase) chemistry, containing relatively high salt in the sequencing reaction, since the salt distorts the bands of the faster moving oligonucleotides, resulting in a cone-like appearance of every track. This effect is not observed with the shark teeth comb with the exception of the two outermost tracks. To eliminate this edge effect it is recommended to apply two additional tracks containing residual sample or a sequencing buffer mixed with an equal volume of the formamide/dye loading buffer. Some sequencing protocols suggest desalting the sequencing reaction by ethanol precipitation, which indeed eliminates band curvature, but this step may become extremely cumbersome with multiple samples. An optimal well width for manual band reading is about 4 mm (Ansorge and Barker 1984); for automatic reading and/or digital image processing the wells can be as small as 1 mm (Middendorf et al. 1992).

Loading Devices

Generally, a sequencing sample with a volume of 1–5 µl can be loaded with a standard piston pipette (e.g. Eppendorf or Gilson) fitted with a flat plastic tip corresponding to the gel thickness. For a long time this loading technique required that the pipette tip was actually inserted into the well. Loading a gel thinner than 0.3 mm in this way is quite slow and requires particular skill. Recently, with the introduction of cycle sequencing chemistry, which produces a relatively large amount of sequencing products, another technique has been gaining popularity. Sample can be loaded with a thin tip (1–10 μl), which is not actually inserted into the well but used to slowly inject a compact bubble of the sample into the well (see "Procedure"). This technique is very rapid and can be performed with multichannel pipettes from microtiter plates, making it possible to load 48–64 tracks with ease. A higher productivity can only be achieved with a robotic device, of which only a few are currently on the market. Apart from a comb sequencing principle, recently introduced by Pharmacia, relatively little has been done to automate sample loading in automated sequencers using slab gels. By contrast, the first commercial CGE sequencer (ABI Prism 310) already features a fully automatic sample loader.

Composition of the Three-Gel System

The system of three different gels (Table 1) allows unambiguous manual sequencing with read lengths equivalent to the best models of automated sequencers currently on the market. It consists of two gel formulations

Table 1. Composition and running parameters for the three-gel system[a]

Type	Composition	T%	C%	Gel buffer	Tank buffer	Dimensions	Power input	Run time	Range (nt)
Standard	Acrylamide/Bis	5	5	2× HTBE	1× HTBE[b]	300×500× 0.4 mm	55 W[c]	3.5 h	35–350[d]
Long run	Long Ranger	5	–[e]	2× HTBE	2× HTBE	300×850× 0.4 mm	70 W[c]	15 h	350– 750[d]
Formamide	Long Ranger	8	–[e]	2× HTBE	1× HTBE	300×500× 0.4 mm	50 W[c]	4 h	20–250[d]

[a] Although gels, containing urea and formamide, electrolyte gradient gels, as well as HTBE buffer have been published previously, this combination is original. All gels contain 7 M urea, unless specifically stated otherwise (see "Gel Stock" Solutions).
[b] Lower tank buffer contains 0.8–1 M sodium acetate on 1× HTBE to create an electrolyte gradient, which results in the reduced distance between the first 100 bands about twofold relative to that in a homogenous buffer system.
[c] Two aluminium plates must be attached to the gel for heat distribution.
[d] Resolution range is listed for the autoradiography of dry gels to a single-sided emulsion, i.e. Kodak Biomax MR. The use of imaging screens will result in reduced effective resolution range.
[e] The gel is prepared from a proprietary 50 % concentrate with unknown cross-linker concentration.

with different gel lengths, providing two overlapping separation ranges, and an auxiliary gel for resolution of compressions. Though we suggest a thickness of 0.4 mm, our recipes can easily be adapted for thinner gels, allowing for an up to 50 % increase in run speed. In the author's practice, which includes teaching the technique to undergraduate students, the use of ultrathin gels for the batch technique with subsequent autoradiography did not provide any advantage apart from an increase in separation speed; but such gels are more difficult to cast and handle for a novice. They are, however, highly recommended for those setups that use active thermostating and/or do not require post-electrophoretic gel handling, such as for certain automated sequencers. Ultrathin gels may be also advantageous for reading in the region between 750 and 1000 bases.

Outline

Outline of the Procedure

1. Prepare gel and buffer stock solutions for storage.

2. Wash and treat gel plates with silane, assemble the gel mold.

3. Prepare the working gel solution. Cast the gel.

4. Mount the gel in the electrophoresis apparatus, perform pre-electrophoresis.

5. Load samples.

6. Perform electrophoretic run.

7. Treat the gel for autoradiography and expose to film.

8. Process data.

9. Regenerate the glass plates and dispose of the waste.

Materials

Equipment

Choosing the Type of Equipment

As discussed in the Introduction, there are three principal systems of sequencing gel electrophoresis available on the market, that for the batch technique, automated sequencers and DBE devices. Within the first group there are many models, with prices (including power supply) ranging from $5000 to $15,000; prices in the second group range from $60,000 to $250,000. There are a few DBE devices on the market and these typically cost around $15,000. The choice of equipment depends mostly on the type of experiment in which it is going to be used. Although an extensive discussion is not possible here, certain important points can be considered:

- Does the experiment require only the separation image, or is subsequent handling of the nucleic acid fragments necessary?

- What type of labelling and detection of the target nucleic acid is used in the experiment?

- How many identical separations per unit time are foreseen?

- Is handling of radioactive materials allowed in the experimental facility?

While the first two issues deal with the design of the experiment, the other two relate to cost/safety considerations. Standard sequencing experiments are indeed only limited by the amount of radioactivity one can handle at a time, but the current trend to cut down on the use of radioactive materials in biological experiments still may come in conflict with the experimental design and/or running cost considerations. Several laboratories usually share an automated sequencer but retain standard batch technique equipment for nonsequencing applications or use a DBE device. DNA sequencing technology (and the equipment market) has reached a complexity comparable to that of imaging or sound recording. Thus, while considerations of a theoretical nature may aid in initial selection, at the final stage it may be a good idea to get the device for a 2 week trial in the lab before making the actual purchase. This is certainly possible with automated sequencers, but may also apply to expensive pieces of batch technique equipment.

Aside from standard sequencing, sizing of DNA fragments is required in techniques, such as transcript mapping or footprinting of DNA-protein complexes, that are still performed using radioactive labelling and batch technique equipment. Most automated sequencers currently on the market come with software that does not allow automated analysis of applications other than sequencing (and even then, sequencing only according to a limited number of sequencing chemistries); they do, however, allow fragment sizing (fragment analysis) with a separate software package. It should be also kept in mind that certain genetic analysis applications which currently involve direct DNA sequencing (such as point mutation detection) may soon be replaced by a different methodology and respective devices, such as DNA chips (Drmanac et al. 1989; Drmanac and Drmanac 1994), which do not require electrophoresis at all.

Safety Considerations

Below a minimal set of electrophoretic equipment, most of which can be made in a workshop for a relatively low cost, is described. In the author's opinion, this set provides the highest feature/cost ratio possible without compromising the safety of the experiment. It should be stressed here that one does not necessarily need expensive equipment to perform DNA sequencing electrophoresis. However, the importance of safety cannot be overemphasised in DNA sequencing, since this technique employs high-voltage power supplies, hazardous chemicals and radioactive materials. In most countries a special licence is required for personnel to carry out these procedures. Safety is particularly important to consider when one decides to build his/her own system at a low cost. While it is certainly possible to make a cheap basic power supply, it may lack such features as constant power output control and leakage detection. When cost is the problem, it is recommended to buy the best stand-alone high voltage power supply one can afford and make the gel holder and glass plates in a local workshop. Nonetheless, expensive sequencing equipment usually features increased safety, so it is certainly worth the money if it is to be used extensively and particularly in teaching institutions. It is also a good idea, not only for safety reasons, but also for stable separation quality, to place the gel holder in an enclosure fitted with a safety lock which would cut off power when opened. Equal attention should be paid to good quality power leads and connectors and proper grounding of the equipment. In the existing equipment these parts should be periodically checked for leaks, burns and other damage and replaced properly.

Note: Insufficient attention to these details can result in lethal electric shock or set a fire (apart from ruining an expensive experiment).

Minimal Equipment Setup

- A pair of glass plates, 4 mm thick (for sizes see Table 1). Gel plates, if they do not come with the gel holder, should be ordered cut from floated soda lime glass. To use the gel casting technique of sliding, one plate has to be 1 cm wider than the other. One plate should have a cut-out ("ears") allowing contact of the gel with the upper buffer compartment.
- A pair of aluminium alloy plates, 2 mm thick, somewhat smaller than the glass plates (e.g. for a 30×50 cm glass plate, a 28×40 cm aluminium plate)

- Plastic spacer and comb set, 0.4 mm thick (e.g. Bio-Rad)
- Gel holder, commercial or home made
- High voltage power supply delivering at least 100 W, 2.5 kV and 50 mA; should allow running in constant power mode
- Glass plate drying rack (e.g. FMC or home made)
- Electronic thermometer with an external sensor, or temperature monitoring strips (FMC)
- Large plastic tray, e.g. 60×80 cm
- 8–10 large binder clamps
- Piston pipette, e.g. Gilson P2 or P20, with micro tips or flat end tips
- Thermoblock or water bath for 85 °C
- Large gel dryer, e.g. BioRad model 583 or Hybaid Gel-Vac
- Plastic wedge spatula and hard toothbrush
- Bubble remover stainless steel wire (optional)
- X-ray film cassette, 33×45 cm or 30×50 cm
- Large hair dryer or a drying oven

Reagents and Solutions

All reagents in this section are prepared from ultrapure or electrophoresis grade chemicals and deionised water. Additional precautions may be necessary for fluorescent sequencers using visible light fluorescence. In this case the use of reagent grade water (e.g. MilliQ system from Millipore Inc.) is recommended, as it effectively counteracts seasonal water quality fluctuations which may seriously affect the performance of the instrument in certain locations.

HTBE buffer, 20× concentrate   **Buffers**

Tris-base 162 g
Boric acid 27 g
EDTA di-sodium salt 10.2 g
Water to 1000 ml

Dissolve on a stirrer. The pH is ~9 (not adjusted). Store at room temperature.

TTE buffer, 20× concentrate

Tris-base 215.6 g
Taurine 71.3 g
EDTA di-sodium salt 4.77 g
Water to about 800 ml

Dissolve on a stirrer, adjust pH to 8.5 with taurine and bring volume to 1000 ml. Store at room temperature (turns yellow over time).

TBE buffer, 20× concentrate

Tris-base 108 g
Boric acid 55 g
EDTA di-sodium salt 7.4 g
Water to about 800 ml

Dissolve on a stirrer, adjust pH to 8.3 with boric acid and bring volume to 1000 ml. On storage a white precipitate forms in the solution, which can be avoided by adjusting pH to 8.9 (Mayeda and Krainer 1991), or by preparing a 5× concentrate instead of a 10×.

Gel Monomer Concentrates

General purpose stock solution: 38 % acrylamide/2 % methylene bis-acrylamide, w/v (T=40 %, C=5 %), BioRad or Gibco-BRL. Store at 4 °C. Long Ranger, 50 % concentrate (FMC). Store at room temperature.

Gel Stock Solutions

Note: Gloves should be worn at all times in preparing and using these solutions!

Standard gel stock solution (T=5 %, C=5 %)

210 g urea (ultrapure)
62.5 ml 40 % monomer concentrate
5 g AG501×8 resin (BioRad), optional
Water to 450 ml

Stir to dissolve (about an hour); filter through a Nalgene 0.45 μm cartridge and add 50 ml 20× buffer. Store in the cold room wrapped in aluminium foil. For daily use, store at room temperature. To polymerise, add 0.5 ml of 10 % ammonium persulfate (see below) and 60 μl TEMED per 100 ml cold gel solution, mix well on a stirrer, and cast the gel immediately.

Note: Acrylamide is a neurotoxin!

Long run gel stock solution (T=5 %)

210 g urea (ultrapure)
50 ml Long Ranger concentrate
50 ml 20× HTBE
Water to 500 ml

Stir to dissolve, filter through a Nalgene cartridge, store as described above. To polymerise, add 0.5 ml of 10 % ammonium persulfate and 60 µl TEMED per 100 ml cold gel solution, mix well on a stirrer and cast the gel immediately.

Note: Long Ranger is a neurotoxin!

Formamide gel stock solution (T=8 %)

210 g urea (ultrapure)
80 ml Long Ranger concentrate
50 ml 20× HTBE
Formamide (deionised) to 500 ml

Warm to 37 °C and stir to dissolve urea (takes a long time), filter through a Nalgene cartridge, store at room temperature (storage longer than a week is not recommended). To polymerise, add 0.5 ml of 10 % ammonium persulfate and 250 µl TEMED per 100 ml gel solution, prewarmed to 37 °C, and mix well on a stirrer. The solution is very viscous and may be difficult to pour into mold.

Note: This solution is toxic and may cause skin cancer!

Gels for automated sequencers are prepared as described above, but with a different T, depending on the WTR distance. As a guide, for 25 cm WTR distance an optimal T=6 %; for 35–45 cm, 4.5–5.5 %; for 50–60 cm, 3.5–4.5 %. Longer reads are achieved with gels having an acrylamide/bis-acrylamide ratio C=2.5–3.5 % (such as in stock solutions marketed for protein electrophoresis by Bio-Rad), as well as with stock solutions having a proprietary cross-linker, e.g. Long Ranger, from FMC, or RapidGel, XL, from Amersham. For DNA sequencers with relatively thick gels (e.g. ALFexpress) it may be necessary also to use TBE buffer instead of HTBE. For other details it is recommended to consult the instrument's manual. **Gels for Automated Sequencers**

Gels for DBE are prepared as described for the standard gel with T=4 % (27 cm WTR) or 3.5 % (50 cm WTR). Longer reads are achieved with gels **Gels for DBE** having an acrylamide/bis-acrylamide C=2.5–3.5% (such as in stock solutions marketed for protein electrophoresis by Bio-Rad). It may be necessary to prepare a gel stock solution and perform a test run, after which the belt speed may be adjusted (±15%). For other details it is recommended to consult the instrument manual.

Auxilliary Gel Preparation Reagents

Note: All reagents in this section (except fixing solution) should be prepared and used in a fume hood; gloves and protective goggles are to be worn at all times!

- Ammonium persulfate (APS) solution: Dissolve 1 g APS (BioRad) in 10 ml of water, dispense in 1 ml aliquots and store frozen at –20 °C.
- TEMED (BioRad), stored at 4 °C.

Note: Different brands may affect polymerisation time. It is recommended to test a new bottle and adjust amount accordingly (more reagent, faster polymerisation).

- Fixing solution: 10% acetic acid; 10% ethanol (v/v). Both can be of technical grade. Store at room temperature.
- Gel binding solution. 1% γ-methacrylpropyloxysilane (Sigma) in acetone. Store at 4 °C.
- Gel repellent solution. 5% dimethyldichlorosilane (Sigma) in carbon tetrachloride or in dichloromethane. Store at 4 °C.
- Gel loading solution: 90% deionized formamide; 25 mM EDTA, pH 8.0; 0.1% bromophenol blue (for manual sequencing) or methyl violet (for fluorescent sequencing); 0.1% xylene cyanol (optional, only for manual sequencing)
  It is recommended to prepare a 2% stock solution of the bromophenol blue (sodium salt) first and filter it through a 0.22 µm cartridge. Deionised formamide can be stored indefinitely in 0.9 ml aliquots at –20 °C. To prepare the working solution, mix this aliquot with 50 µl 2% dye stock and 50 µl 0.5 M EDTA, pH 8.0.

Additional Items for Gel Drying and Auto-radiography

- Thick chromatography paper (i.e. Whatmann 3MM)
- Thin aluminium foil or a polyvinylchloride film (e.g. Saran Wrap, Dow Chemical Co.)
- X-ray film, preferably BioMax MR (Kodak) or Betamax (Amersham) and respective development kit

◾ Procedure

Note: Please read these instructions thoroughly and completely before performing an actual experiment, since appropriate tests may be necessary in advance of a real separation, particularly when using new equipment!

Cleaning and Care of the Glass Plates

Good results in sequencing electrophoresis can only be obtained using special floated soda lime glass plates. The plates should be maintained with great care, particularly avoiding scratches and small fractures. The plates should be stored on a special rack, or in a drawer, interleaved with paper towels. It is also highly recommended to rinse the plates with water immediately after each use, so as not to allow drying of residual buffer and gel; harsh cleaning procedures are not usually required. For cleaning the plates for manual sequencing and DBE we recommend Deconex 11 Universal (Merck). The use of dishwasher detergents for manual sequencing is acceptable, however, they carry the risk of introducing unwanted additives into the gel; a particular detergent must be tested to be determined suitable for manual sequencing. An additional problem with these products is that they are often altered by the manufacturer (e.g. by adding creams for hand protection) without clear notice, and in many countries the regulations do not require a clear listing of the components. For automated sequencers special detergents are recommended by vendors of the instruments. This does not mean that other detergents are unsuitable, but that the user may try those only at his/her own risk. One should use only a few drops of the detergent per plate, distribute them evenly with a soft brush, allow to soak for a few minutes, then rinse with plenty of running water, followed by a rinse with deionised water and finally with a generous amount of ethanol from a squirt bottle. The plates should be dry within a few minutes.

There are several versions of glass plate treatment, depending on the technique and the apparatus used. It is recommended to clearly mark the surfaces of the glass plates that will be in contact with the gel (e.g. with a piece of adhesive tape on the opposite side) and keep to this selection.

Special Treatments of Glass Plates

Glass Plate Treatment Before Gel Casting

Drying Gel on Paper Backing

This technique is the simplest but requires a gel dryer to prepare the gel for autoradiography. The plate with a cutout is treated with a gel-repelling solution (see "Reagents"), so as to ensure that the gel stays flat on the other plate after opening the sandwich.

1. Place the plate on a horizontal support in a fume hood and pour a small amount of the repelling solution in the middle of the plate. Distribute evenly with a Kleenex tissue and let dry.

2. Polish the plate with a lint-free tissue to remove oily spots. Rinse with ethanol and let dry.

If the plates are not meticulously clean, this technique fails even though the repellent layer was present, the gel partially adheres to both plates and is torn upon taking the glass plates apart. There is no safe bet to avoid this other than taking great care of the plates from the very beginning. There are commercial repellents that are less toxic (Sigmacote, Sigma, or Gel-Slick, FMC), however, they are expensive. Recently, a number of inexpensive nontoxic household items for this purpose was suggested, including furniture polish, car windshield wiper fluid and cooking oil spray (Hengen 1995), but they may not be available in all countries.

The repellent should be applied before every run, even though the layer is clearly stable for several runs. While some manuals recommend treating both plates with the repellent solution, in the author's opinion it only complicates gel casting.

Preparing Ultrathin Gels

Ultrathin (<0.3 mm) gels are handled very easily if covalently bound to the glass plate during polymerisation (Garoff and Ansorge 1981). To achieve this, one of the plates is treated with the gel binding solution (see "Reagents") by gently wiping the glass plate **once** with a lint-free tissue wetted in the solution. After the solution has dried, the plate is rinsed with ethanol and polished with a lint-free tissue. The other plate is treated with the repelling agent as described above. After opening of the plates the gel will stay attached to the plate treated with the binding reagent and can only be removed by special treatment (see "Regeneration of the Glass Plates"). This technique can also be applied to 0.4 mm gels; while a gel dryer is not needed to prepare the gel for autoradiography, a large plastic tray (to accommodate the gel) and a shaker will be required. Although this technique indeed yields maximum possible clarity of the autoradiographic image, it is not very practical with long gels, and an improvement in band sharpness it offers over drying the gel on paper is marginal.

Note: Do not use the same pair of glass plates alternatively for the gel dryer technique (as above) and for binding the gels, as this bears a high risk of binding the gel to glass when one does not need it.

In DBE both plates are treated with the binding reagent to ensure good well geometry and a flat lower edge of a low percentage gel. In automated sequencers the treatment may be restricted to the slots area. It should be noted, however, that in the latter case care should be taken to avoid spreading of excess reagent, as upon drying it forms a grease-like substance. This is particularly important with automated sequencers using thin glass spacers (e.g. ALFexpress), as an inadvertent treatment of the spacer makes it impossible to open the gel sandwich without breaking the spacer. Gels sandwiches of this type are easy to cast but are somewhat difficult to open without a special plastic wedge and a certain amount of practice.

DBE and Automated Sequencers

Gel Casting

In the author's opinion, this technique is almost universal and allows easy casting of gels with thicknesses from 0.4 mm down to 0.05 mm and a length of up to 1 meter. However, one needs bench space of about a 1 m², which should be reasonably level. The glass plate without the cutout is placed on an elevated surface, such as a plastic box, in the middle of a large tray (Fig. 2, upper left). Spacers are fixed to the sides with six to eight large binder clamps, so that the clamps protrude about 5 mm into the spacer. The distance between the clamps should allow the other, slightly narrower, glass plate to rest on the spacers and slide in between the clamps. The gel solution is mixed with APS and TEMED and slowly drawn into a 50 ml plastic syringe. A small portion of this solution is spread evenly at the front edge of the glass plate, so that it forms a ~5 cm spread across the entire plate width (Fig. 2, upper right). Using the right hand, the second glass plate is positioned (with the flat end forward) on the front edge of the lower plate with about a 10 cm overlap and at about 45° angle. It is then slowly lowered (like closing a book) until it rests on the spacers. Using the left hand, the gel solution is slowly injected in front of the upper plate, taking care not to introduce bubbles. Enough of the solution is added to form a thick boundary at the edge of the upper plate (Fig. 2, lower right). The upper plate is then slowly moved forward, watching the gel solution being sucked in between the plates. More solution is added and the movement continued until the two plates almost

Casting Gel by the Sliding Technique

Fig. 2. Casting sequencing gel by the sliding technique (*clockwise from upper left*). For explanation, see text

completely align, leaving about 3 mm offset to allow good contact with the lower buffer chamber. The binder clamps are moved so as to fix both plates together; however, the clamps should still reside on the spacer area, otherwise the gel will be distorted and will suck up air bubbles upon removal of the clamps. Any bubbles, if present, may at this point be removed using a V-shaped stainless steel wire (offered by some manufacturers as a "bubble buster"). Finally, the rake comb or the flat part of the shark teeth comb is inserted in the cutout, taking care not to introduce bubbles, and fixed with two binder clamps, which should not protrude too far (Fig. 2, lower left). The flat part of the shark teeth comb or the teeth of the rake comb should only be inserted to a depth of 4–5 mm, otherwise the sample loading may be extremely complicated. This is particularly important with gels thinner than 0.3 mm. Wipe dry all gel solution from the outer surface, as it may save you time cleaning it after polymerisation.

The gel sandwich is completely assembled, using side spacers and either binder clamps or special "rails", fixing the mold at the sides; the bottom side is left open. No taping or greasing is necessary. The sandwich is placed on a horizontal surface (for gels 0.4–0.5 mm thick) or inclined at a small angle (for gels 0.1–0.25 mm thick) and the gel solution is slowly injected from the cutout end, so as to fill the mold completely across its width. The cutout area is then kept filled with the gel solution, as it is being sucked into the mold. It may be necessary to tap the glass plate with a finger to avoid formation of small air bubbles and to maintain an even front of the solution progressing towards the bottom. As the solution reaches bottom, the gel is placed horizontally (where applicable) and the comb is slowly inserted and fixed with two binder clamps. For 0.1–0.2 mm gels a special pressure plate may be necessary to achieve even thickness around the comb area. This technique will certainly not work with less than perfectly flat and clean glass plates; however, it is almost always recommended for automated sequencers, which would have not properly worked without such plates anyway.

Casting a Gel by the Injection Technique

In this technique the gel mold should be assembled and perfectly sealed at the three sides, leaving only the cutout side open. The gel solution is introduced from the cutout along one of the spacers, taking care to maintain an even stream of liquid and avoid bubbles, while holding the mold at about 45–60° or even vertically. As the mold is filled completely, the comb is inserted and the gel is left to polymerise in a horizontal or a vertical position. These gels are notoriously prone to leaks, though there are many ways of sealing the mold at the sides. However, pouring the gels thinner than 0.3 mm and longer than 40 cm by this technique is problematic. Many companies offer various gadgets and tapes to make a reliable seal but none of these comes cheap.

Casting a Gel by the Traditional Technique

Polymerisation

The gel polymerisation can be observed as an appearance of a refractive zone bordering the comb when the gel is viewed at a small angle. Circular refractive zones also form around occasional air bubbles. The gel will typically be set within 30–45 min (for polymerisation problems see "Troubleshooting"). If the gel is to be left overnight (for convenience), it

Checking the Gel for Polymerisation

should be wrapped at both ends with a cling film to prevent drying. Alternatively, it can be placed in a large plastic bag with the end folded over. There is no need to place it in the cold room, doing so may even induce urea crystallisation in the gel.

Running the Gel

Preparing the Gel for the Run

1. Check that there are no unnoticed air bubbles in the gel as well as no unpolymerised areas adjacent to the spacers. These rather often form with the Long Ranger as well as with low percentage gels in general and may distort band migration due to local shunting of the voltage. In addition, they may cause buffer drain in long runs. It is easy to block these channels by pushing the gel a little using a piece of plastic to make the gel adhere to the spacer.

2. Add a little running buffer to the cutout and slowly remove the comb with a gentle rocking motion. Some force and patience may be required, which indicates a good fit of the comb. Immediately flush the well(s) with the running buffer from a small syringe, otherwise unpolymerised acrylamide may distort the wells. At this point it is recommended to examine the wells (particularly of the rake comb) for the absence of gel pieces. Even a small difference in thickness between the comb and the spacers (or insufficient strength of the clamps) may result in gel polymerisation between the comb and the glass plate. Such a gel is almost certainly lost, and it is better to make it again, after having corrected the primary cause (e.g. find a matching pair of spacers and combs). If the well(s) appear to be clean, proceed to the next step.

3. Check that the outer surface of the glass plate is clean and dry, rinse under the tap and dry with paper towels, if necessary. Dirt and gel residues on the outer surface may affect heat distribution and distort band migration. Two aluminium alloy plates are attached at both sides of the gel, and the sandwich is fixed in the gel holder (Fig.3 A). Running buffer is added into both compartments and the upper gel surface is inspected for the absence of air bubbles. These can be removed by flushing with buffer from a small syringe. If the comb area was treated with a binding agent, the gel can be left at this point for a long time, even overnight. In other cases, it is advisable to start the run within 30 min, otherwise the gel surface may swell and the wells become distorted.

Fig. 3a, b. a Electrophoresis setup in the author's laboratory. The power supply, as well as the upper buffer chamber of the 85 cm gel are fixed to the rails, mounted in a fume hood, and equipped with a safety lock. **b** Loading sample into the wells of a shark teeth comb. The *thick black line* at the bottom is the buffer tank gasket

1. Before starting the run, check the gel and the holder for leaks, since the high voltage used in the system can easily lead to a short circuit, sparking and burning of the plastic parts. If small leaks are detected in the gasket of the upper buffer chamber or in the gel itself (between the sides of the gel and the spacers), the buffer should be drained, the cause of the leaks eliminated and the system assembled again.

Setting the Running Parameters

Program the running parameters using the data from Table 1.

Note: These parameters are valid only for a given gel volume and buffer system! If your actual gel is smaller, it may overheat and the gel plates may break. The exact power setting for a new gel holder should be determined by a test run and monitoring of the gel temperature, which should not exceed 45–50 °C. For a test run set the power supply in the constant power mode, using as a guide 0.8 Watt per ml gel volume, set the voltage for 2.5 kV and the current for 50 mA. Start the power supply and monitor the gel temperature after 10 and 30 min. When using a microcomputer controlled power supply that allows for a temperature feedback (from an external probe), set the temperature to 45 °C and the power output to 1 Watt per ml gel. Do not place the probe (or temperature monitoring strips) in the middle of the gel, as these objects may locally affect heat dissipation and cause band distortion. Once a power setting has been determined, it is recommended to always run the gel in the same place, protected from strong air currents.

A leaking upper buffer chamber is not unusual even in an expensive automated sequencer, as some of them still employ a variation of a Studier type holder. In some devices the upper chamber is glued to one of the glass plates. This setup usually does not leak; however, it complicates gel cleaning and casting, particularly if the gel is large. It often helps changing and/or modifying the gasket of the upper chamber to prevent leaks. Modern power supplies can often detect a current leak during the run and shut down the voltage, but the run will nevertheless be lost if the gel run was unattended. In order to anticipate leakage it is recommended to set a constant power mode with a crossover voltage limit only 100–200 V above the starting value. This prevents the power supply from skyrocketing the voltage if the buffer from the upper compartment drains and the current is reduced to almost zero. At voltages above 2 kV such an event could easily start a fire. For the same reason it is risky to run gels at critical temperatures (e.g. above 60 °C, sometimes recommended to increase the denaturing effect) unattended. It is a general safety strategy to run gels relatively cool (45 °C) and to resolve compressions using a separate composite gel, containing urea and formamide, at the end of the sequencing project.

Pre-running the Gel

Previously, it was believed that pre-electrophoresis of 10–20 % sequencing gels for 1–4 h was necessary to remove charged substances formed during polymerisation, which affected the separation of the fastest moving oligonucleotides, as well as to pre-heat the gel to the desired temperature. As sequencing gels with T>10 % are no longer widely used for standard sequencing applications, pre-runs are considered necessary for as long as the system needs to achieve a constant temperature.

Start the pre-run using previously determined or pre-programmed conditions. Control the temperature of the gel when using DBE or an automated sequencer. In the case of the standard gel described here, a short run of about 10 min was found to be sufficient. During the pre-run sample data can be entered into a log file and the samples prepared for loading.

Note: The shark teeth comb is not inserted during the pre-run.

Ideally, a dry desalted sample from an enzymatic reaction should be dissolved in the gel loading buffer. It is also possible to mix loading buffer with the reaction in the ratio of 1:1 or 1:2 if the resultant salt concentration is <50 mM.

Sample Loading

1. The samples should be heated at 85 °C for 3 min and then placed on ice.

2. After the pre-run time has elapsed, the power is shut down and the wells are flushed with running buffer to remove urea leaching out of the gel. The shark teeth comb is inserted at this time and the samples are loaded within 10 min (Fig. 3B). If quick loading is not possible, about half of the gel can be loaded and the run started for about 5 min, during which time the samples compress into a thin line at the bottom of the well. The power is shut down, and the remaining wells can then be flushed again and the loading completed.

Note: Samples containing noncycled polymerase reactions (1:1 template to termination product ratio) should be loaded shortly after denaturation; for repeated loadings they should be denatured again. In contrast, samples from cycle sequencing (about tenfold excess of products over template) require only a single denaturation. This can be done on another day, but the samples have to be heated up to room temperature before loading. Samples containing 7-deaza-GTP should not be excessively denatured, as they are degraded by more than a few minutes incubation at or above 85 °C.

Typical run parameters are shown in Table 1. During the first hour of a run the gel temperature and/or current should be monitored to exclude the possibility of a leak, overheating or sparking. Afterwards the gel can be run without surveillance using a timer. Long runs are usually started at about 16.00 to be finished early the next morning. Formamide gel can also be run overnight and shut down by a timer. The standard gel should not be run longer than recommended, since the separation slows down considerably because of the formation of the field gradient in the bottom part, which in turn promotes overheating of the upper part due to its higher resistance. During running of the standard gel the voltage shortly rises and then slowly creeps down (by about 25 %) because of the build-up of an electrolyte gradient and subsequent current increase. By contrast, in the long run gel the voltage slowly creeps up and then stabilises, because the charged products of polymerisation leave the gel. Under the running conditions used by the author's lab for the formamide gel the voltage slowly creeps up during the entire run.

Running the Gel

Post-electrophoresis Treatment of Gels

Drying on Paper Contrary to earlier recommendations, it is no longer necessary to wash these gels to remove urea before drying.

1. After completion of the run the aluminium plates should be taken off, and the gels should be placed on a horizontal surface to cool down for about 10–15 min. During this time, a sheet of Whatmann 3MM paper should be cut to the size of the film to be used for exposure.

2. A plastic wedge or a spatula is carefully inserted at the lower side of the gel and the upper plate is slowly and gently pried apart. The plates should not be violently taken apart, as it may take 10–20 s for air bubbles to enter. After the sandwich has been opened, the pre-cut sheet of Whatmann paper is immediately placed over the gel and gently rubbed all over to make it stick to the gel. If only a portion of the gel surface contains samples, it is advisable to cut the paper to a size slightly larger than the data area before sticking it to the gel and to trim the surrounding gel by cutting with a spatula.

3. The gel, adhering to the paper, is slowly peeled off the plate and placed, paper side down, on the porous support of a gel dryer. The gel is covered with a cling film or a thin aluminium foil. No effort should be made to achieve a perfectly smooth surface, only large air bubbles between the foil and the gel should be avoided. The gel is covered by the rubber gasket of the dryer and the vacuum pump started; during a few minutes run without heating remaining wrinkles can be smoothed gently by hand. The dryer is programmed for "sequencing cycle" (rapid rise in temperature), and 80 °C for 30–60 min (depending largely on the pump strength).

4. After completion of the drying cycle the gel is allowed to cool on the bench, during which time it loses the initial curl and becomes completely flat.

Note: The covering foil is not removed for exposure. The gel should be securely taped in the cassette to allow tight contact with the film.

Glass-Bound Ultrathin Gels.

1. After completion of the run the aluminium plates should be taken off, and the gels should be placed on a horizontal surface to cool down for about 10–15 min.

2. A plastic wedge or a spatula is carefully inserted at the lower side of the gel and the upper plate is slowly and gently pried apart. The plates

should not be violently taken apart, as it may take 10–20 s for the air bubbles to enter.

3. The gel, adhering to the glass plate, is submerged in a tray with fixing solution and shaken at low speed for 15 min.

4. The gel is then placed under running tap water for 10 min and subsequently air-dried. To speed up drying, a hair dryer or an 80 °C oven can be used. After the gel is dry it should be completely transparent and not sticky. If urea crystals appear, the fixing and washing step should be repeated.

The formamide gel should be washed in acetic acid, even if it is going to be dried on paper, otherwise drying takes a long time.

Gels Containing Formamide

1. The opened gel is placed on a large plastic box in the middle of a large tray (same as used for casting) and sufficient fixing solution is poured on the gel to create a thin layer of liquid without overflow (about 150 ml). The gel is left without shaking for 5 min, and the liquid is drained by a slight tilt of the tray. The procedure is repeated three times with the fixing solution and a fourth time with water. The faster moving dye changes its colour to yellow.

2. The gel is made to adhere to a piece of Whatmann 3MM paper, dried and exposed exactly as described above for the standard gel.

Regeneration

1. If the plates were treated with a gel binding reagent, they should be regenerated before the next use. The wet gel is scraped off as much as possible with a plastic spatula (not razor blade) and the plate is soaked in 1 M NaOH for 1 h, whereupon it is rinsed in running water and the remaining gel debris removed with a hard toothbrush. When the gel binding reagent has been applied (see "DBE and Automated Sequencers") only to the area around the slots, it is not necessary to soak the entire plate, as a small amount of alkali can be applied to a piece of wicking paper, soaked in the solution and sandwiched between the two plates.

Regeneration of the Glass Plates

2. The plate, treated with a repelling agent, can be regenerated by soaking in concentrated sulphuric acid (**corrosive!**) for a few hours or overnight. This sometimes is necessary if the gel has creeped up the glass plate during the run, distorting the wells.

3. Soaking both plates in 1 M NaOH for 1 h is recommended when the gel sticks to the "wrong" plate (or both plates) after opening. With frequent gel preparation such alkali treatment should be done on a regular basis, e.g. once a month. However, prolonged alkali treatment may damage the surface of the plates.

Waste Disposal Sequencing electrophoresis produces radioactive and general chemical wastes that should be disposed of properly. Although sequencing separations typically use less than 50 μCi of radioactive material per run, most of which is distributed between the gel and the lower tank buffer (and does not present a serious radiation hazard), the cumulative effect of such small quantities may well lead to unacceptable levels of laboratory contamination if proper waste disposal procedures are not used. Similar kinds of self-contamination can occur when using fluorescent sequencing: though it does not present a radiation hazard, it may contribute to the "noise" in subsequent experiments. A place should be reserved to prepare and treat sequencing gels after the run; this practice prevents small amounts of radioactive (and, similarly, fluorescent) compounds from being distributed on all benches in the lab. Lower tank buffer should be drained in a container, labelled as containing radioactive material and disposed of according to the regulations of the country. Sequencing gels dried on paper can be stored in a radioactive waste storage room or treated as a low level solid radioactive waste. When the gel was dried on glass, it should be rehydrated in a large tray with water, carefully scraped onto a piece of Saran Wrap (using a plastic spatula!) and discarded into a radioactive waste container. Unused acrylamide solutions should be diluted to 4–5 % and then polymerised in a beaker, after which they can be discarded in a regular waste container.

▨ Results

Typical results of DNA sequencing electrophoresis using the standard gel and the long run gel are presented in Fig. 4. This figure was produced from a sequencing reaction carried out with a Thermosequenase cycle sequencing kit (Amersham) using ^{33}P-labelled primer. The gel was dried on paper and exposed to BioMax MR X-ray film (Kodak). Portions of the actual autoradiograms are presented to show sufficient detail and to demonstrate that the bands should move along nearly straight lines from the original slots (denoted by vertical grids). Close captioned portions of the separation also demonstrate the character of the bands, correspond-

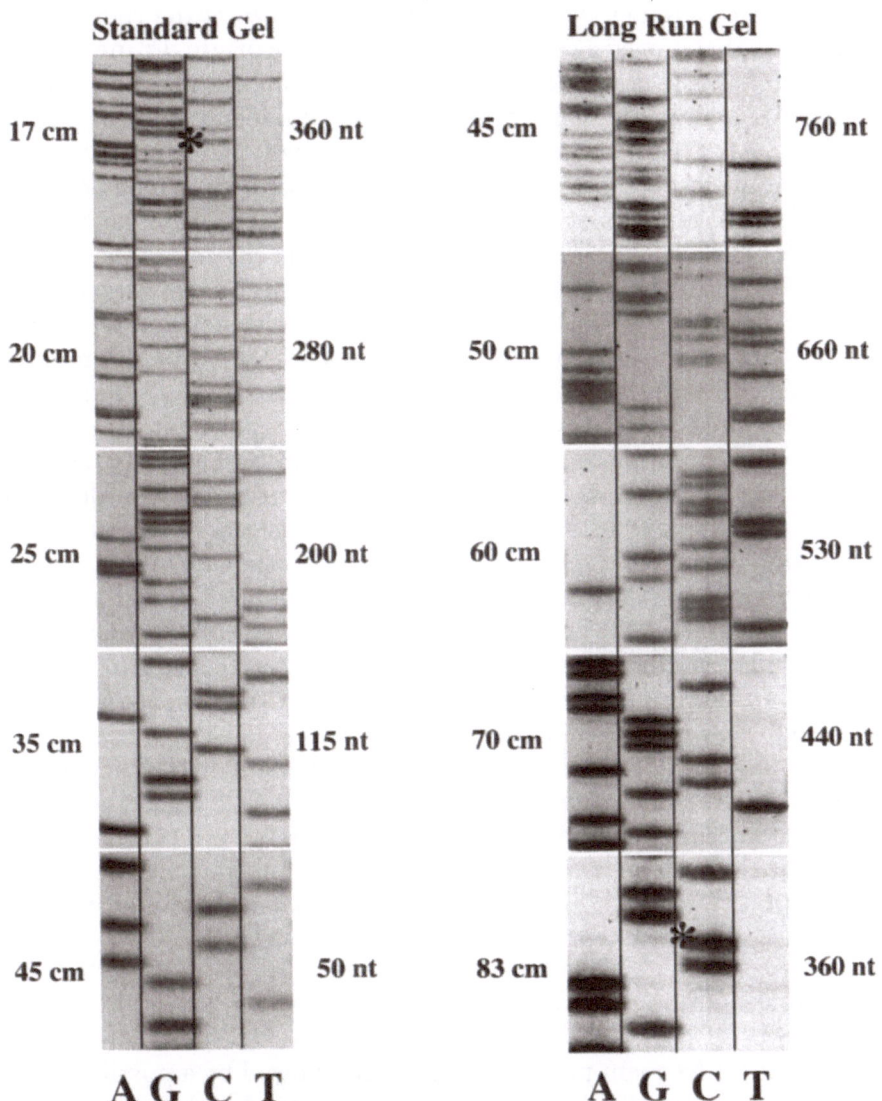

Standard Gel

17 cm — 360 nt
20 cm — 280 nt
25 cm — 200 nt
35 cm — 115 nt
45 cm — 50 nt

A G C T

Long Run Gel

45 cm — 760 nt
50 cm — 660 nt
60 cm — 530 nt
70 cm — 440 nt
83 cm — 360 nt

A G C T

Fig. 4. Typical results of the separations on a standard gel and a long run gel. The band migration distance from the origin is shown at the *left side* of each panel, and the absolute length of oligonucleotides resolved is shown at the *right side*. For other details, text

ing to oligonucleotides with a certain length (shown at the right side of each panel) that moved a certain distance from the origin during electrophoresis (shown at the left of each panel). Thus, the standard gel (on the left panel) resolves oligonucleotides with absolute lengths from ~35 to 360 bases. At WTR<17 cm the distance between the bands does not allow unequivocal estimation of the number of bands in a block of identical bases (above the asterisk). The long run gel resolves the same region at a WTR of 83 cm (the right panel). At WTR of 45 cm on this gel the distance between the bands also becomes very small, but for oligonucleotides of >760 bases. In both gels the band width and the band separation distance increase noticeably with increasing WTR (and separation time). This effect, as discussed earlier, puts an upper limit to the separation range with single-base resolution. Thus, increasing the run time of the long run gel from 15 to 27 h under our conditions (i.e. at the same power/voltage settings as in Table 1) will allow oligonucleotides with lengths >760 bases to be resolved; however, in this case near doubling of the separation time will result in a marginal increase in band separation (not shown). Analysis of the figure also shows that automated sequencers with a typical WTR of 25–35 cm will resolve up to 400 bases, while devices with a WTR of 45–50 cm will resolve up to 700 bases. To read the sequence further, on-line devices must use some sort of software compensation to estimate the number of bands in unresolved blocks and will consequently produce reliable data only with chemistries generating bands with even intensity. However, software development bears the potential of accurate reading further than the human eye can do, particularly if the dependence of the band distance from the local nucleotide composition (Sanger 1975) is included in the base-calling algorithm.

Troubleshooting

Here, the most frequent problems are listed followed by a possible cause and a solution. This list is far from complete (see also the next section, "Getting Help from Colleagues").

- Upon filtration of the gel stock solution the liquid flow is blocked after 100–150 ml have been filtered.
  - Certain brands of acrylamide are known to contain a contaminant which blocks the 0.45 µm filter cartridges.
    - Replace acrylamide by a different brand or purchase a liquid acrylamide-bis concentrate; alternatively, prepare a 40 % stock

solution and filter it first through a Whatmann GF/C glass fibre filter, which removes the bulk of the contaminant, then filter through a 0.45 μm cartridge.

- Crystals form upon storage of the gel stock solution.
  - Some recipes contain 8.3 M urea, which will crystallise upon cooling to 4 °C.
    - In the recipes in this chapter 7 M urea is used, which should not crystallise at 4 °C; the denaturing ability of both gels is similar. Check that the storage temperature is equal or above 4 °C; when using commercial gel formulations, check that they give 7 M and not 8.3 M urea solution.

- Gel polymerises before it is cast.
  - The ambient temperature is above 25 °C; the gel stock solution is too warm.
    - This is one of the most frequent problems. In the recipes in this chapter the gel stock solution is prepared in advance, stored at 4 °C and polymerised while still cold. The old belief that the gel should always be made fresh from solid chemicals usually results in the starting solution being too warm and hence polymerised too quickly; room temperature solution requires about 50 % less TEMED to be polymerised in the same time as the solution brought shortly from the cold room. If the stock solution is freshly made or stored at room temperature, use about 50 % less TEMED than indicated in "Reagents and Solutions". It is advisable to determine the polymerisation time in advance with a small aliquot of the gel stock solution.

- Gel does not polymerise in 30 min; after 2 h most of the gel volume has polymerised, but areas around the spacers and the comb are not polymerised.
  - Gel plates, combs and spacers were not properly cleaned, or incorrect concentrations of TEMED and APS were used because of low quality of one of them or both.
    - This was a typical problem earlier, but is now almost always replaced by the previous one, due to the generally increased quality of chemicals. Replace both APS and TEMED sources; do not use these chemicals when their usability date has expired or is uncertain; do not use unchecked detergents or household cleaning aids to clean the glass plates. Note, however, that a formamide gel always requires fivefold more TEMED for polymerisation than does a standard gel.

- During polymerisation the gel shrinks inside the mold (a reticular appearance of the gel, detached partially from one or both glass plates).
  - Very fast gel polymerisation; a gross mistake in acrylamide and/or APS concentration.
    - Check preparation of the stock solutions; see also "Gel Polymerises Before It Is Cast".

- Upon removing clamps and the comb many air bubbles enter the gel around the comb area; rake comb-formed pockets are visually distorted; shark teeth comb cannot be inserted.
  - Too much or uneven pressure on the comb during polymerisation; glass plates are warped.
    - This problem is more frequent with 0.1–0.2 mm gel setups than with 0.4 mm setups. Clamps should be positioned only above the plastic and not protrude farther into the gel area; use only floated glass plates. Check plate flatness when getting new plates. Use matching spacers and combs and take appropriate care of them so that they do not get warped.

- Upper buffer chamber gasket leaks.
  - A typical problem of many gel holders, not unusual even in expensive automated sequencers; it results from worn out or defective gasket.
    - Examine the chamber gasket and replace with a new one; it often helps to redesign the gasket, if one has such a possibility. Do not use grease to save the gasket.

- During pre-run the current is too low (<10 mA).
  - Gel does not contain buffer.
    - Replace the gel stock solution and cast a new gel.

- Sample does not go to the bottom of the well during loading.
  - Urea leaching out of the gel.
    - Wells should be flushed with running buffer and loaded within 10 min after that. Wells are very difficult to flush when the gel is < 0.2 mm thick and the wells are not square shaped (see "Casting Gels").

- During the run the gel is too hot; one or both glass plates eventually crack.
  - Power applied to the gel is not optimal. See optimisation of this parameter discussed above.

- Gross mistake in the buffer concentration in the gel.
  - Perform a test run with a new equipment to determine the optimal conditions; do not use the parameters given in the instruments manual "out of the box", as there may be a discrepancy between the presumed gel/buffer composition and the one used.

- Upper tank buffer is drained after overnight run; sparking and burning around the slots.
  - Leaking upper chamber gasket or buffer leak between the gel and the spacers.
    - Check for the absence of buffer leak between the gel and the spacers by applying some gel loading solution (power off!); buffer drain should be visible by the movement of the dye; press the gel towards the spacer to block the gap.

- Bubble formation in the slots after >2 h run.
  - Buffer degassing because of elevated temperature.
    - Before filling the tanks pre-warm the running buffer to 50 °C and degas for 10 min under reduced pressure; a better strategy is to remove the shark teeth comb after 15 min run (the loading dye has moved >1 cm from the slots) and not to perform multiple loadings on the same gel. When doing multiple loadings, pre-treat the comb area with the gel binding reagent and change tank buffer with each loading.

- Slots are distorted after >2 h run.
  - Gel swelling and possibly ionic depletion at the boundary between the gel and the tank buffer, on long runs practically unavoidable; or a buildup of repelling agent on the glass.
    - Pre-treat the comb area with the gel binding reagent; do not flush sample wells excessively with running buffer; do not apply gel repelling solution excessively, i.e. more than once per run, rinse with ethanol and polish with a lint-free paper. If the problem persists, regenerate glass plates by soaking in concentrated sulphuric acid overnight, wash and visually check that they have lost repellent coating, then carefully treat again with repelling agent.

- Gel sticks to the "wrong" glass plate or is finally recovered on one plate wrinkled and distorted.
  - Gel overheated during run and/or not cooled down before sandwich opening; or dirty or improperly cleaned glass plates.
    - Control the gel temperature during runs, let the gel cool for 10–15 min before opening the plates; use only proper detergents.

If the problem persists, soak the glass plates in 1 M alkali for 1 h, than rinse with plenty of water.

- The gel is torn or evenly split between the two plates upon opening; the sandwich is very difficult to open.
  - One or both glass plates were improperly treated (possibly both with the binding reagent), excess binding reagent present during polymerisation.
    - Do not use the same glass plates for different plate treatment techniques (e.g. for gel dryer technique and for binding gel to glass); if the problem persists, include additional washing step after treatment with the gel binding agent. If this problem is encountered with intentionally glass bound ultrathin gels, include an additional detergent cleaning step between the treatment with the binding reagent and the gel casting.

- The gel does not stick to paper, instead it remains on the glass plate.
  - The source of this problem is sometimes difficult to locate; generally, the gels with T>8 % do not stick well to paper, while gels described in this chapter usually stick properly, unless a torn gel or difficult to open sandwich is involved. The problem seems to be more common with gels left to polymerise overnight.
    - Use only chromatography paper, not cheap substitutes (e.g. wicking paper), periodically (e.g. once in a month) treat the glass plates with 1 M alkali for 1 h. Treatment of both plates with the gel repelling reagent does not guarantee that the gel will stick well to paper.

- The gel is not dry after 1 h on the gel dryer.
  - Insufficiently low pressure.
    - Check the quality of the low pressure source. Often the in-house central line is inappropriate for this, because the pressure may change significantly during working hours. Check that the gel dryer is clean (one may need to change a porous support plate) and holds the vacuum.

- Crystals are seen on the gel dried on the glass plate.
  - The urea is not removed during washing and fixing step. A typical problem with wedge-shaped gels, a method used to create a field gradient by variable thickness of the gel (Ansorge and Labeit 1984).
    - Use a shaker to treat the gel with the fixing solution, and extend the tap water wash step for a few minutes.

- The film sticks to the gel during exposure.
  - This problem is typical for the technique, requiring direct contact of the gel with the film, and may be particularly pronounced in very humid climates.
  - Do not remove protective foil during exposure to film. Aluminium foil ensures a reliable gel insulation, yet it does not block the signal considerably, while some vinyl films may be penetrable to water. When using gels dried on glass it is particularly important to wash away urea and place the gel in the exposure cassette as fast as possible after removing it from the drying oven.

- Fuzzy bands.
  - Gel overheated during the run or apparatus located in the area with strong air currents, or inadequate contact of the film with the dried gel during exposure.
  - In no occasion should the gel be run near an open window or without thermostating plates! Although chemicals can also be the source of this problem, this is now rarely the case. It is also not uncommon to pay little attention to the quality of the exposure cassette or the position of the gel/film sandwich in it.

Comments

Getting Help from Colleagues

Sequencing gel electrophoresis has evolved into a complex technology, and automated and manual sequencing are already two separate areas with specific problems. Increasingly more researchers use sequencing as an instrument in standard research institutes as well as in relatively unusual settings, such as diagnostic laboratories or field stations. Though the author made every effort to provide an integrated approach and eliminate unnecessary rituals left over from older methods, it was impossible to cover situations that may arise in using this technique, say, in extreme climatic conditions or at remote locations. However, if one has access to the Internet or at least can use an electronic mail system, there is a way to get additional help from colleagues. The system of USENET electronic bulletin boards ("newsgroups") covers thousands of topics, of which two groups discuss DNA sequencing problems, the "bionet.molbio.methds-reagnts" and the "bionet.genome.autosequencing". The Frequently Asked Questions (FAQ) list for bionet.molbio.methds-reagnts, from which the following information was taken, is available by anonymous FTP from ftp.ncifcrf.gov as the file pub/methods/FAQlist. It is also available by anonymous FTP from net.bio.net as pub/BIOSCI/METHDS-REAGNTS/METHODS.FAQ. If one does not know how to use FTP to obtain the FAQ list, a copy may be requested by e-mail to pnh@ncifcrf.gov. New users of BIOSCI/bionet may want to read the BIOSCI FAQ list. This list provides details on how to participate in these forums and is available by anonymous FTP from net.bio.net in pub/BIOSCI/biosci.FAQ. It may also be requested by sending e-mail to biosci-help@net.bio.net. The FAQ sheet is posted on the first of each month to the newsgroup BIONEWS/bionet.announce immediately following the posting of the BIOSCI information sheet. The best way to search for past articles is by using the new WWW service from BIONET. The service is located at net.bio.net and all one will need to do in order to read and/or post to any of the newsgroups is point a World Wide Web browser (such as Netscape Navigator) to the URL http://www.bio.net and then click on the "Access the BIOSCI/bionet Newsgroups" hyperlink.

Acknowledgements. The author is grateful to M. Woznica and U. Rosenberg (MWG Biotech AG), A. Tauer, K. Nohava and A. Zikopoulos (Pharmacia Biotech), M. Mueller (Perkin-Elmer Switzerland), U. Zuber and R. Junet (Amersham Switzerland) for providing technical information and the possibility to get hands-on experience on various automated sequencers. The help of P. Gazzotti in preparing the figures, as well as his valuable advice on the procedure descriptions is also highly appreciated. Finally, the author is grateful to Prof. E. Carafoli for hospitality and a productive atmosphere in the lab.

References

Ahmed A (1987) A simple and rapid procedure for sequencing long (40 kb) DNA fragments. Gene 61: 363–372

Anderson S (1981) Shotgun sequencing strategy using cloned DNase I–generated fragments. Nucleic Acids Res 9: 3015–3027

Ansorge W, Barker R (1984) System for DNA sequencing with resolution of up to 600 base pairs. J Biochem Biophys Methods 9: 33–47

Ansorge W, Labeit S (1984) Field gradients improve resolution on DNA sequencing gels. J Biochem Biophys Methods 10: 237–243

Ansorge W, Sproat BS, Stegemann J, Schwager C (1986) A non-radioactive automated method for DNA sequencing determination. J Biochem Biophys Methods 13: 315–323

Bankier AT, Weston KM, Barrell BG (1988) Random cloning and sequencing by the M13/dideoxynucleotide chain termination method. Methods Enzymol 155: 52–93

Barnes WM, Bevan M, Son PH (1983) Kilo-sequencing: creation of an ordered nest of asymmetric deletions across a large target sequence carried on phage M13. Methods Enzymol 101: 98–122

Beck S, Pohl FM (1984) DNA sequencing with direct blotting electrophoresis. EMBO J. 3: 2905–2909

Biggin MD, Gibson TJ, Hong GF (1983) Buffer gradient gels and 35S-label as an aid to rapid DNA sequence determination. Proc Natl Acad Sci USA 80: 3963–3965

Brassard E, Turmel C, Noolandi J (1992) Pulsed field sequencing gel electrophoresis. Electrophoresis 13: 529–535

Brumley RL, Smith LM (1991) Rapid DNA sequencing by horizontal ultrathin gel electrophoresis. Nucleic Acids Res 19: 4121–4126

Carninci P, Gustuncich S, Bottega S, Patrosso C, Del Sal G, Manfioletti G, Schneider C (1990) A simple discontinuous buffer system for increased resolution and speed in gel electrophoretic analysis of DNA sequence. Nucleic Acids Res 18: 204

Cherry JL, Young H, Di Sera LJ, Ferguson FM, Kimball AW, Dunn DM, Gesteland RF, Weiss RB (1994) Enzyme-linked fluorescent detection for automated multiplex DNA sequencing. Genomics 20: 68–74

Church G, Gilbert W (1984) Genomic sequencing. Proc Natl Acad Sci USA 81: 1991–1995

Church GM, Kieffer-Higgins S (1988) Multiplex DNA sequencing. Science 240: 185–188

Deininger P (1983) Random subcloning of sonicated DNA: application to shotgun DNA sequencing. Anal Biochem 129: 216–223

Drmanac R, Labat I, Brukner I, Crkvenjakov (1989) Sequencing of megabase plus DNA by hybridization: theory of the method. Genomics 4: 114–128

Drmanac S, Drmanac R (1994) Processing of cDNA and genomic kilobase-size clones for massive screening, mapping and sequencing by hybridization. Biotechniques 17: 328–329, 332–336

Drossman H, Luckey JA, Kostichka AJ, D'Cunha J, Smith LM (1990) High-speed separations of DNA sequencing reactions by capillary electrophoresis. Anal Chem 62: 900–903

Favello A, Hillier L, Wilson RK (1995) Genomic DNA sequencing methods. Methods Cell Biol 48: 551–569

Fulton LL, Hillier LD, Wilson RK (1995) Large-scale complementary DNA sequencing methods. Methods Cell Biol 48: 571–582

Fulton LL, Wilson RK (1994) Variations on cycle sequencing. Biotechniques 17: 298–301

Garoff H, Ansorge W (1981) Improvements of DNA sequencing gels. Anal Biochem 115: 450–457

Hawkins TL, Du Z, Halloran ND, Wilson RK (1992) Fluorescence chemistries for automated primer-directed DNA sequencing. Electrophoresis 13: 552–559

Hengen PN (1995) Frequently asked question (FAQ) list for bionet.molbio.methds-reagnts version number NN.DD.MM.19YY, available via anonymous FTP from ftp.ncifcrf.gov as file pub/methods/FAQlist (a WWW publication, updated monthly)

Hengen PN (1995). Mini-prep plasmid DNA isolation and purification using silica-based resins. In: Griffin and Griffin (eds) Molecular biology: current innovations and future trends. Wymondham, UK, Horizon Scientific Press, pp 39–50

Henikoff S (1984) Unidirectional digestion with exonuclease III creates targeted breakpoints for DNA sequencing. Gene 28: 351–359

Kieleczawa J, Dunn JJ, Studier FW (1992) DNA sequencing by primer walking with strings of contiguous hexamers. Science 258: 1787–1791

Kostichka AJ, Marchbanks ML, Brumley RL, Drossman H, Smith LM (1992) High speed automated DNA sequencing in ultrathin slab gels. Biotechnology 10: 78–81

Krishnan BR, Chaplin DD (1994) Fluorescent automated sequencing of supercoiled high molecular weight double-stranded DNA. BioTechniques 17: 854–857

Kutateladze TV, Axelrod VD, Gorbulev VG, Belzhelarskaya SN, Vartikyan RM (1979) New procedure of high-voltage electrophoresis in polyacrylamide gel and its application to the sequencing of nucleic acids. Anal Biochem 100: 129–135

Maxam AM, Gilbert W (1977) A new method for sequencing DNA. Proc Natl Acad Sci USA 74: 560–564

Mayeda A, Krainer AR (1991) Long-term storage of concentrated tris borate-EDTA electrophoresis buffers without precipitation. Biotechniques 10: 182

McCombie WR, Heiner C, Kelley JM, Fitzgerald MG, Gocayne JD (1992) Rapid and reliable fluorescent cycle sequencing of double-stranded templates. DNA Seq 2: 289–296

Middendorf LR, Bruce JC, Bruce RC, Eckles RD, Grone DL, Roemer SC, Sloniker GD, Steffens DL, Suffer SL, Brumbaugh JA, Patonay G (1992) Continuous, on-line DNA sequencing using a versatile infrared laser scanner/electrophoresis apparatus. Electrophoresis 13: 487–494

Mouradian S, Brumley RL, Smith LM (1994) Separating field strength, temperature, and pulsing effects in pulsed field electrophoresis. Electrophoresis 15: 1084–1090

Murray V (1989) Improved double-stranded DNA sequencing using the linear polymerase chain reaction. Nucleic Acids Res 17: 8889

Nishikawa T, Kambara H (1991) Analysis of limiting factors of DNA band separation by a DNA sequencer using fluorescence detection. Electrophoresis 12: 623–631

Nishikawa T, Kambara H (1992) High resolution-separation of DNA bands by electrophoresis with a long gel in a fluorescence-detection DNA sequencer. Electrophoresis 13: 500–505

Olsson A, Moks T, Muhlen J, Gaal AB (1984) Uniformly spaced banding patterns in DNA sequencing by the use of a field-strength gradient. J Biochem Biophys Methods 10: 83–90

Peacock AC, Dingman CW (1967) Resolution of multiple ribonucleic acid species by polyacrylamide gel electrophoresis. Biochemistry 6: 1818–1827

Pentoney SL, Konrad KD, Kaye W (1992) A single-fluor approach to DNA sequence determination using high performance capillary electrophoresis. Electrophoresis 13: 467–474

Pogue RR, Cook ME, Livingstone LR, Hunt III SW (1993) Preparation of template for automated sequencing using QIAGEN resin. BioTechniques 17: 377–378

Prober JM, Trainor GL, Dam RJ, Hobbs FW, Robertson CW, Zagursky RJ, Cocuzza AJ, Jensen MA, Baumeister K (1987) A system for rapid DNA sequencing with fluorescent chain-terminating dideoxynucleotides. Science 238: 336–341

Richterich P, Heller C, Wurst H, Pohl F (1989) DNA sequencing with direct blotting electrophoresis and colorimetric detection. Biotechniques 7: 52–59

Ruiz-Martinez MC, Berka J, Belenkii A, Foret F, Miller AW, Karger BL (1993) DNA sequencing by capillary electrophoresis with replaceable linear polyacrylamide and laser-induced fluorescence detection. Anal Chem 65: 2851—2858

Sanger F, Coulson AR (1978) The use of thin acrylamide gels for DNA sequencing. FEBS Lett 87: 107–110

Sanger F, Coulson RA (1975) A rapid method for determining sequence in DNA by primed synthesis with DNA polymerase. J Mol Biol 94: 441–448

Sheen J-Y, Seed B (1988) Electrolyte gradient gels for DNA sequencing. BioTechniques 6: 942–944

Slater GW, Drouin G (1992) Why cannot we sequence thousands of DNA bases on a polyacrylamide gel? Electrophoresis 13: 574–582

Smith LM, Sanders JZ, Kaiser RJ, Hughes P, Dodd C, Connel CR, Heiner C, Kent SBH, Hood LE (1986) Fluorescence detection in automated DNA sequence analysis. Nature 321: 674–679

Studier FW (1973) Analysis of bacteriophage T7 early RNAs and proteins on slab gels. J Mol Biol 79: 237–248

Sulston J, Du Z, Thomas K, Wilson R, Hillier L, Staden R, Halloran N, Green P, Thierry-Mieg J, Qui L, Dear S, Coulson A, Craxton M, Durbin R, Berks M, Metzstein M, Hawkins T, Ainscough R, Waterston R (1992) The C.elegans genome sequencing project: a beginning. Nature 356: 37–41

Swerdlow H, Dew-Jager KE, Brady K, Grey R, Dovichi NJ, Gesteland R (1992) Stability of capillary gels for automated sequencing of DNA. Electrophoresis 13: 475–483

Tietz D, Chrambach A (1993) DNA shape and separation efficiency in polymer media: A computerized method based on electrophoretic mobility data. Electrophoresis 14: 185–190

Voss H, Wiemann S, Grothues D, Sensen C, Zimmermann J, Schwager C, Stegemann J, Erfle H, Rupp T, Ansorge W (1993) Automated low-redundancy large-scale DNA sequencing by primer walking. BioTechniques 15: 714–721

Voss H, Wiemann S, Wirkner U, Schwager C, Zimmermann J, Stegemann J, Erfle H, Hewitt NA, Rupp T, Ansorge W (1992) Automated DNA sequencing system resolving 1000 bases with fluorescein-15-dATP as internal label. Meth Molec Cell Biol 3: 153–155

Voss H, Zimmermann J, Schwager C, Erfle H, Stegemann J, Stucky K, Ansorge W (1990) Automated fluorescent sequencing of cosmid DNA. Nucleic Acids Res 18: 1066

Wilson RK (1993) High-throughput purification of M13 templates for DNA sequencing. BioTechniques 15: 418–420, 422

Wilson RK, Chen C, Hood L (1990) Optimization of asymmetric polymerase chain reaction for rapid fluorescent DNA sequencing. BioTechniques 8: 184–189

Wilson RK, Koop BF, Chen C, Halloran N, Sciammis R, Hood L (1992) Nucleotide sequence analysis of 95 kb near the 3'-end of the murine T-cell receptor alpha/delta chain locus: strategy and methodology. Genomics 13: 1198–1208

Suppliers

A list of major suppliers of equipment and reagents for DNA sequencing with USA addresses was compiled using largely data available at http://www.biosupplynet.com/alpha-supps.htm. This list is not intended to be complete and it is also recommended to check the WWW directory for up-to-date addresses.

Amersham-Pharmacia Biotech, Inc.
800 Centennial Avenue
P.O. Box 1327, Piscataway, NJ 08855-1327, USA
Web Site: http://www.apbiotech.com

Bertan High Voltage Corporation
121 New South Rd, Hicksville, New York 11801, USA
Phone: +1-516-433-3110
Fax: +1-516-935-1766
e-mail: info@bertan.com (tech Support)
Web Site: http://www.bertan.com

Bio-Rad Laboratories
2000 Alfred Nobel Drive, Hercules, California 94547, USA
Phone: +1-510-741-6781, 1-800-424-6723
Fax: +1-510-741-5800
e-mail: crickey@haley.genetics.bio-rad.com
Web Site: www.bio-rad.com

FMC BioProducts
191 Thomaston Street, Rockland, Maine 04841, USA
Phone: +1-207-594-3495
Fax: +1-207-594-3491
e-mail: biotechserv@fmc.com
Web site: http://www.bioproducts.com

GATC GmbH
Fritz-Arnold-Str. 23, D-78467 Constance, Germany
Phone: +49-7531-57209
Fax: +49-7531-57313
e-mail: gatc@lakelink.cl.sub.de

Genomyx Corp.
353 Hatch Drive, Foster City, California 94404, USA
Phone: +1-415-572-8800
Fax: +1-415-572-8096
e-mail: info@genomyx.com
Web site: http://www.genomix.com

Genesys Technologies Inc.
1100 Dallas Street, Sauk City, Wisconsin 53583, USA
Phone: +1-608-643-4166
Fax: +1-608-643-5048
e-mail: info@genesys-tech.com
Web site: http://www.genesys-tech.com

Jordan Scientific Co., Inc.
4315 S. State Road, Bloomington, Indiana 47401, USA
Phone: +1-812-334-1543
Fax: +1-812-334-1509
e-mail: jordan@intersource.com

Eastman Kodak Company (includes IBI)
25 Science Park, New Haven, Connecticut 06511, USA
Phone: 1-800-243-2555
Fax: +1-203-786-5674
e-mail: SiS-support@kodak.com
Web site: http://www.kodak.com/go/scientific

LI-COR Inc.
Biotechnology Div.
4308 Progressive Ave.
Lincoln, NE 68504, USA
Phone: +1-402-467-0700
Fax: +1-402-467-0819
e-mail: biohelp@bio.licor.com
Web site: http://www.licor.com

Life Technologies/Gibco BRL
8400 Helgerman Court, Gaithesburg, Maryland 20877, USA
9800 Medical Centre Dr. Rockville, MD 20849-6482
Phone: +1-301-840-8000
Fax: +1-301-670-1394 + 1-800-331-2286
e-mail: info@lifetech.com, molbio@lifetech.com
Web site: http://www.lifetech.com

Molecular Dynamics
928 East Arques Avenue, Sunnyvale, Virginia 94086, USA
Phone: +1-408-773-1222
Fax: +1-408-773-1493
e-mail: info@mdyn.com
Web site: http://www.mdyn.com/

National Diagnostics
305 Patton Drive, Atlanta, Georgia 30336, USA
Phone: +1-404-699-2121
Fax: +1-404-699-2077

Perkin-Elmer Corporation/Applied Biosystems
850 Lincoln Centre Drive, Foster City, California 94404, USA
Phone: 1-800-327-3002, 1-800-545-7547
Fax: +1-415-638-5998, +1-415-638-5875
e-mail: info@perkin-elmer.com
Web site: http://www.perkin-elmer.com

Pharmacia Biotech, Inc.
800 Centennial Avenue
P.O. Box 1327, Piscataway, NJ 08855-1327, USA
Web Site: http://www.apbiotech.com

Promega Corporation
2800 Woods Hollow Road, Madison, Wisconsin 53711–5399, USA
Phone: +1-608-274-4330
Fax: +1-608-277-2516
e-mail: custserv@promega.com (Promega Technical Support)
Web site: http://www.promega.com

Stratagene
11011 North Torrey Pines Road, La Jolla, California 92037, USA
Phone: 1-800-424-5444
Fax: +1-619-535-0034
e-mail: TechServices@stratagene.com (Technical Services)
Web site: http://www.stratagene.com

Whatman, Inc.
5285 NE Elam Young Pkwy, Bldg A-400, Hillsboro, Oregon 97124, USA
Phone: 503-549-0762
Fax: 201-472-6949
e-mail: info@whatman.com
Web site: http://www.whatman.com

Chapter 3

An Introduction to DNA Fingerprinting Using RFLP and RAPD Techniques

R.J. Snowdon and A. Langsdorf

Introduction

All individual organisms, provided they are not clones or identical twins, have a unique genetic makeup. Thus, individuals can be genetically distinguished from one another using processes collectively described as DNA fingerprinting. Multilocus DNA fingerprinting was first described by Jeffreys et al. (1985), the popular name describing its application in identifying individuals on the basis of a unique genetic pattern. Just as conventional fingerprints are compared by analysis of a set of distinguishable characteristics in their patterns, a DNA „fingerprint" also comprises a set of distinguishable characteristics: a set of bands, resembling a bar-code, representing numerous different-sized fragments of the genomic DNA of an individual, separated by electrophoresis. When two individuals are genetically different, differences between their banding patterns, or DNA fingerprints, can be identified. The degree to which these differences, known as DNA polymorphism, allow one to distinguish between individuals (or to determine similarity or identity between individuals), is statistically dependent on the number of bands investigated.

This chapter provides an introduction to the two most common established methods for generating DNA fingerprints: the RFLP (restriction fragment length polymorphism) method of Jeffreys et al. (1985), and the RAPD (randomly amplified polymorphic DNA) technique developed by Williams et al. (1990). Both of these methods elucidate genetic differences at the DNA sequence level which result from evolutionary mechanisms such as DNA deletions, additions, substitutions, repetitions and

R.J. Snowdon, A. Langsdorf, Department of Biometry and Population Genetics, Justus-Liebig Universität, Ludwigstrasse 27, 35390 Giessen, Germany)
Correspondence to R.J. Snowdon: *phone* +49-641-99-37542; *fax* +49-641-9937549; *e-mail* Rod.Snowdon@agrar.uni-giessen.de)

translocations. Although each method results in a banding pattern and band polymorphism which can be analyzed in much the same way, the two techniques are based on quite different principles:

- RFLP analysis relies on digestion of DNA samples with restriction enzymes. Differences arise between individuals in sizes of corresponding fragments, resulting from sequence differences at DNA restriction sites or within restriction fragments. The fragments are then separated by agarose gel electrophoresis, which generates a characteristic smear pattern owing to the continuous distribution of fragment sizes. DNA is transferred from the gel and fixed on a solid membrane, and band polymorphism is visualized, after hybridization of a labeled probe to homologous sequences on the separated fragments, by detection of hybridization sites. Thus, a set of probe-specific bands is produced from the smear of DNA fragments. Different types of probe allow investigation of varying numbers of bands and different degrees of polymorphism.

- RAPD investigation uses the polymerase chain reaction (PCR; Mullis et al. 1986; Mullis and Faloona 1987) to amplify DNA samples with short oligonucleotide primers that anneal randomly throughout the genome. The result is a distinctive set of amplification products, whose sizes vary where differences exist at or between primer annealing sites. Differences are visualized by staining gels after agarose gel electrophoresis of amplification products. Use of different oligonucleotide sequences as primers brings different banding patterns, and individuals or populations are characterized by the set of banding patterns they produce for a number of different primers.

DNA fingerprinting techniques are used for an extremely wide variety of applications. Some applications require much more intricate analyses than others, but the basic RFLP and RAPD methods are broadly applicable – in principle, eukaryotic DNA samples can always be analyzed in much the same way, regardless of the organism from which they come. The method of choice for DNA fingerprinting, however, greatly depends on the aims of the comparison and particularly on the degree to which the individuals of interest differ. In forensic comparisons of crime-scene samples with those from suspects, for example, an extremely high degree of accuracy is demanded, requiring genetic fingerprints with the maximum possible number of useful bands. This is best achieved with RFLP analysis. For other applications, for example differentiating between crop varieties in agricultural breeding programs, sufficient information may be gained from considerably fewer bands. Identification of even a single, cultivar-specific band may be sufficient for such purposes.

RFLP and RAPD methodologies differ widely in terms of simplicity, time and cost considerations. RAPD analysis is in general a much more straight-forward process and the time-frame for screening genetic variation within and among large population samples, for example, is normally considerably shorter than using RFLP technology. In terms of per-analysis costs, RFLP investigation also tends to be more expensive than RAPD analysis. Nonetheless, there are instances in which the level of genetic differentiation provided by RAPD fingerprinting is insufficient for the desired application, and time and cost concerns are made redundant by the fact that a single, multi-copy RFLP probe may provide the same level of information as a large number of RAPD runs with different primers. In addition, whereas a RAPD is just that – a randomly amplified sequence – the RFLP method is able, when appropriate probes are used, to provide more control over which sequences are identified. A wide variety of different RFLP probes can be applied – with little alteration in the methodology – from gene-specific, low- or single-copy sequences, through to minisatellite sequences with very high copy-numbers throughout the genome.

Sometimes, the amount of genomic DNA that can be obtained from a sample must be taken into account when choosing which DNA-fingerprinting method is more appropriate. Whereas RAPD fragments can theoretically be amplified from a single DNA strand, and thus require minimal amounts of DNA, the amount of DNA required for successful RFLP analysis is normally considerably greater. When sufficient DNA cannot be obtained, for example in population studies involving small or rare organisms in which nondestructive sampling is important, RFLP techniques must often be ruled out. In this case, RAPD analysis provides the simplest fingerprinting option. A new method of generating DNA fingerprints, the AFLP (amplified fragment length polymorphism) technique, may also be considered. This method, developed by Zabeau and Vos (1993), combines the ideas of RFLP and RAPD fingerprinting in the random PCR-amplification of fragments generated by restriction digestion. The AFLP method and its application to DNA fingerprinting are described in detail by Vos et al. (1995).

Two other considerations when deciding between RFLP and RAPD fingerprinting are safety and reliability. The first applies specifically to RFLP analysis, for which, until recently, the standard detection method relied on radioactively labeled probes. Today, however, detection sensitivity of widely available nonisotopic labeling compounds is very much improved and often as good as the traditional isotopes. For this reason, and because of the strict controls, high costs of handling facilities and disposal difficulties involved with isotope usage, not to mention personal safety hazards, we concentrate here on nonisotopic RFLP analysis. The question of reliability particularly concerns RAPD investigations. The PCR process is extremely sensitive to contamination and to amplification conditions, which can introduce amplification artifacts (e.g. Ellsworth et al. 1993). Using the same template (genomic) DNA, the number of amplification products for a given oligonucleotide primer can alter drastically according to differences in the composition of the reaction mix and in cycle conditions. It is thus important, when carrying out population analyses for example, that DNA samples are uniformly contamination-free and that PCR cycling is identical for all samples. In our experience, even differences between thermal cyclers can potentially affect results. Thus, in order to make assumptions about genetic similarity or differences between individuals, it is vital that standards are always included whose amplification products can be shown to be consistently identical in different RAPD experiments.

We present here simple protocols which, although their purpose in this context is to introduce new users to the RFLP and RAPD fingerprinting techniques, are actually used by us in our routine analyses. The first requirement for both RAPD and RFLP fingerprinting is high-quality genomic DNA, since activity of the thermostable polymerase enzyme used to amplify PCR products for RAPDs and the restriction enzymes used to digest genomic DNA for RFLP experiments can both be adversely affected by impurities in the DNA. We thus include a reliable and robust DNA-extraction method, which we have successfully applied to a very diverse range of plant species representing many taxonomic groups. We have also tested this method for DNA extraction from various tissues in fish and other animal species, where it provides similar excellent yields of high-quality genomic DNA.

A flow-diagram describing the major steps in RFLP and RAPD fingerprinting analysis is given in Fig. 1. Protocols for RFLP analysis are outlined in Sect. 3.2, and a general protocol for RAPD amplification is given in Sect. 3.3. In "Results" we briefly discuss basic approaches to data analysis from fingerprinting results. In our laboratories, RFLP and RAPD fingerprinting applications generally focus on investigation of levels of molecular genetic variability in and among populations of various plant species and prokaryotic plant pathogens. Exactly the same methods can however be applied to comparisons of populations and individuals in animal species.

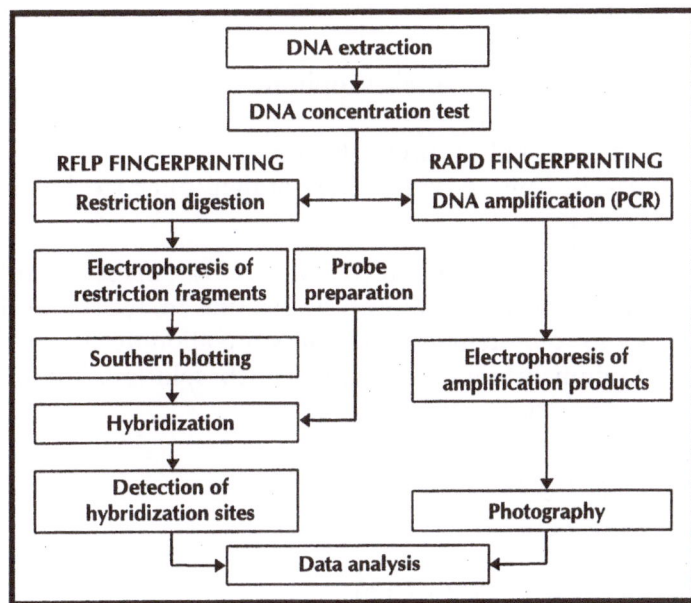

Fig. 1. Flow diagram describing the key steps in RFLP and RAPD fingerprinting

3.1
DNA Extraction

This is a protocol for DNA extraction from plant material, adapted from Doyle and Doyle (1990). It is a robust method that gives high yields of good-quality DNA from leaf and root tissue in a surprisingly wide range of plant species covering diverse taxonomic groups. Best yields are obtained from young leaves, either fresh or frozen at $-70\,^{\circ}C$, but we also extract DNA routinely from freeze-dried or dehydrated plant material. In addition, we have found the method to also give excellent DNA yields when applied to various tissues from fish and other animals. Bear in mind that for RAPD analysis, the amount of DNA required is considerably less, thus the volumes and quantities shown may be scaled down by a factor of ten for mini-preparations if DNA is not also required for RFLP analysis.

The concentration of genomic DNA extracted using CTAB-based buffers cannot be accurately estimated by spectrophotometry, as residual CTAB causes high interference in absorbance measurements. Instead, approximation of DNA concentration is easily achieved by rapid electrophoresis and comparison with concentration standards.

■ Materials

Equipment   – Mini-gel horizontal electrophoresis chamber (gel size approximately 10 cm×8 cm)
– Ultraviolet light source (302 nm)
– Polaroid camera with yellow-orange filter (to block UV rays under 545 nm)

Chemicals   – RNAase-A: 10 mg/ml in 10 mM Tris-HCl, pH 7.5; 15 mM NaCl. DNAase activity must be removed by heating to 100 °C for 15 min. Allow to cool to room temperature and store aliquots at –20 °C.
– Ethidium bromide: 10 mg/ml stock solution. Dissolve in water and store in a darkened bottle at 4 °C.

Buffers   TAE buffer (50× stock solution)

242 g Tris base
57.1 ml glacial acetic acid
100 ml 0.5 M EDTA (pH 8.0)
Dilute 1:50 in deionized water for use.

Gel loading buffer

0.25 % bromophenol blue
15 % Ficoll (Type 400)
Dissolve in distilled water and store at room temperature.

Variations of this recipe can be found in Sambrook et al. 1989.

DNA extraction buffer

4 % CTAB (hexadecyltrimethylammonium bromide)
1.4 M NaCl
0.2 % (v/v) 2-mercaptoethanol
20 mM EDTA
100 mM Tris-HCl (pH 8.0)

TE buffer

10 mM Tris-HCl (pH 8.0)
1 mM EDTA (pH 8.0)
Sterilize by autoclaving.

▨ Procedure

1. Grind up to 1 g of fresh or deep-frozen plant material (for animal tissue or dehydrated plants, use 0.1–0.5 g) to a fine powder in liquid nitrogen in a pre-chilled mortar and pestle. A small amount of sterile sand may be added to assist grinding.

DNA Extraction

2. Transfer powder to 4 ml DNA extraction buffer in a 10 ml centrifuge tube, pre-heated to 65 °C in a water bath. Incubate 30–45 min with occasional gentle mixing.

3. Remove from water bath and add 4 ml chloroform:isoamyl alcohol (24:1). After gentle but thorough mixing, spin 10 min at 4000 rpm in a swinging-rotor centrifuge to separate the phases. The chloroform phase extracts to the bottom, the DNA remains in the aqueous phase. A thick layer of organic matter normally separates the two phases at this point.

Note: Chloroform is toxic by skin contact and inhalation. Wear rubber gloves and work in a fume hood.

4. Using a wide-bore (e.g. Pasteur) pipette, remove the aqueous (upper) phase to a clean tube. If particles of organic matter are still visible, repeat steps 3 and 4.

5. Add two volumes of ice-cold ethanol (or one volume of ice-cold isopropanol). Mix gently to precipitate the nucleic acids. Often, the precipitate will become visible immediately; if not, store the samples for 30 min to several hours at –20 °C.

6. Wash the nucleic acid 30 min to overnight as follows:
   - If strands of DNA are visible, spool the DNA with a glass hook and transfer to 70 % ethanol to wash.
   - If spooling is impossible, centrifuge at 1000 rpm for 5–10 min to produce a soft pellet. If a pellet is not seen, centrifuge at 4000 rpm for 5 min, after which the nucleic acids should be visible as a hard white pellet. Remove the supernatant carefully and add 70 % ethanol to wash. Harder pellets are more difficult to wash.

7. Dislodge the pellet using a glass hook and transfer to a sterile 1.5 ml tube (if DNA was spooled, obtain pellet by centrifugation for 5 min at 4000 rpm). Leave tube open and allow to dry thoroughly (a vacuum-centrifuge, or alternatively a 37 °C incubator, may be used to speed up drying, though we generally simply leave samples 15–20 min at room temperature).

8. Resuspend the nucleic acids in 200 µl TE buffer. The pellet may be quite difficult to dissolve, especially if overdried. We generally leave samples overnight at 4 °C before tapping tubes gently to dissolve the DNA. Do not vortex or repeatedly pipette, as genomic DNA comprises very long strands that are easily damaged.

9. Add RNAase-A to a final concentration of 10 µg/ml and incubate 60 min at room temperature. DNA samples in TE buffer can be stored at 4 °C, or deep-frozen for long-term storage.

Testing DNA Concentration

1. Prepare a small agarose gel by boiling 1× TAE buffer containing 0.8 % (w/v) agarose. This is easiest in a microwave oven, but can also be done on a hotplate with constant stirring. The gel should boil vigorously to minimize air bubbles, thus a boiling flask at least five times the volume of the gel is recommended to avoid spillage. Water lost during boiling should be replaced (weigh the flask before and after boiling, add distilled water to recover the difference).

2. After cooling until the flask is comfortable to touch (45–50 °C), pour the gel to a depth of about 5 mm into a mini-gel casting tray with an appropriate well-comb.

3. When the gel has set, cover it in the gel chamber with 1× TAE buffer and remove the well-comb.

4. To 1 µl of each DNA sample add 7 µl sterile distilled water and 2 µl gel loading buffer. Mix by gently pipetting and load the 10 µl samples into the gel slots. For estimates of DNA concentrations, load three or four slots with a series of DNA concentration standards. We generally use 50, 100, 150 and 200 ng bacteriophage-λ DNA as concentration standards. A DNA size standard can also be added, but is not required for a genomic DNA test gel.

5. Mini-gels to test DNA concentration can be electrophoresed quite quickly, since separation of bands is unimportant. Electrophorese at 100 V (6 V/cm) for 20–30 min or until the bromophenol blue marker dye has traveled approximately half the length of the gel. DNA is negatively charged, thus the gel loading wells must be at the anodal end of the gel chamber!

6. Stain the nucleic acids by transferring the gel, using disposable rubber gloves, to a dish containing ethidium bromide diluted in deionized water to a concentration of 0.5 µg/ml and staining 15 min at room temperature.

Note: Ethidium bromide is a strong mutagen. Use disposable rubber gloves when handling ethidium bromide -containing gels and solutions. The staining solution can be used a number of times and should be disposed of only after appropriate decontamination (see Sambrook et al. 1989). After pouring off the staining solution, wash the gel in water for a minimum of 15 min (leave under gently running water, or change water at least once during washing). Although ethidium bromide bound to DNA fluoresces brighter under UV light than unbound ethidium bromide, and is thus visible even before washing, the wash step removes unbound ethidium bromide from the gel and thus reduces the possibility of contamination of work areas, at the same time reducing background fluorescence.

7. In a darkroom, lay the gel over a UV light source emitting at 302 nm. Pink-fluorescing bands of genomic DNA will be seen near the wells. Photograph the gel with a Polaroid camera mounted with a yellow-orange filter over the UV source.

Results

An example of a DNA test gel is shown in Fig. 2. Typical yields for this method range between 100 and 200 ng/μl. DNA samples should be compared against concentration standards and the concentrations altered

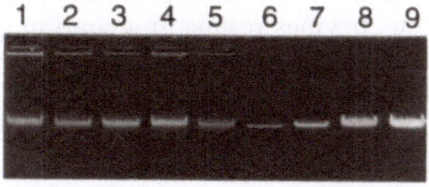

Fig. 2. Test gel to measure yields from genomic DNA extraction. Concentrations of DNA samples for fingerprinting should be measured and equalized by appropriate dilution, in order to prevent between-sample variation in fingerprint banding intensity. In *lanes 1–5* are 1 μl DNA samples extracted from different plant species; *lanes 6–9* are concentration standards for double-stranded DNA (50 ng, 100 ng, 150 ng and 200 ng bacteriophage-λ DNA, respectively). Genomic DNA was electrophoresed for 20–30 min at 120 V with unlimited current, in a 0.8 % (w/v) agarose minigel with TAE gel and electrode buffer. After electrophoresis the gel was stained for 15 min with 0.5 μg/ml ethidium bromide, destained in water for at least 15 min, and DNA was visualized on a UV light box emitting at 302 nm. Using a Polaroid camera mounted above the light box, bands were photographed through a yellow-orange filter onto Polaroid 667 black and white instant film

accordingly depending on the desired application. For RFLP fingerprinting a DNA concentration of approximately 50–100 ng/µl is ideal, whereas for best results in RAPD analysis, DNA should be diluted with TE buffer to a concentration of 10–25 ng/µl.

3.2
RFLP Fingerprinting

Southern (1975) described a method for detection of specific DNA sequences by hybridization of labeled probes to DNA fragments separated by electrophoresis and transferred from the gel to a solid matrix. The name Southern is today widely used to refer not only to this DNA-DNA hybridization technique, but also to the technique of DNA transfer from an agarose gel to a solid filter matrix ("Southern blotting").

DNA for subsequent RFLP analysis is first digested with a DNA restriction enzyme. Restriction enzymes recognize and cleave at specific sequences along the DNA and thus produce, for each unique DNA strand, a particular set of so-called restriction fragments. Restriction enzymes with a six-base recognition sequence ("six-cutters") are most commonly used for RFLP analysis. These enzymes, which are generally less sensitive to DNA impurities than four-cutter restriction enzymes, cut on average once every 4096 bp, yielding fragments of a size easily resolved by agarose gel electrophoresis. Because eukaryotic genomic DNA is many, many millions of base pairs long, its cleavage with restriction enzymes results in a continuous distribution of fragments ranging in size from a few base pairs to many thousands of base pairs long. When the restriction fragments obtained by digestion of eukaryotic genomic DNA are fractionated by electrophoresis, the result is thus a characteristic smear.

■ Materials

Equipment
- Large horizontal electrophoresis chamber (gel length at least 15 cm)
- DNA transfer membrane (e.g. Amersham, Boehringer Mannheim, Qiagen; we recommend positively-charged nylon membrane)
- Sheets of thick filter paper
- Stacks of paper towels
- X-ray film and autoradiography cassette
- Film developing facilities

- DNA restriction enzyme (e.g. *Hin*dIII, *Eco*RI or another six-cutter) with appropriate enzyme reaction buffer (e.g. Promega Corporation, Gibco-BRL)
- Digoxigenin random-primed labeling kit (e.g. Boehringer Mannheim, Amersham Life Sciences, Promega Corporation)
- Antidigoxigenin-AP FAB-fragments
- 25 mM CSPD chemiluminescence substrate (e.g. Boehringer Mannheim, Promega Corporation)
- EDTA stock solution: 2 M, pH 8.0
- Na-acetate stock: 3.0 M, pH 5.2

Chemicals

Buffers

20× SSC (saline sodium citrate, stock solution)

3 M NaCl
3 M Na-citrate (pH 7.0)

Sterilize by autoclaving.
- SDS (10 % stock solution): Dissolve 10 g SDS (sodium dodecyl-sulfate) in 100 ml distilled water by heating to 65 °C. Adjust pH to 7.2 before use.

Maleic acid buffer

0.1 M maleic acid
0.15 M NaCl (pH 8.0)

- **Either** Dig blocking solution (10 % stock solution) **or** BLOTTO
  - Dig blocking solution (10 % stock solution): Dissolve 10 g Dig blocking reagent (Boehringer Mannheim) in 100 ml maleic acid buffer by gentle heating with constant stirring. Sterilize by autoclaving.
  - BLOTTO: 5 % (w/v) non-fat dried milk; 0.02 % Na-azide. Dissolve in water and sterilize by autoclaving.

Hybridization buffer

5× SSC
0.1 % N-laurylsarcosine
0.02 % SDS
1 % blocking reagent (1:10 dilution of blocking solution)

Post-hybridization wash buffer A (for minisatellite probes)

2× SSC
0.1 % SDS

Post-hybridization wash buffer B (for oligonucleotide probes)

6× SSC
0.1 % SDS

Detection buffer

0.1 M Tris-HCl (pH 9.5)
0.1 M NaCl
50 mM $MgCl_2$

■ Procedure

DNA Restriction Digestion

1. To a sterile 1.5 ml reaction tube add the following in the given order:

2–5 μg DNA
20 μl 10× enzyme reaction buffer (supplied with enzyme)
Sterile distilled water to a total volume of 195 μl
5 μl restriction enzyme (2–5 enzyme units per 1 μg genomic DNA)

Note: If DNA is particularly unclean or proves difficult to cut, 1 μl 50 mM spermidine can be added to the reaction to facilitate digestion.

Mix thoroughly, spin tubes briefly to bring contents to bottom, and incubate overnight at 37 °C (depending on concentration and cleanliness of DNA, and the particular enzyme used, shorter incubation times normally suffice; however overnight incubation generally ensures complete digestion).

2. Add 20 μl sodium acetate stock and precipitate DNA by mixing thoroughly with 660 μl ice-cold ethanol and leaving for 15 min at −20 °C.

3. Spin 15 min at 12,000 rpm in a bench-top centrifuge, remove supernatant and wash the DNA pellet briefly in 70 % ethanol. Remove ethanol and dry DNA at room temperature for 10 min. (Droplets adhering

to the tube walls can be carefully removed with a sterile pipette tip attached to a vacuum pump, but take care not to suck up the pellet!)

4. Resuspend the DNA in 20 µl TE buffer by gently tapping the tubes with the index finger. Resuspension can be accelerated by incubating samples 30 min at 37 °C in a water bath. When the pellet is fully dissolved, add 5 µl gel loading buffer, mix thoroughly and spin samples for a few seconds in a microfuge to bring contents to the bottom of the tubes.

Electrophoresis of Restriction Fragments

For RFLP analysis, DNA fragments must be well separated and gel distortion must be minimized. Electrophoresis is thus carried out at low voltage to avoid overheating of the gel, and maximum separation of fragments is achieved by electrophoresis for extended time periods in a long gel. Note that, for RFLP analysis, ethidium bromide should not be added to DNA samples before electrophoresis, as is the practice in some laboratories for normal nucleic acid electrophoresis. Ethidium bromide can cause DNA concentration-dependent differences in migration rate, which can lead to band-shifting problems.

1. Prepare a large gel, using the same method as Sect. 3.1, containing 0.8 % agarose in 1× TAE buffer. About 150 ml is required for a 15×20 cm gel tray. After cooling the gel to 45–50 °C, pour the gel to a depth of approximately 5 mm into the gel tray, containing the appropriate well comb, on a level surface. Take care not to generate bubbles in the gel while pouring (if a few bubbles do occur, these should be removed immediately by touching them gently, before the gel sets, with the corner of a paper tissue).

2. When the gel has fully set, transfer it to the electrophoresis chamber, cover with 1× TAE buffer and gently remove the well comb (removing the comb in the absence of buffer can cause collapse of or subsequent air bubbles in the wells, which cause problems when loading the gel).

3. Load the 25 µl samples of digested DNA and gel loading buffer to the gel wells using a micropipetter with a fine tip. Load a DNA size standard appropriate for fragment sizes of 500–20,000 bp (e.g. *Hind*III/ *Eco*RI digested bacteriophage-λ DNA).

4. Electrophorese for 16–20 h at 45 V (1.5 V/cm).

5. Stop electrophoresis when the bromophenol blue marker dye begins to travel out of the cathodal end of the gel. Under the conditions described here, the rate of electrophoresis of bromophenol blue is approximately equivalent to that of double-stranded DNA fragments

Fig. 3. Restriction digest of genomic DNA from rapeseed, showing typical smear patterns containing faint satellite-DNA bands. *Lanes 1–3* are digested with *Hind*III, *lanes 4–6* with *Eco*RI. The smears confirm that each sample is completely digested. The size marker (*m*) is produced by *Eco*RI/*Hind*III double-digestion of phage-λ DNA, and is necessary after Southern hybridization for estimation of the sizes of RFLP bands. Restriction digested genomic DNA was electrophoresed for 16–20 h (until the bromophenol blue marker dye began to migrate out of the gel) at 50 V with unlimited current, in a 0.8 % (w/v) agarose gel with TAE gel and electrode buffer. After electrophoresis the gel was stained for 15 min with 0.5 µg/ml ethidium bromide, destained in water for at least 15 min, and DNA was visualized on a UV light box emitting at 302 nm. Stained DNA was photographed through a yellow-orange filter onto Polaroid 667 black and white instant film using a Polaroid camera mounted above the light box

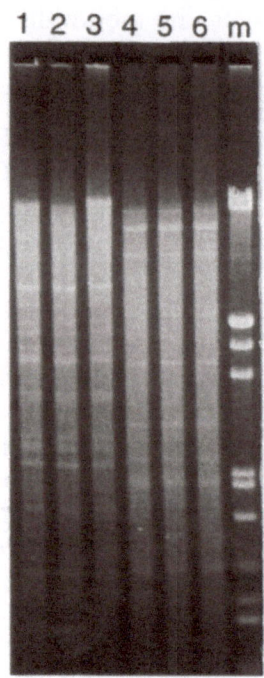

about 400 bp long. The fragments of interest for RFLP analysis are generally larger than 400 bp.

6. Stain and photograph the gel as in Sect. 3.1. If the gel casting tray used is not UV-transparent, carefully transfer the gel onto the UV light source using a large acetate sheet (e.g. an overhead transparency sheet) for support to avoid gel breakage.

A typical restriction digest gel is shown in Fig. 3. Each lane shows a smear representing the continuous distribution of fragment sizes produced by digestion of genomic DNA. Highly repetitive minisatellite sequences are seen as light bands within the smear.

Southern Blotting   Digested DNA is transferred from the gel to a solid matrix (membrane), where it is hybridized with labeled probes that allow visualization of bands corresponding to particular DNA sequences. In this way, a DNA fingerprint is produced from the smear of DNA fragments. Earlier, the matrix used was generally nitrocellulose, however today most laboratories use nylon membranes, which are more robust and retain DNA more efficiently for re-probing. We normally use positively charged nylon membrane, which as well as allowing even more frequent reprobing than uncharged nylon has the added advantage that DNA does not require fixing after alkaline blotting.

The two most widely used transfer techniques are vacuum-blotting and capillary transfer. While vacuum blotting is very rapid, the capillary transfer method we describe here, adapted from Molenaar and Wilkins (1991), is also relatively quick and has the advantage that no special equipment is necessary. We routinely use capillary transfer for all our Southern hybridization experiments.

1. After photographing the gel, transfer it (in the gel casting tray or on a plastic sheet) to a plastic dish. If the gel is not left in the casting tray, we recommend turning it upside-down at this point, since imperfections or scratches in the washing dish can damage the gel underside during the following wash steps. The washes are carried out at room temperature, on a slowly shaking platform if available.

2. Wash 10 min in 0.25 M HCl. (Acid treatment depurinates DNA, which after subsequent denaturation results in cleavage of DNA strands, shortening longer fragments and allowing for more efficient transfer. Over-depurination results in fragments which are too small to be efficiently retained by the membrane.)

3. Remove HCl (by carefully pouring off liquid while supporting the gel with a gloved hand), rinse briefly with distilled water, and denature by washing 2×15 min in 0.4 M NaOH (DNA strands denature under alkaline conditions. DNA must be denatured to permit subsequent hybridization).

4. During the depurination and denaturation steps, cut a piece of nylon membrane to the same size as the gel. Cut one corner of the membrane for later reference. To the same dimensions as the membrane, cut four or five pieces of thick filter paper (1–2 mm; the card used by breweries for manufacture of beer-mats is ideal for this purpose!) and a stack of paper towels about 5 cm high. Thoroughly wet the membrane and all but one filter card in 0.4 M NaOH.

Note: Always handle the membrane and filters at the corners and only with clean rubber gloves or forceps. The membrane is very sensitive to oils or proteins from the fingers which can inhibit blotting and cause high background hybridization.

5. Construct the blot as follows, taking care to avoid bubbles between the different layers: Using rubber gloves, layer three or four pieces of

the pre-wetted filter card on a sheet of glass on the workbench. Carefully overturn the gel (to avoid disaster, this is best done between two overhead-transparency sheets!) and lay it, well-side down, on the layer of filter papers (the lower surface of the gel is perfectly flat and provides better contact with the membrane, whereas the upper gel surface exhibits imperfections from pouring that can inhibit blotting when the gel is not inverted beforehand). Lay the pre-wetted membrane on top of the gel, with the cut corner corresponding to the first well. Removal of air bubbles, which block the capillary action, is vital here (air bubbles are best removed by rolling a glass rod or pipette over the filter). Lay a further pre-wetted filter card over the membrane and remove bubbles again. Lay a dry filter paper on top and complete the construction with the stack of pre-cut paper towels, weighed down with the gel casting tray on top. Blotting is accelerated with a 200–300 g weight (e.g. a bottle of water) on top of the stack.

6. Leave for 3–6 h. DNA is drawn out of the gel by capillary action and retained by the membrane. The layer of paper towels will be damp and the gel will be reduced in thickness by about 80 %.

7. Wearing gloves, deconstruct the blot apparatus and transfer the membrane to a clean dish. Wash briefly in 5× SSC, making sure that no residual agarose is stuck to the membrane.

8. If positively charged membrane was used, air-dry the filter between two sheets of filter paper. If uncharged membrane was used, the DNA must be fixed to the membrane before hybridization. This can be achieved by UV cross-linking, but we find the easiest and most reliable method is baking at 80 °C for 30 min.

RFLP Probes A wide range of different probes are suitable for RFLP fingerprinting. These range from short, simple-sequence oligonucleotide repeats, e.g. $(GATA)_4$, $(CAC)_5$, $(TG)_n$), to minisatellite probes that directly detect specific groups of tandem repeat sequences. In principle, the same methods can be adapted – with alterations in hybridization and wash temperatures – for use with all types of probe.

One very useful fingerprinting probe is genomic DNA from the bacteriophage M13, which has been found to identify a high degree of minisatellite sequence polymorphism in an incredibly wide range of species encompassing humans, fish and other animals, plants and microorganisms (e.g. Vassart et al. 1987; Ryskov et al. 1988; Fields et al. 1989; Zimmerman et al. 1989). Simple oligonucleotide repeat sequences are also very useful as general RFLP fingerprinting probes (e.g. Weising et al. 1991). Other options include the use of highly repetitive DNA sequences like ribosomal DNA or species-specific minisatellite sequences isolated from satellite bands seen after electrophoresis of restriction digest products.

Until recent years, the method of choice for RFLP fingerprinting was radioactive labeling of probes, owing to the high sensitivity of isotopic labels in comparison with nonradioactive labeling and detection. Improved chemiluminescent detection methods have now been developed, however, which approach the sensitivities achieved with radioactive systems and avoid the associated safety hazards. Reliable RFLP fingerprinting techniques are thus now accessible to laboratories without facilities for usage and disposal of radioactive chemicals.

Probe Preparation

Probe DNA is labeled by the random priming technique developed by Feinberg and Vogelstein (1983). This procedure is greatly simplified by use of a commercial labeling kit, which are widely available, reliable and generally cost-effective. We normally label probes with digoxigenin, but the same nonradioactive techniques can be applied with biotin (and subsequently the appropriate biotin detection antibodies). For oligonucleotide probes we find it most convenient to purchase pre-labeled DNA.

1. Label the DNA according to the instructions of the kit manufacturer.

2. Stop the reaction by adding EDTA to a concentration of 0.02 M. Dig-labeled probe DNA can be used without further purification, since unincorporated nucleotides are removed from the membrane during the subsequent hybridization and detection washes.

3. Denature probe DNA by incubating in a boiling water bath for 10 min (Dig-labeled probes cannot be alkaline-denatured) and chilling immediately on ice. Nonhomologous single-stranded probes like $(GATA)_4$ can be used without denaturing.

Hybridization

Labeled, denatured probes are hybridized to their complementary sequences fixed on the Southern blot membrane. Before hybridization, membranes must be blocked in a pre-hybridization step, to prevent the probe binding directly to the membrane surface. Pre-hybridization is carried out in hybridization solution **without** the labeled probe.

The hybridization solution described here is a suitable for most applications. Many variations are possible, but most importantly the solution must contain a blocking reagent which binds to the membrane and prevents nonspecific hybridization of the labeled probe. We use the Dig-blocking reagent supplied by Boehringer Mannheim, but many laboratories substitute this with BLOTTO, which is simply 5 % non-fat dried milk containing 0.02 % Na-azide as a preservative, or occasionally with Denhardts reagent.

A large number of protocol variations are possible for Southern blot hybridization, but the simple protocol given here has proved to be extremely versatile and robust in our hands. Pre-hybridization, hybridization and the subsequent washing and detection steps are best performed in hybridization bottles rotating in a hybridization oven. If a hybridization oven is not available, use a sealable plastic dish in a shaking water bath; in this case solution volumes should be increased according to the size of the dish.

1. Wet the Southern blot membrane in 2× SSC containing 0.1 % SDS. Wearing gloves, roll the membrane and place it, DNA-side up, in a hybridization bottle. Unroll by adding a small amount of 2× SSC and rolling until the membrane adheres to the tube walls. Replace the SSC with 20 ml of pre-warmed hybridization solution and pre-hybridize 30 min at the appropriate hybridization temperature.

Note: For minisatellite probes, hybridize at 65 °C; for oligonucleotide probes the normal hybridization temperature is 38 °C.

2. Add labeled, denatured probe DNA directly to the pre-hybridization solution and mix well (take care not to pipette concentrated probe directly onto the membrane, as this can result in an area of high background hybridization).

3. Hybridize 2 h to overnight at hybridization temperature. (Highly repetitive oligonucleotide probes require lower hybridization times, minisatellite probes give best results with overnight hybridization. It is best to begin with longer hybridization times when testing new probes; if hybridization is very strong then times may be reduced.)

4. Pour off the hybridization solution (Dig-labeled probes can be reused a number of times when stored in hybridization solution at −20 °C and re-denatured before each use).
   - For **minisatellite probes**, wash the membrane 15 min at room temperature, followed by 2×15 min at 65 °C, in wash buffer A. The washes remove the probe from partially homologous sequences, depending on the SSC concentration and temperature. Increased stringency (i.e. lower SSC concentration or higher temperature)

reduces the number of bands present in the RFLP fingerprint, and is necessary in cases of high nonspecific background hybridization in sample lanes.
- For **oligonucleotide probes**, wash 5 min in wash buffer B at room temperature and 5 min at 38 °C.

We routinely use a chemiluminescence-based system for detection of Dig-labeled probes, giving sensitivities approaching those obtained by radioactive detection. As with DNA-labeling, reasonably cost-effective commercial kits are also available for chemiluminescent detection of Dig-labeled probes. The chemiluminescence substrate we use is CSPD, which has succeeded the somewhat less sensitive AMPPD. Variations of CSPD can provide even higher sensitivity, though may result in higher backgrounds with certain membrane types. Unless otherwise stated, the detection steps are carried out at room temperature in rotating hybridization bottles.

Detection of Hybridization Sites

1. Without letting the membrane dry out (if detection is not carried out immediately after washing, store the membrane damp at 4 °C in a sealed plastic bag), wash the membrane briefly in detection wash buffer.

2. Dilute 5 ml Dig-blocking solution (or BLOTTO) with 45 ml maleic acid buffer and block membrane with this blocking buffer for 30 min.

3. Add 5 µl anti-digoxigenin-AP (FAB-fragments) directly to blocking buffer (taking care not to pipette directly onto the membrane) and incubate 30 min.

4. Wash 2×15 min in maleic acid buffer containing 0.02 % Tween 20.

5. Rinse membrane briefly in detection buffer.

6. Dilute 10 µl CSPD in 1 ml detection buffer. Lay the membrane on a sheet of clear plastic and, without allowing it to dry out, pipette the diluted CSPD solution onto one end of the membrane. Roll a second plastic sheet over the first, spreading the CSPD solution over the entire membrane. Incubate the membrane, between the two plastic sheets, in the dark at 37 °C for 5–15 min. After incubation, squeeze out the excess fluid by wiping over the top plastic sheet with a paper towel.

7. In a photographic darkroom, expose the filter to X-ray film for 20–30 min. The membrane, still inside the plastic sheets, should be

placed DNA-side up in an autoradiography cassette, with the film laid on top directly before sealing the cassette (if an autoradiography cassette is unavailable, cardboard filmholders can be constructed by sandwiching two sheets of thick black card between perspex sheets and sealing tightly with binding clamps). Develop, fix and wash the film. If the hybridization signal is too weak, re-expose for a longer time. If the film is overexposed, shorten the exposure time.

8. Proceed with data analysis (see "Results").

3.3
RAPD Fingerprinting

Randomly amplified polymorphic DNA (RAPD) fingerprints, first described by Williams et al. (1990), are generated by the random amplification of genomic DNA segments using single primers of arbitrary nucleotide sequence. A primer is mixed with genomic DNA in the presence of a thermostable DNA polymerase, and DNA fragments are exponentially amplified using the polymerase chain reaction (PCR; Mullis et al. 1986; Mullis and Faloona 1987). Amplification products are generated when the primer anneals to homologous sites on opposite strands of the genomic DNA and within an amplifiable distance of each other (a few thousand nucleotides). A RAPD fingerprint is generated by agarose gel electrophoresis of the amplification products of a single primer, and the presence or absence of DNA fragments is diagnostic for oligonucleotide binding sites on the genomic DNA. To generate sufficient RAPD fingerprint information for comparisons between individuals and populations, DNA amplification of a set of samples is repeated with many different primers, under conditions resulting in several amplified bands from each primer. Generally, oligonucleotide primers of ten bases are used, with a base composition ranging from 50–70% G+C. Palindromes greater than six bases in length and complementary at the 3'-end should be avoided. Various random primers are available for purchase (e.g. Operon Technologies Inc.).

We present here a basic protocol that can be easily applied for initial RAPD screening. In practice, different oligonucleotide primers require different PCR conditions for optimal amplification. Since the optimal conditions for any particular primer (and DNA from any particular species) are best determined by simple testing, we include a brief outline of the most important factors in the optimization of amplification conditions.

▨ Materials

– Thermocycler
– Large horizontal electrophoresis chamber (gel length 10–20 cm)

– Stock solution dNTPs: dATP, dCTP, dTTP and dGTP (e.g. Gibco-BRL, Promega Corporation), each diluted to 1 mM in sterile double-distilled water and stored in small aliquots at –20 °C
– Primer stock solution: Each oligonucleotide primer (e.g. Operon Technologies Inc.) is diluted to 20 µM in 10 mM Tris-HCl, pH 7.4; 1 mM EDTA and stored at 4 °C.
– *Taq* DNA polymerase (e.g. Gibco-BRL, Promega Corporation; normally supplied at 5 U/µl, dilute to 1 U/µl before use)
– Filter-sterilized mineral oil
– Nu-sieve agarose (FMC Bioproducts)

PCR buffer (10× stock solution)

100 mM Tris-HCl (pH 8.3)
500 mM KCl
15 mM $MgCl_2$
0.01 % (w/v) gelatin

TBE buffer (5× stock solution)

54 g Tris base
27.5 g boric acid
20 ml 0.5 M EDTA (pH 8.0)
Dilute 1:10 in deionized water before use.

▨ Procedure

Depending on the thermocycler used, amplification reactions are carried out in sterile 96-well microtiter plates or PCR tubes (0.2 or 0.5 ml volume). To minimize pipetting errors and assist automation, the components of the amplification reaction are mixed together directly before use and this reaction mixture is added to each DNA sample.

1. For each primer, prepare the following reaction mix by adding the components, in the order given, to a sterile tube. Mix gently and spin to bring tube contents to bottom of tube.

Note: A 24 µl reaction mix is required for each amplification reaction – the given volumes should therefore be multiplied by the number of DNA samples to be amplified. Volumes for a single reaction mix are:

2.5 µl 10× PCR buffer
1 µl dNTP stock solution
1 µl primer stock solution
18.5 µl sterile double-distilled water
1 µl *Taq* polymerase (1 U/µl)

---

2. Pipette 1 µl (10–25 ng/µl) of each DNA sample into the PCR tubes or microtiter plate wells. Add 24 µl of pre-prepared reaction mix.

3. Overlay with 50 µl sterile mineral oil (not needed when using a thermocycler with heated lid).

4. Carry out RAPD PCR using the following standard cycle conditions: 40 cycles of 1 min at 94 °C (denaturation of DNA), 1 min at 36 °C (primer annealing), 2 min at 72 °C (primer extension); followed by one cycle of 10 min at 72 °C. Cool and hold at 5 °C.

5. Add 5 µl gel loading buffer (Sect. 3.1) to each sample by inserting the pipette tip through the oil overlay. Store samples at 4 °C until electrophoresis.

Electrophoresis of Amplification Products

1. Prepare a 1.4 % Nu-sieve agarose gel in 0.5× TBE buffer using the method described in Sect. 3.1, "Procedure."

2. After cooling the gel to 45–50 °C, pour the gel to a depth of approximately 5 mm in a gel casting tray, containing the appropriate well comb, on a level surface.

3. When the gel has fully set, transfer it to the electrophoresis chamber, cover with 0.5× TBE buffer and gently remove the well comb.

4. Taking care to pipette from underneath the oil overlay, load 10 µl of each RAPD PCR sample. Load a DNA size marker appropriate for double-stranded DNA between 200 and 2000 bp (e.g. Gibco-BRL 100 bp DNA-ladder).

5. Electrophorese approximately 2.5 h at 120 V (4 V/cm). The bromophenol blue marker dye should run at least 10–15 cm. Bromophenol blue migrates in 1.4 % Nu-sieve agarose in TBE buffer at approximately the same rate as amplification fragments of around 200 bp.

6. Stain and photograph gel as in Sect. 3.1, "Procedure".

7. Proceed with data analysis (see "Results" below).

Optimizing RAPD Analysis

The standard protocol given will generally produce RAPD products with all genomic DNA samples for most primers tested. Although cycling conditions may be altered, we have found that the cycle described produces excellent results with genomic DNA from a wide range of species, and also in different styles of thermocycler from different manufacturers. Normally, unsatisfactory amplification can be rectified by modification of the PCR reaction conditions. Altering the components of the reaction mixture tends to improve results in cases of high background or unreliable bands caused by unspecific priming and amplification. Amplification conditions should be optimized for each primer, to obtain a maximal number of clear, reproducible RAPD bands. In general, the given concentrations of dNTPs and *Taq* polymerase allow efficient amplification without mispriming or non-specific amplification (although we recommend the use of a single brand of *Taq* polymerase for all comparative RAPD tests, due to variation in activity among different brands). Under constant cycling conditions, the most important factors in optimization of amplification conditions are the concentrations of template DNA and Mg^{2+}, respectively. More extensive information regarding optimization of PCR reactions can be obtained from Innis and Gelfand (1990).

DNA Concentration

Differences in DNA template concentration can result in large variation in amplification products. In our experience, this has more to do with the concentration of contaminants in the genomic DNA than the DNA concentration itself. While amplification can theoretically proceed from a single DNA strand, contaminants in the genomic DNA sample may inhibit amplification until they are sufficiently diluted. If amplification is unsatisfactory or varies over a number of similar samples, we recommend testing DNA amounts from 0.1 ng to 50 ng in the standard RAPD reaction. In our hands, best results are obtained with 10–20 ng of the DNA extracted using the method given here. Excessive DNA can result in nonspecific amplification (evident as a background smear in the gel lane), while too little DNA may give variable or nonreproducible bands.

Mg²⁺ Concentration

Magnesium concentration has a large influence on the efficiency and accuracy of the RAPD reaction. The standard reaction (1.5 mM $MgCl_2$) provides a slight surplus of Mg^{2+} which is generally sufficient for reliable amplification. Excessive concentrations can result in nonreliable banding patterns, in which unexpected bands are observed and others are missing. It is important to remember, however, that the presence of EDTA or other chelators in the primer stocks or template DNA may disturb the apparent magnesium optimum. On the other hand, DNA samples may also contain some residual Mg^{2+}. Thus, if amplification is unsatisfactory, it is often worthwhile to repeat amplification with different Mg^{2+} concentrations.

Results

Typical RFLP and RAPD fingerprint results are shown in Figs. 4 and 5, respectively. In general, a single RFLP fingerprint probe can give a large amount of information about the genetic variation (or similarity) among even closely related individuals. To obtain a sufficient number of useful (polymorphic) bands using RAPD fingerprinting, a number of different primers must be assayed; however single primers can be analytical in comparisons between less closely related individuals, for example samples from genetically isolated populations or plant cultivars.

In principle, data obtained from RAPD and RFLP fingerprinting techniques can be treated in much the same way. Although the possibility exists that two RAPD bands of identical migration might comprise a different DNA sequence (only the primed ends must be identical), for the purposes of data analysis it is assumed that fingerprint bands (RFLP fragments or RAPD amplification products) of equal size represent the same sections of genomic DNA. Differences in fragment lengths are thus interpreted as genetic differences. For each individual investigated, bands are scored as present or absent and differences between individuals are calculated according to the variability between their genotypes (the complete set of fingerprint data for an individual, pooled from all RFLP or RAPD markers). For analysis of large numbers of individuals, for example in extensive analysis of populations, it is recommended to digitalize band patterns and analyze using specialized software (e.g. Scanalytics RFLPscan).

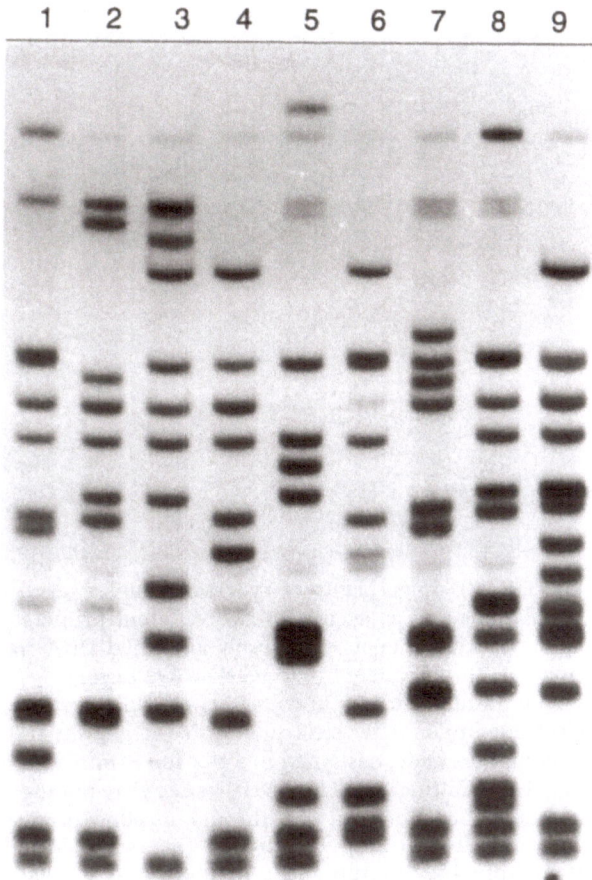

Fig. 4. Multilocus RFLP fingerprints of barley powdery mildew isolates, produced by hybridization of digested, electrophoresed and blotted genomic DNA with a highly polymorphic repetitive minisatellite probe. Each lane represents a different individual. Multilocus minisatellite fingerprints typically produce a large number of polymorphic bands, thus allowing distinction of even closely related individuals. The nine individuals shown here all belong to the same species, however no two have the same RFLP fingerprint. Such comparisons can be useful in the search for molecular markers linked to important traits like disease or pest resistance. A common strategy in plant breeding programs, for example, is to search for RFLP bands which are present more frequently in individuals known to express desired genes, since such bands are likely to be linked to the character of interest and thus be useful molecular markers

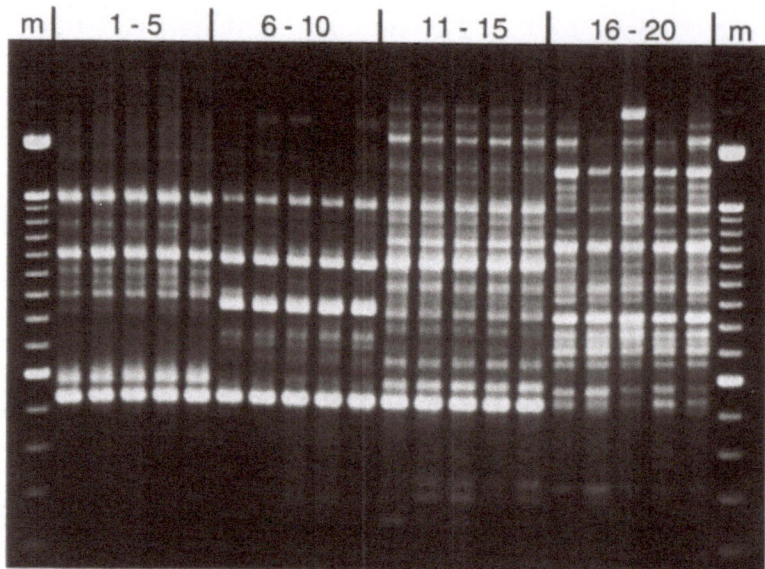

Fig. 5. RAPD fingerprints of ryegrass samples taken from two genetically isolated populations and assayed with two different primers producing population-specific amplification products. Randomly amplified DNA was electrophoresed about 2.5 h (until the bromophenol blue marker dye began to migrate out of the gel) at 120 V with unlimited current, in a 1.4 % (w/v) Nu-sieve agarose gel with 0.5× TBE gel and electrode buffer. After electrophoresis the gel was stained for 20 min with 0.5 µg/ml ethidium bromide, destained in water for 30 min, and bands were visualized on a UV light box emitting at 302 nm. DNA bands were photographed, with a Polaroid camera mounted above the light box, through a yellow-orange filter onto Polaroid 667 black and white instant film. The marker (*m*) is a DNA ladder (Gibco-BRL) with 100 bp steps from 100–1500 bp and a further band at 2000 bp. The first primer (OPA-10, Operon Technologies Inc.) shows little or no variation within the respective populations, but the two populations can be distinguished from one another by their specific RAPD patterns (*lanes 1–5* and *6–10*, respectively). The second primer (OPB-15, Operon Technologies Inc.) also shows no variation within the first population sample (*lanes 11–15*), but again this sample is clearly distinguishable from the second (*lanes 16–20*), in which within-population variability is also apparent. On the basis of RAPD fingerprint patterns with different primers, an unknown individual could be identified with respect to which population it originated from. In addition, population-specific fingerprint bands often allow identification of genotypes containing important characters, and screening of RAPD fingerprints then enables identification for breeding purposes of individuals or populations exhibiting desired traits

The degree to which DNA fingerprinting allows discrimination between genetically different individuals depends not only on the degree to which the individuals differ, but also particularly on the number of informative markers investigated. Understandably, investigation of a large number of fingerprint bands thus allows more reliable determination of genetic variation or similarity. A detailed description of factors contributing to the statistical significance of fingerprinting results, and suggestions for analysis strategies, are given by Weir (1990).

Since this chapter is intended as a simple introduction to practical RFLP and RAPD methods, we will only provide a few basic statistical methods for calculation of genetic similarity between individuals assayed by DNA fingerprinting. For further introductory information about statistical methods involved in RFLP and RAPD fingerprinting, we recommend Bruford et al. (1992). In-depth information on genetic data analysis can be found in Weir (1990); however it should be mentioned that, when more than a few individuals are investigated, statistical analysis is best undertaken using appropriate computer software (e.g. SPSS, NTSYS-PC).

For the purposes of comparing individuals or populations on the basis of genetic fingerprints, the first step is calculation of genetic similarity between and among the samples assayed. The most commonly used method for calculating similarity coefficients between two genotypes is the Dice coefficient, as described by Rohlf (1992): **Similarity Coefficients**

$$s = 2N_{ab}/(2N_{ab} + N_a + N_b)$$

where N_{ab} is the number of bands shared by both genotypes in the comparison, and N_a and N_b are the numbers of bands present in the respective genotypes.

An alternative is the similarity coefficient of Jeffreys (1985):

$$x = ((N_{ab}/N_a) + (N_{ab}/N_b))/2$$

Other variations are sometimes used, but in principle the similarity measurement adopted does not greatly influence the outcome. In each case the most important factor is the relative genetic difference among the samples investigated. Pairwise comparisons of genetic similarity among all individuals investigated provide information for subsequent cluster analysis to determine genetic relationships within and among populations.

Cluster Analysis   For analyzing relationships within or between groups of individuals (populations or population samples) a cluster analysis is normally applied. The UPGMA (unweighted pair group method with arithmetical averages) algorithm of Sneath and Sokal (1993) is the most common method, whereby similarity indices within or among population samples are compared with each other using the squared Euclidean distances between paired genotypes. The UPGMA method groups related individuals and calculates relative genetic distances within and among groups. These distances are used to create a so-called cluster diagram, or dendrogram, which illustrates the genetic relationships calculated from the DNA fingerprint data.

References

Bruford MW, Hanotte O, Brookfield JFY, Burke T (1992) Single-locus and multi-locus DNA fingerprinting. In: Hoelzel AR (ed) Molecular genetic analysis of populations. Oxford University, Oxford, pp 225–269

Doyle JJ, Doyle JL (1990) Isolation of plant DNA from fresh tissue. Focus 12:13–15

Ellsworth DL, Rittenhouse KD, Honeycutt RL (1993) Artifactual variation in randomly amplified polymorphic DNA banding patterns. Biotechniques 14:214–217

Feinberg AP, Vogelstein B (1983) A technique for radiolabelling DNA restriction endonuclease fragments to high specific activity. Anal Biochem 132:6–13

Fields RD, Johnson KR, Thorgaard GH (1989) DNA fingerprints in rainbow trout detected by hybridization with DNA of bacteriophage M13. Trans Am Fish Soc 118:78–81

Innis MA, Gelfand DH (1990) Optimization of PCRs. In: Innis MA, Gelfand DH, Sninsky JJ, White TJ (eds) PCR Protocols. Academic, San Diego, California, pp 3–12

Jeffreys AJ, Wilson V, Thein SL (1895) Individual-specific 'fingerprints' of human DNA. Nature 31:76–79

Molenaar AJ, Wilkins RJ (1991) A simple and convenient way of blotting nucleic acids. Biotechniques 10:334–335

Mullis K, Faloona F, Scharf S, Saiki R, Horn G, Erlich H (1986) Specific enzymatic amplification of DNA in vitro: The polymerase chain reaction. Cold Spring Harbor Symposia on Quantitative Biology, Volume LI. Cold Spring Harbor Laboratory, New York. pp 263–273

Mullis KB, Faloona (1987) Specific synthesis of DNA in vitro via a polymerase-catalyzed chain reaction. Methods Enzymol 155:335–350

Rohlf FJ (1992) NTSYS-pc numerical taxonomy and multivariate analysis system, Version 1.7. Owner manual

Ryskov AP, Jincharadze AG, Prosnyak NI, Ivanov PL, Limborska SA (1988) M13 phage DNA as a universal marker for DNA fingerprinting of animals, plants and microorganisms. FEBS Letters 233:388–392

Sambrook J, Fritsch EF, Maniatis T (1989) Molecular cloning: a laboratory manual. 2nd edition. Cold Spring Harbor Laboratory, New York

Sneath PHA, Sokal RR (1973) Numerical taxonomy. WH Freeman, San Francisco, California

Southern EM (1975) Detection of specific sequences among DNA fragments separated by gel electrophoresis. J Mol Biol 98:503–517

Vassart GM, Georges R, Monsieur H, Brocas AS, Lequarre AS, Christophe D (1987) A sequence in M13 phage detects hypervariable minisatellites in human and animal DNA. Science 235:683–684

Vos P, Hogers R, Bleeker M, Reijans M, van de Lee T, Hornes M, Frijters A, Pot J, Peleman J, Kuiper M, Zabeau M (1995) AFLP: a new technique for DNA fingerprinting. Nucl Acids Res 23:4407–4414

Weir BS (1990) Genetic data analysis. Sinauer Associates, Sunderland, Massachusetts

Weising K, Beyermann B, Ramser J, Kahl G (1991) Plant DNA-fingerprinting with radioactive and digoxygenated oligonucleotide probes complementary to simple repetitive DNA sequences. Electrophoresis 12:159–169

Williams JGK, Kubelik AR, Livak KJ, Rafalski JA, Tingey SV (1990) DNA polymorphism amplified by arbitrary primers are useful as genetic markers. Nucl Acids Res 13:6531–6533

Zabeau M, Vos P (1993) Selective restriction fragment amplification: a general method for DNA fingerprinting. European Patent Application, Publication #0534858-A1, Office europeén des brevets, Paris

Zimmerman PA, Lang-Unnasch N, Cullis CA (1989) Polymorphic regions in plant genomes detected by an M13 probe. Genome 32:824–828

Information sources in the Internet

Internet newsgroups: bionet.molbio.methds-reagnts
Other useful Internet sites:
Many very good links to molecular biology, genetics, and electrophoresis resources and protocols: http://www-lmmb.ncifcrf.gov/~pnh/index.html

Links to molecular biology protocols: http://www.horizonpress.com/gateway/protocols.html
Molecular biology protocols, methods discussion forum: http://research.nwfsc.noaa.gov/protocols.html

▪ Suppliers

Amersham Life Sciences
2636 South Clearbrook Drive, Arlington Heights, Illinois 60005, USA
Phone: 1-800-323-9750
Fax:1-800-223-8735
e-mail: alssah@ixx.netcom.com

Boehringer Mannheim
9115 Hague Road, Indianapolis, Indiana 46250-0414, USA
Phone: 1-800-262-1640
Fax: 1-800-262-4911
e-mail: biochemts–us@bmc.boehringer-mannheim.com

FMC Bioproducts
1919 Thomaston Street, Rockland, Maine 04841, USA
Phone: 1-800-341-1574
Fax: 1-800-362-5552
e-mail: biotexhserv@fmc.com

Gibco-BRL Life Technologies
8400 Helgerman Court, Gaithersburg, Maryland 20884-9980, USA
Phone: +1-301-840-8000
Fax: +1-301-670-8599
e-mail: gibco-brl@aol.com

Operon Technologies, Inc.
1000 Atlantic Avenue, Alameda, California, 94501, USA
Phone: 1-800-688-2248
Fax: +1-510-865-5255
e-mail: dann@operon.com

Promega Corporation
2800 Woods Hollow Road, Madison, Wisconsin 53711-5399, USA
Phone: 1-800-356-9526
Fax: +1-608-277-2576
e-mail: techserv@promega.com

Qiagen
9600 De Soto Avenue, Chatsworth, California 91311, USA
Phone: 1-800-426-8157
Fax: 1-800-718-2056
e-mail: qiagen@kaiwan.com

Pulsed-Field Gel Electrophoresis for Genomic Fingerprinting of Pathogenic Bacteria

MICHAEL B. COULTHART, WENDY M. JOHNSON, FRASER E. ASHTON

Introduction

Before techniques for direct analysis of sequence variation in DNA were available, genetic polymorphisms used to type intraspecific strains of pathogenic bacteria were inferred by comparing biochemical and/or serological phenotypes. Now, clinical laboratories can analyze DNA simultaneously from many bacterial strains, comparing them for presence or absence of specific genes or for varying nucleotide sequences at particular homologous loci. Interstrain discrimination is thus limited in principle only by the amount of genetic polymorphism naturally present within a bacterial species. Also, with an appropriate theoretical framework it should be possible to use such data to accurately reconstruct phylogenetic (evolutionary) relationships among those lineages. This approach, commonly called "molecular epidemiology" (although this term has much broader connotations – see Schulte and Perera 1993), is actively developing as a methodological framework for studying disease transmission in human populations.

One widely used laboratory technique in bacterial molecular epidemiology relies on electrophoretic separation of large (50–500 kilobase pairs, kbp) DNA fragments produced by cleavage with a restriction endonuclease having few (10–20) genomic recognition sites. The resulting fragment mobility patterns can reveal interstrain differences, called restriction-fragment length polymorphisms (RFLP), which can be used as genetic markers for strain tracing. The specialized electrophoretic methods required to separate such populations of large DNA fragments into countable numbers of gel bands are collectively called pulsed-field

Correspondence to Michael B. Coulthart, Bureau of Microbiology, Federal Laboratories for Human and Animal Health, 1015 Arlington Avenue, Winnipeg, Manitoba, Canada R3E 3R2, Phone: +01-204-789-6026, Fax: +01-204-789-2097, E-mail: mike_coulthart@hc-sc.gc.ca

gel electrophoresis, or PFGE (Schwartz and Cantor 1984; Carle et al. 1985; Chu et al. 1986). During standard, single orientation electrophoresis, the varying molecular sieving effects of the gel matrix on DNA molecules of different sizes leads to their separation. However, above a certain size, differently sized DNA molecules tend to migrate at indistinguishable rates in a unidirectional field, probably because of molecular "tunneling" through the porous gel matrix and the consequent loss of the sieving effect. PFGE techniques all depend on periodic switching of the orientation of the electric field during electrophoresis. The size-dependent rate of reorientation of DNA molecules within this variable electric field causes the molecular species to be separable over a much wider size range. Furthermore, varying the parameters of the electric field (e.g., voltage; time interval between successive reorientations; angle of reorientation) allows optimization of the size range of DNA separated, according to the needs of the experimenter.

Since the original applications of PFGE in molecular epidemiology (Grothues et al. 1988; Arbeit et al. 1990; Goering and Duensing 1990), the method has been shown repeatedly to discriminate between even closely related strains of pathogenic bacteria, often succeeding in detecting polymorphism when other methods fail. But despite its widespread application in bacterial strain tracing, PFGE remains somewhat technically demanding, and results can be difficult to reproduce between laboratories. Because of this, and because there is currently no universally accepted set of technical standards for PFGE, it is not always obvious how best to interpret the resulting genetic data. The main purpose of this chapter is to provide experimental protocols that we have found in our laboratory to work reproducibly and produce high-quality RFLP data for a variety of pathogenic bacterial species (e.g., *Escherichia coli*; *Salmonella* spp.; *Campylobacter* spp.; *Neisseria meningitidis* and *N. gonorrheae*; *Burkholderia cepacia*; *Pseudomonas aeruginosa*; *Legionella pneumophila* and *L. mcdadei*; *Bordetella pertussis*; and *Staphylococcus aureus*). For excellent general reviews of PFGE theory and techniques, see for example Burmeister and Ulanovsky (1992) and Gemmill (1991).

Outline

Short Protocol
- Day 0: Inoculate 180 mm agar plate or broth; grow overnight.
- Day 1: Prepare bacterial suspensions in agarose blocks; begin lysis.
- Continue lysis.
- Complete lysis; inhibit or remove proteinase from lysis step.

- Day 3 or 4: Begin restriction enzyme cleavage.
- Day 4 or 5: Complete restriction enzyme cleavage; set up gel; begin electrophoresis.
- Day 5 or 6: Complete electrophoresis; stain gel; record results.

Materials

Chemicals

- Agarose, low melting-point (Boehringer-Mannheim LM-MP, cat. no. 1-441-353)
- Agarose, multipurpose (Boehringer-Mannheim MP, cat. no. 1–88-991)

Note: An important physical parameter in judging the suitability of agarose for PFGE is gel tensile strength, measured in units of g/cm^2 for a standard 1% (w/v) gel. Boehringer-Mannheim MP agarose has extremely high gel strength (more than $1800\,g/cm^2$) and yields gels of excellent quality.

- Pefabloc (Boehringer-Mannheim, cat. no. 1-429-876)
- Phenylmethyl sulfonyl fluoride (Sigma, cat. no. P-7626)
- n-Lauroyl sarcosine, sodium salt (Sigma, cat. no. L-5777)
- Brij 58 (polyoxyethylene 20 cetyl ether) (Sigma, cat. no. P-5884)
- Deoxycholic acid, sodium salt (Sigma, cat. no. D-6750)

Enzymes and Markers

- Lysostaphin (Sigma, cat. no. L-7386)
- Proteinase K (Boehringer-Mannheim, cat. no. 1-092-766)
- Various restriction endonucleases (e.g., NotI, BglII, SpeI, SmaI; various suppliers, e.g., New England Biolabs, Boehringer-Mannheim, Promega)

Note: Restriction endonucleases used in preparing DNA for PFGE should be of the highest possible quality, with maximal activity and free of contaminating endonucleases. Many companies now supply "PFGE-certified" restriction enzymes for which appropriate application-specific quality control is maintained, and these are usually to be preferred over similar products lacking this explicit quality control. PFGE certification does not, however, constitute a guarantee that the preparation will always function up to the user's standards, and if difficulties are encountered there may be no option but to experiment with different enzymes or lots of a given enzyme from a particular supplier, or to switch suppliers.

No one restriction enzyme is necessarily suitable for PFGE-based RFLP studies in all bacterial species. See further discussion on this topic in "Choice of Restriction Enzyme".

- High-molecular-weight DNA markers (concatemers of bacteriophage λ, strain c1857Sam7) (Bio-Rad, cat. no. 170-3605)

Equipment, Other Supplies

- Contour-clamped homogeneous-electric-field (CHEF) system (e.g., Bio-Rad CHEF DRIII, including electrophoresis chamber, buffer circulator/cooler and control unit)
- Photography equipment (ultraviolet transilluminator, photography stand, Polaroid camera)
- Kodak Type 57 Polaroid film

Stock Solutions

Note: After adjustment to final volumes as stated, the following eight solutions should be autoclaved and stored at room temperature.

- Tris-HCl buffer: 1 M, pH 8.0. Dissolve 121.12 g Tris in 750 ml distilled H_2O (dH_2O) and 50 ml 1 M HCl. Adjust pH to 8.0 with 1 M HCl. Add dH_2O to 1 l.
- Tris-HCl buffer: 1 M, pH 7.4. As above, but adjust pH to 7.4 with 1 M HCl.
- EDTA: 0.5 M, pH 8.0. Add 186.1 g $Na_2EDTA.2H_2O$ (EDTA) to 700 ml dH_2O. With stirring, add 10 M NaOH (~50 ml) to pH 8.0. Add dH_2O to 1 l.

Note: EDTA will begin visibly to dissolve as pH approaches 8.0 with addition of NaOH.

- EDTA: 0.5 M, pH 9.5. Prepared and stored as with previous stock solution, but continue adding 10 M NaOH to pH 9.5.
- $MgCl_2$: 1 M. Dissolve 20.3 g $MgCl_2.6H2O$ in 80 ml dH_2O. Add dH_2O to 100 ml.
- KCl: 1 M. Dissolve 7.46 g KCl in 80 ml H_2O. Add dH_2O to 100 ml.
- NaCl: 5 M. Dissolve 292 g NaCl in 800 ml dH_2O. Add dH_2O to 1 l.
- TBE buffer: 20×. Dissolve 181.6 g Tris, 30.9 g boric acid and 0.74 g EDTA in 800 ml dH_2O. Add dH_2O to 1 l.

Note: There is no need to autoclave the following three solutions, if presterilized dH_2O and glassware are used to prepare and store them. All should be stored at room temperature.

- Sarkosyl: 10 % (w/v). Dissolve 10 g n-lauroyl sarcosine, sodium salt, in ~70 ml sterile dH_2O in a glass beaker. Add sterile dH_2O to 100 ml.
- Brij 58: 10 % (w/v) Dissolve 10 g Brij 58 in ~70 ml sterile dH_2O. Add sterile dH_2O to 100 ml.
- Deoxycholate: 10 % (w/v). Dissolve 10 g sodium deoxycholate in ~70 ml sterile dH_2O. Add sterile dH_2O to 100 ml.

Note: The following four solutions should be stored at −20 °C in small aliquots (e.g., 0.5–1 ml).

- Lysostaphin: 2 mg/ml. Dissolve 20 mg lysostaphin in 10 ml sterile TE (see below).
- Proteinase K: 10 mg/ml. Dissolve 100 mg proteinase K in 10 ml sterile dH$_2$O.
- PMSF: 100 mM. Dissolve 0.174 g phenylmethyl sulfonyl fluoride in 10 ml 95 % ethanol.

Note: PMSF is extremely toxic, and volatile. Exercise extreme care when weighing the dry chemical, and wherever possible, solutions containing it (especially the concentrated stock) should be handled in a fume hood. Pefabloc [4-(2-aminoethyl)-benzene sulfonyl fluoride, supplied as hydrochloride by Boehringer-Mannheim] is being marketed as a less toxic and more stable substitute for PMSF. We have used Pefabloc successfully to prepare high-molecular-weight DNA from *Neisseria meningitidis* (see below).

- Pefabloc: 150 mM. Dissolve 0.359 g Pefabloc in 10 ml sterile dH$_2$O.
- 10× restriction enzyme reaction buffer concentrates: Supplied by enzyme manufacturer, or prepared in the user's laboratory.

Note: The small aliquots of reaction buffer concentrate usually provided with restriction enzymes may not be sufficient to prepare the large volumes of working buffer needed for gel-slice equilibration (see below). We have found that home-made preparations of these 10× buffer concentrates (two examples are given below), which ideally should be made to match exactly the compositions of those supplied by the manufacturer, can replace the commercial solutions. If in doubt about the recommended reaction buffer composition for a particular enzyme, contact the manufacturer's technical support department, or consult published reviews on the subject (e.g., Bhagwat 1992).

10× *Not*I/*Bgl*II reaction buffer

500 mM Tris-HCl, pH 8.0
1 M NaCl
100 mM MgCl$_2$

10× *Spe*I reaction buffer

---

200 mM Tris-HCl, pH 7.4
500 mM KCl
50 mM $MgCl_2$

- PFGE dye mixture: Dissolve 20 ml glycerol and 0.125 g bromophenol blue in 50 ml dH_2O and 5 ml 20× electrophoresis buffer. Add sterile dH_2O to 100 ml and store at 4 °C.
- Ethidium bromide: 10 mg/ml. Dissolve 100 mg of ethidium bromide in 10 ml dH_2O. Store at 4 °C in a plastic tube or brown glass bottle wrapped with aluminum foil.

Note: Ethidium bromide is a potent mutagen, moderately toxic, and a highly suspect carcinogen. Wear gloves and exercise extreme care when handling the dry powder. Dispose of it responsibly by treating used gel-staining solution with charcoal or ion-exchange resin (Sambrook et al. 1989), then incinerating the used adsorbent.

Working Solutions

Unless otherwise stated, the following working solutions are prepared by diluting the above stock solutions with sterile dH_2O to give the specified concentrations.

- $T_{10}E_1$ buffer: 10 mM Tris-HCl, pH 8.0; 1 mM EDTA
- $T_{10}E_{50}$ buffer: 10 mM Tris-HCl, pH 8.0; 50 mM EDTA
- Low-melt agarose: 1.2 % (w/v): Immediately before use, dissolve 1.2 g low-melt agarose in 100 ml $T_{10}E_1$ buffer by boiling (e.g., in a microwave oven). To keep molten during use, incubate at 40 °C. Can be stored in the gelled state between uses in a tightly closed bottle at 4 °C and remelted as needed.
- Proteinase K lysis buffer: 0.25 M EDTA, pH 9.5; 0.5 % (w/v) sarkosyl; 0.5 mg/ml proteinase K

Note: Prepare the proteinase K lysis buffer and the following four working solutions from stocks immediately before use.

Lysostaphin lysis buffer

---

10 mM Tris-HCl, pH 8.0
1 M NaCl
0.2 % (w/v) deoxycholate
0.5 % Brij 58
0.5 % sarkosyl
50 µg/ml lysostaphin

Note: While the proteinase K lysis buffer described above is effective when used alone in lysis of gram-negative species, the cell walls of *Staphylococcus aureus* are more effectively disrupted by an initial treatment with lysostaphin, deoxycholate and Brij 58 as in the above buffer, followed by a standard treatment with proteinase K (see below).

- PMSF buffer: 1 mM PMSF in $T_{10}E_1$ buffer
- Pefabloc buffer: 1.5 mM Pefabloc in $T_{10}E_1$ buffer

Note: The dH_2O used to prepare the following two working solutions need not be sterile.
- Electrophoresis buffer: Dilute 150 ml of 20× TBE stock to 3 l with dH_2O.
- Gel-staining solution: 0.5–1.0 µg/ml ethidium bromide

▉ Procedure

Note: At least until step 6 is completed (and prudently, beyond), one should assume that the sample contains a large number of viable, infectious bacteria. Observe appropriate biosafety regulations for handling live bacterial pathogens by working in a biosafety cabinet, using barrier precautions such as gloves and masks, and applying effective disinfection and disposal procedures.

Day 1

1. Prepare sufficient 1.2 % low-melt agarose solution to accommodate the number of bacterial strains being processed (see next step for volumes required). Incubate at 40 °C to keep molten until needed.

2. **Thoroughly** suspend 10^{10}–10^{11} bacterial cells, for example from a freshly grown overnight broth culture or 180 mm agar plate, by vortex mixing or repeated pipetting in a 5 ml tube containing 1 ml $T_{10}E_1$ buffer. For suspension cultures, sediment cells first by a brief low-speed centrifugation, then resuspend as above.

Note: When a laboratory begins work on a new species, the optimal input of bacterial culture per loaded sample volume should be determined empirically. Input cell numbers in the range 10^9–10^{10} per gel lane (ideally counted using a hemocytometer on a small aliquot of a suspension such as that prepared in step 2 above) should suffice. To elaborate: most pathogenic bacteria contain on the order of 2–5×10^{-15} g of DNA per cell, so that 10^9 cells contain a total DNA mass of about 2–5 µg. The range of DNA amounts per gel band that are clearly resolved by PFGE and conveniently detected with ethidium bromide staining, transillumination with 260 nm ultraviolet light, and photography with Polaroid Type 57 film is 10–500 ng. Thus, one can predict that, for a bacterial sample originally containing 10^9 cells each with a genome size of 2×10^{-15} g, a 50 kbp gel band (representing 2.5 % of the genome) will ideally contain up to ~25 ng of DNA, while a 500 kbp band will contain up to ~250 ng (neglecting sample losses during preparation).

Also, the total amount of culture processed can be scaled down from the above-stated quantity, depending on availability of material and need for multiple restriction digests.

3. Add 1 ml of 1.2 % low-melt agarose to the bacterial suspension (final agarose concentration 0.6 %) and mix thoroughly by inversion or gentle vortex mixing.

4. Aspirate agarose-cell suspension into a 3 ml disposable syringe, leaving an air gap at the lower end of the syringe barrel. Invert in a rack and incubate at 4 °C for 15–30 min.

5. Remove the agarose block from the syringe by drawing the plunger out, and slide the block into a disposable petri dish. Using a glass coverslip, slice the block into sections approximately 3 mm thick.

6. Transfer gel slices to 5 ml freshly prepared proteinase K lysis buffer (for gram-negative organisms such N. meningitidis, Escherichia coli or Legionella pneumophila), or lysostaphin lysis buffer (for S. aureus) in a 50 ml sterile tube. Incubate at 50 °C for 24 h, agitating gently occasionally. Replace the initial volume of lysis buffer with a freshly prepared 5 ml volume of proteinase K lysis buffer and reincubate at 50 °C for an additional 24 h.

Note: This incubation is specified to take place over a 48 h period, mainly because we have found it to be most reliable when performed this way (perhaps due to a requirement for diffusion of proteolysis products out of the agarose gel). However, it may be possible, with experimentation, to shorten the length of this incubation period. We have also had success with "weekend digestions" (~60 h with no addition of a second volume of proteinase K lysis buffer).

Day 3   1. Remove tubes from 50 °C waterbath and allow to cool to room temperature. Agarose slices will regain firmness as they cool, and should appear clear.

2. Remove lysis buffer **as completely as possible** with a Pasteur pipette and soak slices for 30 min, twice, at room temperature in 5 ml of freshly prepared PMSF buffer. Agitate gently by hand once or twice during each 30 min incubation.

Note: When diluting 100 mM PMSF stock solution into $T_{10}E_1$ buffer to make the working solution, take care to mix thoroughly until the cloudiness of the initial suspension disappears.

3. Remove PMSF buffer completely and soak slices for 30 min, four times, at room temperature in 5 ml of $T_{10}E_1$ buffer. Agitate gently by hand once or twice during each incubation, taking care to rinse sides of tubes to remove residual PMSF buffer clinging there.

Note: If carefully prepared, the DNA in these gel slices can often be stored at 4 °C for up to 1 year with little or no apparent degradation, at least for some organisms (e.g., *N. meningitidis*). However, with other organisms (e.g., *E. coli*) we have had less success with storage. One should verify storage stability before assuming that degradation will not occur, and it is recommended at least in the first analysis that DNA for PFGE be freshly prepared.

Alternate Procedures for Proteinase K Inactivation/Removal

Due to the toxicity and instability of PMSF, it may be desirable to use alternative methods for inhibition or removal of residual proteinase K, which obviously must be accomplished before addition of restriction enzyme to cleave the DNA. At least with *N. meningitidis*, we have found either of the following two alternative procedures also to be satisfactory.

1. Remove tubes containing agarose slices in lysis buffer from the 50 °C water bath and allow to cool to room temperature, as above.

Inhibition with Pefabloc

2. Remove lysis buffer completely and incubate slices for 2–3 h or overnight at 37 °C in 5 ml of Pefabloc buffer.

3. Remove Pefabloc buffer completely and wash slices for 30 min, three times, at 37 °C in 5 ml of $T_{10}E_{50}$ buffer. Agitate by hand once or twice during each incubation, and take care to rinse sides of tubes to remove residual Pefabloc buffer.

4. Remove $T_{10}E_{50}$ buffer and store slices at 4 °C in 5 ml $T_{10}E_1$ buffer, if desired.

Dilution by Repeated Buffer Washes

1. Remove the tubes containing agarose slices in lysis buffer from the 50 °C water bath and allow to cool to room temperature, as above.

2. Remove lysis buffer completely and incubate slices for 2 h, three times, at 37 °C in 10 ml of $T_{10}E_1$ buffer. Agitate by hand once or twice during each incubation. Perform the same wash a fourth time, overnight.

3. Next day, add an additional 5 ml of $T_{10}E_1$ buffer and store slices at 4 °C, if desired.

Day 4

Note: Although this portion of the procedure is described as part of day 4, if time permits it may be started on day 3, thus saving a day.

1. For each restriction enzyme cleavage reaction to be carried out, place 1 ml of the appropriate 1× reaction buffer into a 1.5 ml plastic microcentrifuge tube. To each tube, add one half a DNA-containing gel slice. Individual slices, retrieved from the storage tube with a sterile Pasteur pipette that has been flamed into a hooked shape, may be sliced into halves with a glass coverslip. Take care to transfer as little of the storage buffer as possible. Incubate at room temperature for 1 h with occasional inversion of the tubes to equilibrate the gel slices with reaction buffer.

2. Remove the reaction buffer completely, and replace with 100 µl of reaction buffer containing 20–40 Units of restriction enzyme. Incubate overnight at the recommended temperature (usually 37 °C, but note that some enzymes, such as *Sma*I, work best when incubated at room temperature).

Day 5

1. Dissolve suitable pulsed-field grade agarose (e.g., Boehringer-Mannheim MP) in 1× electrophoresis buffer to a final concentration of 1 % (w/v), by boiling. Agitate flask intermittently by hand during this process, to ensure that all residual clumps of agarose are fully dispersed and dissolved. Place agarose solution in a 50 °C water bath to equilibrate for 30–60 min before pouring gel.

2. Meanwhile, assemble and level the gel casting tray, and place the well-forming comb with the tips of the teeth approximately 1.5 mm from the tray.

3. Pour the 1 % agarose solution smoothly into the gel casting tray, taking care not to create excessive numbers of bubbles and checking to ensure that no bubbles become trapped under the teeth of the well-

forming comb. Any residual bubbles remaining on the gel surface after pouring can be removed by first using a Pasteur pipette to push them to one side of the tray, and then aspirating them. Once the gel has solidified (judged by a uniform milky appearance), place at 4 °C for at least 30 min to complete gelation.

4. Meanwhile, set the thermostatic circulator of the electrophoresis apparatus to a temperature that will maintain a buffer temperature of 14 °C **in the electrophoresis chamber.**

Note: It should be determined empirically at what temperature the circulator thermostat should be set to maintain the buffer in the electrophoresis chamber at the desired temperature, which because of imperfect cooling efficiency is normally a few degrees above that set at the circulator.

5. Place 3 l of electrophoresis buffer into the electrophoresis chamber and level the apparatus.

6. Slowly remove the well-forming comb from the gel, pulling it straight out and taking great care not to tear the bottoms or sides of the wells as a result of suction. Place gel, on the casting tray, on the platform in the chamber. Begin circulating the electrophoresis buffer to bring it to 14 °C.

7. Meanwhile, add 150 µl of PFGE dye mixture to a gel slice containing DNA that has been digested with restriction enzyme. Incubate the slice at 65 °C for at least 15 min, until the agarose has completely melted. Similarly incubate gel slices containing sufficient molecular-weight markers for at least two gel lanes. Mix melted samples **very gently** by tapping the bottom of the tube, or by gentle inversion followed by a brief centrifugation to collect all liquid to the bottom of the tube.

8. Stop circulation of the electrophoresis buffer. Take one sample out of the 65 °C water bath and, using an automatic pipettor fitted with a cut-off tip, **promptly** load 20–50 µl of sample mixture in a sample well. Repeat this procedure, one tube at a time, until all samples are loaded. Load at least two wells with molecular-weight markers (e.g., in the two outside lanes), using the same procedure.

9. Visually verify that the agarose in the sample wells has solidified. Complete assembly of the electrophoresis chamber, according to the manufacturer's instructions in the user's manual, and resume buffer

circulation. Turn on power supply, and set to regulate field strength at 200 V. Adjust the pulse-ramping interval to an empirically determined optimal value (e.g., from 1 to 30 s over the course of the run) for the size range of DNA fragments expected. Apply power, and run the electrophoresis for 18–24 h or until the bromophenol blue marker has migrated 10–15 cm from the sample wells.

Note: A discussion of optimization of electrophoresis parameters (e.g., gel concentration; temperature; voltage; run time; field-switching intervals and ramping regimes; field reorientation angles) is beyond the scope of this chapter. One widely applicable observation, however, is that the ability of a single PFGE run to separate DNA species in different size ranges can be maximized by gradually increasing ("ramping") the length of the time interval between field reorientations during the course of the separation. In this way, smaller molecules capable of more rapid reorientation are separated optimally first, followed later by larger molecules that require longer times to respond. More extensive theoretical and experimental consideration of the effects of these factors is given, for example, by Birren et al. (1988), Clark et al. (1988), and Birren et al. (1989).

Day 6

1. Turn off power supply, and detach power cables from the electrophoresis chamber. Carefully lift the gel out of the chamber on the supporting gel tray, and place into 500 ml of 0.5–1 µg/ml ethidium bromide stain solution. Agitate gently for 30–60 min.

2. Carefully pour off the ethidium bromide stain solution, holding the gel in place in the tray with a gloved hand. Replace the stain solution with 500 ml of dH$_2$O, and agitate gently for another 30–60 min.

3. Lift the gel out of the tray and place directly on the surface of an ultraviolet transilluminator. Photograph the gel using visible fluorescence excited at 260 nm (for highest sensitivity) or 300–310 nm (for highest personal safety and less sample DNA damage, for instance if band recovery is planned). A commonly used configuration for photography of UV-induced fluorescence of ethidium bromide-stained DNA uses Kodak Type 57 or 667 film and an orange or red filter (e.g., Wratten #23 A). Since UV transilluminator filters tend to become UV-opaque with use, it is not possible to specify a single set of film-exposure parameters that will suffice in all situations. However, a good "rule of thumb" for newer UV filters is to use a $1/4$–1 s exposure at f4.5.

4. Record and analyze banding patterns.

Comments

Data Quality

An aspect of PFGE seldom discussed in molecular epidemiology is visual quality of the banding patterns obtained (band sharpness, flatness and across-gel registration). These can have a large influence on an investigator's ability to score RFLPs for an epidemiological investigation and on the feasibility of comparing results among laboratories, which may eventually be done through electronic databases. Some of the more obvious factors here are: loading optimal amounts of nondegraded, completely restricted DNA on the gel; restriction digests that yield appropriate fragment sizes and numbers (see McClelland et al. 1987); and a homogeneous field PFGE method (such as CHEF) that can produce sharp, flat, well-aligned bands across the gel. However, we have also found that configuration (shape and size) of the sample-loading zone plays a role. Flatter, sharper bands can be obtained by using a 1.5 mm-thick well-forming comb instead of one 3 mm-thick, and by pipetting melted samples into the well after covering the gel with buffer instead of before, to avoid meniscus formation in the well. Despite the increased difficulty of loading the often viscous sample under these conditions, with practice it may be mastered and generally leads to higher quality primary data.

Also important is minimization of sample **volume**, independent of **mass** of DNA present. Figure 1 illustrates the effect of loading equal amounts of DNA in different volumes. With increasing volume (in this case approximately 20, 40 and 60 µl), the quality of the banding pattern as normally viewed from above progressively deteriorates, until the pattern becomes difficult to read and visible bands are even lost from the high and low ends of the molecular-size range (Fig. 1 A, *lanes 3–5, 8–10*). This is somewhat unexpected, since with a perfectly vertical DNA band the total UV-stimulated fluorescence emitted by the ethidium-DNA complexes and detected from a vantage point above the gel should not depend on the band's depth. A partial explanation of this observation may be the phenomenon of band "tilting" that occurs during conventional gel electrophoresis of macromolecules in agarose and polyacrylamide gel matrices (Kozulic 1994), and apparently also during PFGE. When the stained gel lanes were excised, rotated and viewed from the side (Fig. 1B), pronounced band tilting was observed despite no drastic loss of sharpness, suggesting that the main reason for loss of band clarity when viewed from above is simply the dispersion of fluorescent signal from the DNA in the band over a wider gel surface area. Thus, for PFGE

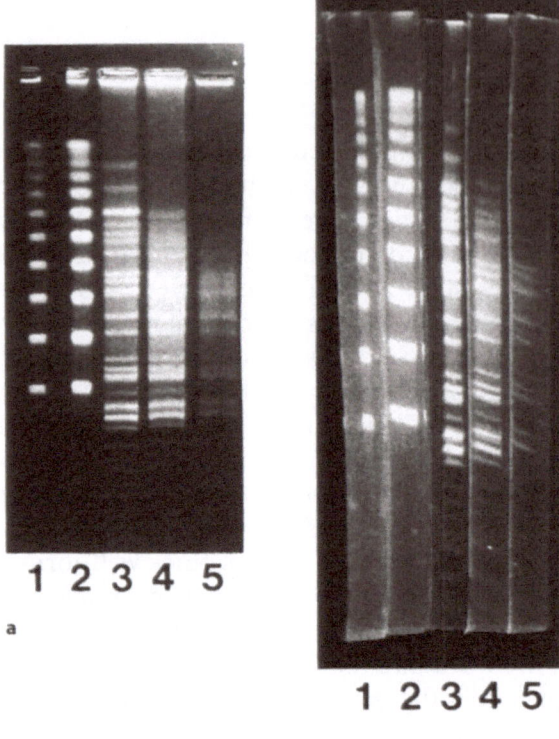

Fig. 1A, B. A PFGE gel, illustrating the effect of varying sample loading volume for a constant amount of DNA. Approximately 10^{10} cells of *E. coli* were suspended in an agarose block, lysed and their DNA cleaved with *Xba*I as described in the text. Six 90 μl portions of the agarose block were mixed with varying volumes of PFGE dye mixture so that the equivalent of 12.5 μl of the original block were loaded on the gel in total volumes of 21 μl (*lane 3*), 43 μl (*lane 4*), and 56 μl (*lane 5*). Gel composition was 1 % (w/v) Boehringer-Mannheim multipurpose agarose cast and run in 1× TBE electrophoresis buffer. Electrophoresis was carried out at 200 V for 22 h, with linear ramping of pulse times from 1 to 35 s at 14 °C. Gel was stained post-electrophoretically for 1 h in 500 ml of 1 μg/ml ethidium bromide. Photograph was taken with Kodak Type 57 Polaroid film, with UV transillumination at 260 nm and an orange filter (Wratten #23 A). *Lanes 1, 2,* high-molecular-weight DNA standards (concatemers of λ-DNA). Note the deterioration in band sharpness when progressing from smaller to larger sample-loading volumes (*lanes 3–5*). **B** Side view of lanes *1–5* in **A**, after slicing the lanes out of the original gel and rotating them 90°. *Top* of original gel is oriented to *left* for each lane. Note that, with increasing loading volumes (*lanes 3–5*), although the angle of slanting of the DNA bands does not change substantially, the total distance between the top and bottom of a band (*rightward* and *leftward* in figure) increases, thus contributing to a diffused appearance for the band (see **A**)

we recommend the use of the minimal sample volume (e.g., 20–25 μl) that will completely cover the bottom of the sample well (so as to form a regularly shaped starting zone) while still containing sufficient DNA to permit efficient band detection (see discussion above, under Day 1 of "Procedure," pertaining to optimization of numbers of bacterial cells per sample).

Choice of Restriction Enzyme

Another important area concerns choice of restriction enzyme(s) that will both: (1) cleave the genomic DNA into a manageable number (usually 10–20) of well-separated fragments and (2) reveal RFLPs having enough diversity and the appropriate phylogenetic behavior to be epidemiologically informative. The first of these requirements is relatively easily met; the seminal paper by McClelland et al. (1987), where rules were defined for predicting restriction enzyme cleavage frequency in bacterial genomes, is worth reading. Conceptual discussion of the second is largely beyond the scope of this chapter, and currently choices must be based on experience. Fortunately however, for some better-studied bacterial species restriction enzymes have been identified that both cleave the genome infrequently and reveal polymorphism. In such cases it is perhaps best to begin by using the established strategy; some examples are given in Table 1. However, it should not be assumed that there is a single optimal enzyme that will serve for all investigations in a given bacterial species, or that broad evaluation of different enzymes was carried out before the data in a published paper was collected. Furthermore, the use of several different enzymes can sometimes reveal patterns of polymorphism that would not be observed if only one enzyme were used.

Interpretation of RFLP Patterns

A real-world illustration of the application of PFGE techniques can be taken from our work in Canada on the molecular epidemiology of the important community-acquired pathogen, *N. meningitidis*. Since 1986, a single genotype of *N. meningitidis*, named ET15 for its multilocus allozyme genotype, has spread and now accounts for nearly all serogroup C meningococcal disease in this country, making the tracing of outbreaks with conventional markers a challenge (Ashton et al 1991; Whalen et al.

Table 1. Restriction enzymes used in PFGE for bacterial molecular epidemiology, with selected references

Species	Enzyme(s) Used	Strains	References
Escherichia coli	*Not*I *Xba*I	Diverse Serotype O157:H7	Arbeit et al. (1990) Bohm and Karch (1992)
Shigella sonnei	*Xba*I	Day-care outbreak	Brian et al. (1993)
Shigella dysenteriae, flexneri, sonnei	*Not*I	Sporadic, epidemic	Soldati and Piffaretti (1991)
Vibrio cholerae	*Not*I	Serotype O1	Cameron et al. (1994)
Salmonella enteritidis	*Avr*II, *Spe*I, *Xba*I	Sporadic, outbreak	Thong et al. (1995)
Salmonella typhi	*Xba*I, *Spe*I	Sporadic	Thong et al. (1994)
Campylobacter coli, jejuni	*Sma*I	Diverse	Yan et al. (1991)
Neisseria meningitidis	*Bgl*II, *Not*I, *Spe*I	Serotype C/ET15	Strathdee et al. (1993)
Neisseria gonorrheae	*Nhe*I, *Spe*I		Poh et al. (1993)
Legionella pneumophila	*Not*I, *Sfi*I *Bss*HII, *Sal*I, *Spe*I	Hospital isolates	Ott et al. (1991) This laboratory (unpublished)
Legionella mcdadei	*Bss*HII		This laboratory (unpublished)
Bordetella pertussis	*Xba*I	Diverse	Khattak et al. (1992)
Burkholderia cepacia	*Spe*I *Nhe*I, *Xba*I	Hospital outbreak	Anderson et al. (1991) This laboratory (unpublished)
Pseudomonas aeruginosa	*Dra*I, *Nhe*I, *Spe*I, *Ssp*I, *Xba*I *Hpa*I	CF (siblings)	Grothues et al. (1988) This laboratory (unpublished)
Staphylococcus aureus	*Sma*I *Sma*I	Diverse, MRSA MRSA	Goering and Duensing (1990) El-Hadami et al. (1991)
Streptococcus pneumoniae	*Apa*I, *Sma*I	Diverse	Lefevre et al. (1993)
Enterococcus faecalis	*Sma*I	Diverse	Murray et al. (1990)
Enterococcus faecium	*Sma*I	Diverse	Miranda et al. (1991)

1995). For example, between November 30, 1993 and February 15, 1994, eight cases of meningococcal disease were reported during a period of 11 weeks from the community of Saanich and environs, located on Vancouver Island, British Columbia. The strains recovered were all sero-

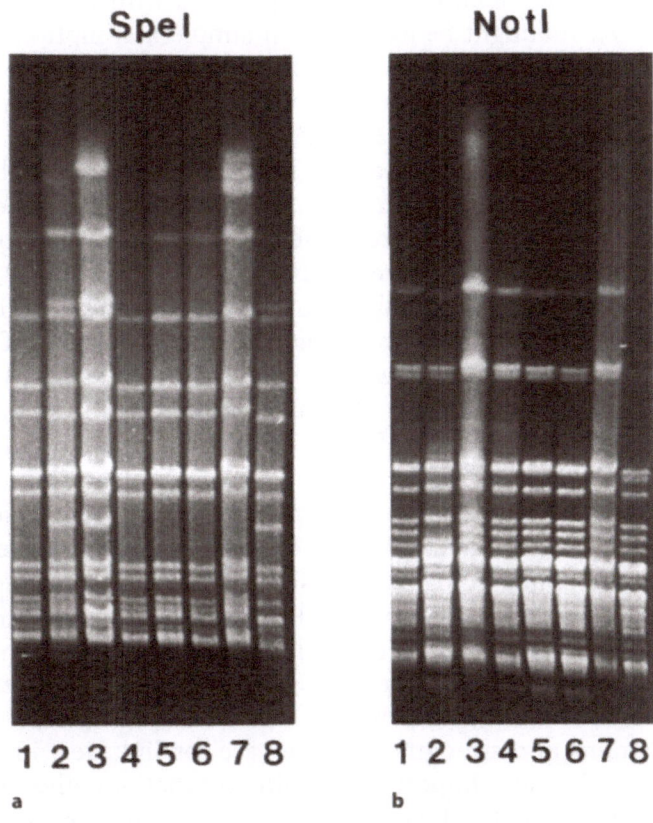

Fig. 2A, B. An example of *Neisseria meningitidis* strain relationships as revealed by PFGE. Sample preparation, restriction enzyme digestion, gel electrophoresis and staining were carried out as described in the text. **a** *Spe*I cleavage patterns of eight strains of *N. meningitidis* recovered from Vancouver Island, British Columbia, Canada during an 11-week period. *Lanes 4–7* Cluster in Saanich, among neighbors in a condominium community; *lane 1* case that occurred 2 months earlier in Saanich (no known link to cluster); *lanes 2, 8* nonlinked cases in Victoria; *lane 3* nonlinked case in Saanich. **b** Same strains as in **a**, *Not*I cleavage patterns. These data illustrate that PFGE has the ability to discriminate between very closely related genotypes of *N. meningitidis*, and can verify genetic similarities between strains suspected of epidemiological linkage (e.g., *lanes 4–7*). However, as discussed in more detail in the text, they also illustrate that PFGE data are subject to the difficulties of inference caused by small samples, and to the need for appropriate local controls

group C, serotype 2a:P1.2,5 or 2a:P1.2, and ET15. When these strains were analyzed by PFGE with *Spe*I and *Not*I, various banding patterns were seen (Fig. 2). It was clear that a spatiotemporal cluster of four of these cases, occurring between January 30, and February 7, 1994 in patients who lived as neighbors in a condominium community in Saanich, shared the identical banding patterns with both enzymes (Fig. 2, *lanes 4–7*). As might be expected on simple assumptions, two approximately contemporaneous cases (within a month either side of the cluster) from the nearby city of Victoria were caused by strains with quite distinct PFGE patterns (Fig. 2, *lanes 2, 8*), as was a third case that occurred within Saanich but had no known epidemiological link to the condominium cluster (Fig. 2, *lane 3*). The latter case was distinguishable with both *Spe*I and *Not*I from the two Victoria cases (compare Fig. 2 *lane 3* with *lanes 2, 8*). Interestingly, an eighth case (Fig. 2, *lane 1*), which had occurred in Saanich two months prior to the focal cluster, showed *Spe*I and *Not*I patterns indistinguishable from that of the cluster strain, despite a lack of known epidemiological linkages to the cluster.

A number of points can be illustrated with results such as these. First, PFGE was capable of splitting strains that were indistinguishable by serogrouping, serotyping and allozyme electrophoresis, demonstrating the discriminatory power of the technique. Second, strains from a cluster of cases (the "condominium cluster") that were suspected but not known to be epidemiologically linked on the basis of nongenetic information displayed indistinguishable PFGE patterns, suggesting that they were in fact linked. Third, three strains for which there was no prior evidence of linkage to the cluster had PFGE genotypes different from the cluster strains, sharpening the contrast between "cluster" and "background" strains and strengthening the case for linkage within the cluster. However, it is also important to point out that not all of the simplest expectations were fulfilled in this investigation. For instance, as might be expected from the general biology of *N. meningitidis* as a respiratory pathogen readily transmitted by casual contact, meningococcal disease in the community of Saanich in 1993–1994 was clearly caused by more than one meningococcal strain. One of these strains gave rise to the condominium cluster, but also to a "sporadic" case in the same community with no known links of transmission. If the intention had been to investigate pathways of transmission over a longer time scale, or with the benefit of less prior information on case contacts, one might be forced to consider possible linkages between this sporadic case and the focal cluster. In this case, the question arises whether the PFGE data on strain similarity would be a sufficient basis on which to undertake such an investigation. Furthermore, if we had by chance only acquired this matching sporadic strain, and not the other Saanich strain whose PFGE genotype did not match that of the suspected cluster strain (Fig. 2, *lane 3*), the nature of the "local-control" strain background against which the genetic similarities among the cluster strains were clearly discernible would perhaps have been quite different. This example highlights the fact that when using a molecular approach, as in other types of epidemiological investigations, sample size and selection of adequate numbers of appropriate local controls can play a key role in the inferences possible from the data.

When Your Best Efforts Don't Succeed

Despite utilization of an excellent technical manual, scrupulous attention to the execution of experimental details, and repeated attempts at troubleshooting, one may still fail to obtain the quality of primary data required for an epidemiological investigation by PFGE. In this case, one is well-advised to draw on the collective expertise of experienced practitioners in the field, perhaps by directly contacting specific workers in the field, for example the authors of published articles reporting the use of PFGE in infectious disease epidemiology.

Such direct consultation may not adequately answer your questions. For those with access to electronic-mail services, one approach that has become extremely popular in recent years is to post an electronic-mail message on the Internet. This can be done for instance in a Usenet news group, or in one of the many more specialized mailing list servers ("LISTSERVS") which "rebroadcast" electronic-mail messages to a group of subscribers who are devoted to a more narrowly defined subject area. In general, proper etiquette involves posing the question as briefly yet as precisely as possible (with a descriptive subject line), humbly asking the general group of subscribers for any help available, and then waiting for replies; the range and generosity of helpful responses from other subscribers is frequently astonishing!

Some examples of Usenet fora in which to post such messages are: bionet.molbio.methds.reagnts; bionet.microbiology; bionet.diagnostics; and sci.microbiology. An excellent example of a mailing-list server is Pro-Med, to which one can subscribe by sending an electronic-mail message containing the command **subscribe promed** – where "–" indicates the electronic-mail address to which you want the Pro-Med mail server to send messages (ideally, the same one from which you are sending the request to subscribe). An even more sophisticated approach that has developed at a remarkable rate within the past few years is interactive access to the Internet via the World Wide Web. Here, dynamic hyperlinks between remote computers are used to simultaneously access widely distributed resources by cross-references, and sophisticated graphical Web browsers (e.g., Netscape Navigator) render a powerful, appealing user interface that greatly simplifies the complex underlying electronic communication processes. In many subject areas, listings of topically related Web sites are available by accessing a centralized resource site. An example of this type in the area of microbiology is the "Microbial Underground", which can be viewed at http://www.ch.ic.ac.uk/medbact/.

Summary

We have presented laboratory protocols to support the application of PFGE in bacterial molecular epidemiology. These protocols have been tested widely by us and found to yield high-quality primary data for a wide range of pathogen species, and it is hoped that they will contribute to a broadly based effort to improve and standardize basic methodology in this important, growing area of clinical microbiology. Another area in need of development, that of the integration of data on molecular genetic variation among pathogen strains into the statistically sophisticated framework of epidemiology, is beyond the scope of this chapter but was touched upon very briefly and intuitively by way of an example. It is in this latter area that we expect the most significant growth in new approaches and understanding to occur over the coming years.

Acknowledgements. We gratefully acknowledge the expertise and skill of Nancy Bigelow, Charles Dendy, Russell Easy, Linda Mancino, Esther Ofori, Alan Ryan, Shaun Tyler, and Gehua Wang, who helped to refine PFGE protocols for various pathogenic bacterial species. Special thanks are due to Gehua Wang for providing the results in Fig. 1, and to Linda Mancino for the alternate protocols for proteinase K removal.

References

Anderson DJ, Kuhns JS, Vasil ML, Gerding DL, Janoff EN (1991) DNA fingerprinting by pulsed field gel electrophoresis and ribotyping to distinguish *Pseudomonas cepacia* isolates from a nosocomial outbreak. J Clin Microbiol 29:648–649
Arbeit RD, Arthur M, Dunn R, Kim C, Selander RK, Goldstein R (1990) Resolution of recent evolutionary divergence among *Escherichia coli* from related lineages: the

application of pulsed field electrophoresis to molecular epidemiology. J Inf Dis 161:230–235

Ashton F, Ryan A, Borczyk A, Caugant D, Mancino L, Huang D (1991) Emergence of a virulent clone of *Neisseria meningitidis* serotype 2a that is associated with meningococcal group C disease in Canada. Can J Microbiol 29:2489–2493

Bhagwat AS (1992) Restriction enzymes: properties and use. Meth Enzymol 216:199–224

Bohm H, Karch H (1992) DNA fingerprinting of *Escherichia coli* O157:H7 strains by pulsed-field gel electrophoresis. J Clin Microbiol 30:2169–2172

Birren BW, Lai E, Clark SM, Hood L, Simon MI (1988) Optimized conditions for pulsed field gel electrophoretic separation of DNA. Nucleic Acids Res 16:7563–7582

Birren BW, Lai E, Hood L, Simon MI (1989) Pulsed field gel electrophoresis techniques for separating 1- to 50-kilobase DNA fragments. Anal Biochem 177:181–286

Brian MJ, Van R, Townsend I, Murray BE, Cleary TG, Pickering LK (1993) Evaluation of the molecular epidemiology of an outbreak of multiply resistant *Shigella sonnei* in a day-care center by using pulsed-field gel electrophoresis and plasmid DNA analysis. J Clin Microbiol 31:2152–2156

Burmeister M, Ulanovsky L (eds) (1992) Pulsed-field gel electrophoresis. Methods in molecular biology, vol 12, Humana, Totawa, New Jersey

Cameron DN, Khambaty FM, Wachsmuth IK, Tauze RV, Barrett TJ (1994) Molecular characterization of *Vibrio cholerae* O1 strains by pulsed-field gel electrophoresis. J Clin Microbiol 32:1685–1690

Carle GF, Frank M, Olson MV (1986) Electrophoretic separation of large DNA molecules by periodic inversion of the electric field. Science 232:65–68

Chu G, Vollrath D, Davis RW (1986) Separation of large DNA molecules by contour-clamped homogeneous electric fields. Science 234:1582–1585

Clark SM, Lai E, Birren BW, Hood L (1988) A novel instrument for separating large DNA molecules with pulsed homogeneous electric fields. Science 241-1203–1205

El-Hadami W, Roberts L, Vickery A, Inglis B, Gibbs A, Stewart PR (1991) Epidemiological analysis of a methicillin-resistant *Staphylococcus aureus* outbreak using restriction fragment length polymorphisms of genomic DNA. J Gen Microbiol 137:2713–2720

Gemmill RM (1991) Pulsed field gel electrophoresis. In: Chrambach A, Dunn MJ, Radola BJ (eds) Advances in electrophoresis, vol 4. VCH, Weinheim, pp 1–48

Goering RV, Duensing TD (1990) Rapid field inversion gel electorphoresis in combination with an rRNA gene probe in the epidemiological evaluation of Staphylococci. J Clin Microbiol 28:426–429

Grothues D, Koopmann U, von der Hardt H, Tümmler B (1988) Genome fingerprinting of *Pseudomonas aeruginosa* indicates colonization of cystic fibrosis siblings with closely related strains. J Clin Microbiol 26:1973–1977

Khattak MN, Matthews RC, Burnie JP (1992) Is *Bordetella pertussis* clonal? Br Med J 304:813–815

Kozulic B (1994) Looking at bands from another side. Anal Biochem 216:253–261

Lefevre JC, Faucon G, Sicard AM, Gasc AM (1993) DNA fingerprinting of *Streptococcus pneumoniae* strains by pulsed-field gel electrophoresis. J Clin Microbiol 31:2724–2728

Murray BE, Singh KV, Heath JD, Sharma BR, Weinstock GM (1990) Comparison of genomic DNAs of different enterococcal isolates using restriction endonucleases with infrequent recognition sites. J Clin Microbiol 28:2059–2063

Miranda AG, Singh KV, Murray BE (1991) DNA fingerprinting of *Enterococcus fae-cium* by pulsed-field gel electrophoresis may be a useful epidemiologic tool. J Clin Microbiol 29:2752–2757

Ott M, Bender L, Marre R, Hacker J (1991) Pulsed field electrophoresis of genomic restriction fragments for the detection of nosocomial *Legionella pneumophila* in hospital water supplies. J Clin Microbiol 29:813–815

Poh CL, Lau QC (1993) Subtyping of *Neisseria gonorrhoeae* auxotype-serovar groups by pulsed-field gel electrophoresis. J Med Microbiol 38:366–370

Prager R, Streckel W, Stephan R, Bockemühl J, Shimada T, Tschäpe H (1994) Genomic fingerprinting of *Vibrio cholerae* O139 from Germany and South Asia in comparison with strains of *Vibrio cholerae* O1 and other serogroups. Med Microbiol Lett 3:219–227

Sambrook J, Fritsch EF, Maniatis T (1989) Molecular cloning (2nd ed). Cold Spring Harbor Press, Cold Spring Harbor, New York

Schulte PA, Perera FP (1993) Molecular epidemiology: principles and practices. Academic Press, San Diego

Schwartz DC, Cantor CR (1984) Separation of yeast chromosome-sized DNAs by pulsed field gradient gel electrophoresis. Cell 37:67–75

Soldati L, Piffaretti JC (1991) Molecular typing of Shigella strains using pulsed field gel electrophoresis and genome hybridization with insertion sequences. Res Microbiol 142:489–498

Strathdee CA, Tyler SD, Ryan JA, Johnson WM, Ashton FE (1993) Genomic fingerprinting of *Neisseria meningitidis* associated with Group C meningococcal disease in Canada. J Clin Microbiol 31:2506–2508

Thong K-L, Cheong Y-M, Puthucheary S, Koh C-L, Pang T (1994) Epidemiologic analysis of sporadic *Salmonella typhi* isolates and those from outbreaks by pulsed-field gel electrophoresis. J Clin Microbiol 32:1135–1141

Thong K-L, Ngeow Y-F, Altwegg M, Navaratnam P, Pang T (1995) Molecular analysis of *Salmonella enteritidis* by pulsed-field gel electrophoresis and ribotyping. J Clin Microbiol 33:1070–1074

Whalen CM, Hockin JC, Ryan A, Ashton F (1995) The changing epidemiology of invasive meningococcal disease in Canada, 1985 through 1992. JAMA 273:390–394

Yan W, Chang N, Taylor DE (1991) Pulsed-field gel electrophoresis of *Campylobacter jejuni* and *Campylobacter coli* genomic DNA and its epidemiologic application. J Inf Dis 163:1068—1072

▧ Suppliers

Bio-Rad Laboratories
2000 Alfred Nobel Drive, Hercules, California 94547, USA
Phone: +1-510-741-1000; 1-800-424-6723
Fax: 510 741-5800; 1-800 879-2289

Boehringer-Mannheim GmbH, Biochemica
Sandhofer Strasse 116
68305 Mannheim, Germany
Phone: +49-621-759 8568
Fax: +49-621-759-4083
Web site: http://biochem.boehringer-mannheim.com

New England Biolabs
32 Tozer Road, Beverly, Massachusetts 01915-5599, USA
Phone: +1-508-927-5054; 1-800 632-5227
Fax: +1-508-921-1350
Web site: http://vent.neb.com

Promega Corporation
2800 Woods Hollow Road, Madison, Wisconsin 53711-5399, USA
Phone:+1-608-274-4330; 1-800-356-9526
Fax: +1-608-277-2516
Web site: http://www.promega.com

Sigma Chemical Company
P.O. Box 14508; St. Louis, Missouri 63178, USA
Phone: +1-314-771-5750; 1-800-521-8956
Fax: +1-314-771-5757; 1-800-325-5052
Web site: http://www.sigma.sial.com

Chapter 5

Ultrasensitive Silver Based Stains for Nucleic Acid Detection

Carl R. Merril, Karen M. Washart, Robert C. Allen

Introduction

The visualization of nucleic acids separated by gel electrophoresis has been frequently accomplished by the use of radioisotopic markers and autoradiography. However, the use of radioisotopes is time consuming and expensive due to the use of photographic films and supplies and the extensive requirements for the disposal of radioactive materials. These factors have stimulated the development of alternative, non-isotopic visualization procedures such as fluorescence, chemiluminescence, enzyme techniques and silver staining, the latter of which is relatively inexpensive, sensitive and rapid.

Silver stains have proven to be more than 100-fold more sensitive than the commonly used fluorescent stain ethidium bromide for the detection of nucleic acids separated on gels. Detection of nucleic acids by silver staining depends on the reduction of silver ions to form metallic silver images (Merril 1986). This use of a silver reduction reaction to create an image of the nucleic acid distribution in a gel depends on oxidation-reduction potential differences between the sites occupied by the nucleic acid molecules and adjacent sites in the gel which are devoid of such molecules. Initially some researchers suggested that silver staining might be due to differential binding of silver ions. However, by employing a silver ion solution with a concentration of silver comparable to that employed in normal silver staining protocols, containing radioactive sil-

R. Merril, Karen M. Washart, Laboratory of Biochemical Genetics, National Institute of Mental Health, NIH, Building 10, Room 2D54, Bethesda Maryland 20892, USA
Robert C. Allen, Department of Pathology and Laboratory Medicine, Medical School of the University of South Carolina, 171 Ashley Ave., Charleston, South Carolina, 29425, USA
Correspondence to Carl R. Merril, 2 Winder Court, Rockville, MD 20850, USA (*phone* +1-301-435-3583; *fax* +1-301-480-9862; *e-mail* merrilc@helix.nih.gov)

ver, it was possible to demonstrate that these ions were uniformly distributed throughout the gel and not concentrated at the site of a biopolymer (Merril et al. 1984). The reduction of ionic to metallic silver was demonstrated through the use of electron microscopy. On development of the gel silver image, it was possible to visualize the formation of silver grains, or particles, associated with the positions of biopolymers in the gel (Merril et al. 1988).

Silver stain protocols for nucleic acid detection (and protein detection) can be divided into three general categories: diamine or ammoniacal stains, non-diamine chemical reduction stains, and stains based on photodevelopment. Both the diamine and non-diamine chemical reduction stains rely on an alkaline environment for the reduction of ionic silver while photodevelopment may occur in an acidic environment.

The silver ions in the diamine stains are complexed as diamines by mixing silver nitrate with ammonium hydroxide. In these stains few silver ions are free for reduction until the ammonium hydroxide is neutralized in a controlled manner by a weak acid, such as citric acid. Image development is initiated by the acidification of the ammoniacal silver solution, usually with citric acid in the presence of formaldehyde. The addition of citric acid lowers the concentration of free ammonium ions which results in the liberation of silver ions for reduction by formaldehyde to metallic silver (Merril 1986, 1990). Diamine silver stains generally have high sensitivity and they have proven to be particularly useful for the staining of proteins separated on polyacrylamide gels which are thicker than 1 mm, although diamine staining is also a highly sensitive method also for ultrathin layer gels (Allen 1980).

Non-diamine chemical reduction stains use silver nitrate to provide silver ions for reaction with protein and nucleic acid sites under acidic conditions (Merril 1986, 1990). Image development is initiated by placing the gel in an alkaline solution, utilizing sodium carbonate and/or hydroxide and other bases to maintain the alkaline pH, while the silver ions are selectively reduced to metallic silver with formaldehyde. The formic acid, produced by the oxidation of formaldehyde, is buffered by the sodium carbonate. This type of stain is relatively rapid and simple to perform. The non-diamine stains generally work best with gels 1 mm or less in thickness.

Photodevelopment silver stains depend on the energy from photons of light to reduce ionic silver to metallic silver (Merril 1986 and Merril et al. 1986). Nucleic acids enhance the photoreduction of ionic to metallic silver in gels impregnated with silver chloride. While the photodevelopment stains are very rapid and can permit the visualization of nucleic acid patterns within 10 min after an electrophoretic separation, they lack the sensitivity of the other silver staining methods. Furthermore, image preservation, which is very good with chemical reduction methods, is poor with photodevelopment stains. This type of silver stain should be reserved for studies of dense nucleic acid bands or spots.

It is possible to combine the chemical and photodevelopment methods. For example, one silver stain protocol, which was developed to detect proteins and nucleic acids on membranes, uses light to initiate the formation of silver nucleation centers with silver halides, followed by the use of chemical development to deposit additional silver on the silver nucleation centers. This protocol provides a stain which is rapid and sensitive. It can detect proteins and nucleic acids in the nanogram range on thin membranes in under 15 min (Merril et al. 1986).

While a complete understanding of the mechanisms underlying the visualization of nucleic acids by silver staining has not yet been achieved, silver staining experiments utilizing nucleic acid polymers, nucleosides and nucleotides indicate that the purine bases adenine and guanine may be of primary importance (Merril et al. 1986). In addition, nucleic acid polymers smaller than 300 bp display an increased density of staining in relationship to their concentration compared to larger nucleic acid polymers (Goldman et al. 1982).

Given the complex nature of silver staining and the numerous gel formulations and applications, there are almost an infinite number of possible silver staining protocols. We have selected two of the protocols that we have found to be particularly useful for the detection of nucleic acid separated on gels, described below.

Materials

Silver Staining Gels Not Bound to Membranes

The solutions are easy to prepare and all but the developer may be kept for at least 6 months. The chemicals utilized in these stains are common, high-purity laboratory reagents which may be purchased from any reputable chemical supply company.

- Fixative: 10 % (v/v) ethanol in deionized water
- Oxidizer: 0.0034 M potassium dichromate (1 g/l) and 0.0032 M nitric acid (or 0.2 ml of concentrated nitric acid, 16 M).

Note: Avoid direct contact with the skin since this solution is an irritant. It is a strong oxidizer and should be kept away from reducing agents. It is also photosensitive and should be stored in an amber or dark bottle.

- Silver solution: 2 % silver nitrate (2 g/l).

Note: Silver nitrate solution is caustic to eyes and mucous membranes, and it will stain both skin and clothing. Store in an amber or dark bottle.

- Developer: 0.28 M sodium carbonate (30 g/l) plus 0.5 ml/l of 37 % formalin, added just prior to use. This solution should be used at 4 °C. It may be stored at that temperature for up to 24 h.

Note: Since formaldehyde vapor may be a carcinogen, this solution should be made and used in a fume hood.

- Stop solution: 5 % acetic acid (v/v) in deionized water

Silver Staining DNA on Ultrathin-Layer Gels Backed on Polyester

- Oxidizer: 2 % (v/v) nitric acid in deionized water
- Silver solution: 0.012 M silver nitrate (1.5 g/300 ml).

Note: Silver nitrate solution is caustic to eyes and mucous membranes, and it will stain both skin and clothing. Store in an amber or dark bottle.

- Developer: 0.28 M sodium carbonate (30 g/l) plus 0.5 ml/l of 37 % formalin, added just prior to use. This solution should be used at 4 °C. It may be stored at that temperature for up to 24 h.

Note: Since formaldehyde vapor may be a carcinogen, this solution should be made and used in a fume hood.

- Stop solution: 10 % (v/v) acetic acid in deionized water
- Storage solution: 10 % (v/v) glycerol in deionized water

Procedure

Silver Staining Gels Not Bound To Membranes

Following the electrophoretic separation of nucleic acids, each polyacrylamide gel is placed in a separate clean glass or plastic tray. The use of clean vinyl or latex gloves, with the powder washed off, and a lab coat will reduce the risk of contaminating the gel and the staining of the investigator. All solutions should be at room temperature except the developer, which is pre-cooled at 4 °C. Subdued laboratory lighting will help to decrease background staining. Development should not be done on a light box. The gel should be handled gently to avoid crushing that may produce artifacts. Solutions should not be poured directly on the gel; instead, tilt the tray and pour solutions into a corner or use individual trays for each solution and move the gels from tray to tray. Solution volumes should be sufficient to completely immerse the gels. The trays should be gently agitated during staining.

1. Following electrophoretic separation, soak the gel in fixative for 10 min. Use a volume of fixative double that needed to cover the gel. The gel may be stored indefinitely in the fixative at this point if development needs to be delayed.

2. The gel is then placed in the dichromate oxidizer for 3–5 min. Then decant the oxidizer. Wash the gel three times with deionized water for 3–5 min each.

3. Place the gel in the silver nitrate solution for 15–20 min and then briefly wash the gel (for 20–30 s) in deionized water to remove silver nitrate from the gel surface.

4. Form the image by washing the gel with pre-cooled developer for approximately 30 s with constant agitation. Observe the gel closely during this process and discard the solution as soon as it loses clarity. Replace with fresh developer and again discard as soon as it loses clarity. Repeat this process until the gel image has formed or until the background staining becomes objectionable.

5. When the DNA bands reach the desired intensity relative to the background, stop development by placing the gel in the stop solution. The DNA bands normally are dark brown or black on a faint yellow/brown background.

Silver Staining DNA on Ultrathin-Layer Gels Backed on Polyester

1. Following electrophoresis, place the gel into a glass or plastic tray and cover with 2% nitric acid until the tracking dye boundary is completely yellow. This usually takes 1–2 min depending on ionic strength of the leading ion and the gel thickness. When a continuous buffer system is used and no tracking dye line is present, 2 min in the oxidizing nitric acid is sufficient for 300-μm-thick gels.

2. Rinse the surface once with deionized water, drain and add the silver solution for a period of 3–5 min (3 min for 200-μm-thick and 5 min for 350-μm-thick gels) at room temperature on a shaker. Reduction in reaction time may be accomplished by reacting the silver nitrate for 2–3 min at an elevated temperature (achieved by placing the gel and stain into a microwave oven for 30 s at maximum power).

Note: Use plastic or glass tray only! In both cases, following the reaction with silver, wash the gel twice for 30 s each with deionized water.

3. Reduce the silver ions to metallic silver with the developer solution. When the solution turns brown, pour it off and add new solution. Continue to change every few minutes until the DNA bands show up well. Best results are obtained with fresh reducing reagent and background staining can be reduced by using the solutions cold, although the development time will be increased. During the development step, a cover may be placed over the staining dish to keep direct light off the gel in order to reduce background staining of the surface. If there is very little DNA in the sample tracks, then staining can be continued to a light yellow background to obtain greater sensitivity.

4. Staining ultrathin-layer gels backed on a polyester backing such as Gel-Bond PAG (FMC, Rockland, ME) requires much shorter reaction times in silver nitrate. Excessive silver staining may lead to unacceptable background darkening due to the staining of the "AcrylAide" coating on the backing polyester film.

5. Stop the silver reduction in water for 5 min, add stop solution for 5 min, and place in distilled water. Stopping the silver reduction reaction with acetic acid directly is not recommended. In the presence of the sodium carbonate, addition of acetic acid may lead to bubble formation in ultrathin-layer gels. These later can cause holes due to evolution of carbon dioxide on drying the gel and will interfere with photography and especially densitometry. A 1 min wash in water prior to addition of the stop solution will prevent bubble formation.

6. Wipe any remaining reduced silver from the surface with a damp cotton ball until no more blackening of the cotton is apparent and then dry the gel in a microwave oven in short 1 min bursts at full power. This process can be repeated.

7. For long-term storage, soak the developed, dried gels 5 min in the storage solution, cover with a clear plastic sheet and blot off excess moisture on a paper towel.

Results

Figure 1 demonstrates a typical silver stained gel, with a 20 cm separation on a 7 %T, 3 %C rehydratable polyacrylamide gel using 30 mM formate as the leading ion and serine as the trailing ion. Glycerol (1.34 M) is used as a moisturizing agent. The separation was carried out in 2 h of constant pulsed power at 20 pulses per second. The gel was stained with silver for 5 min.

Troubleshooting

- For a permanent record, the gel should be photographed on a light box since gels silver- stained with this procedure may darken over time.

- Thinner gels can be stained faster, but they are more fragile and difficult to work with whereas gels thicker than 0.7 mm require longer to stain. To increase the durability of gels that are thinner than 0.7 mm, we recommend casting them on a polyester backing (such as GelBond PAG, FMC, Rockland, ME or Gel-Fix, Serva, Heidelberg, Germany). Alternatively, the gel may be cast and bonded to silanized glass (Bind-Silane, Pharmacia-LKB, Piscataway, New Jersey). Gels that are attached to polyester sheets have an added advantage over unbound gels in that after electrophoresis they can be trimmed to remove the regions that do not contain DNA, thereby decreasing the amount of staining solutions required.

- If gels bound to polyester sheets are utilized, the polyester sheets must be flattened and held tightly by capillary attraction against the backing glass plate of the gel apparatus prior to casting. To accomplish this, a few milliliters of distilled water or 5 % glycerol is pipetted onto the backing gel plate, then the polyester sheet is placed on the plate. One must be careful to place the hydrophobic surface against the backing plate. The sheet is then flattened with a rubber print roller or a wooden rolling pin. It is important to assure that no bubbles or lint is under the backing, otherwise irregularities in the gel will result with attendant disturbances in the electric field.

- Flap or capillary casting (Kruchiniana et al. 1994) are simple methods of preparing backed gels either for direct use or for the preparation of rehydratable gels. To prevent the polyester sheet from moving while

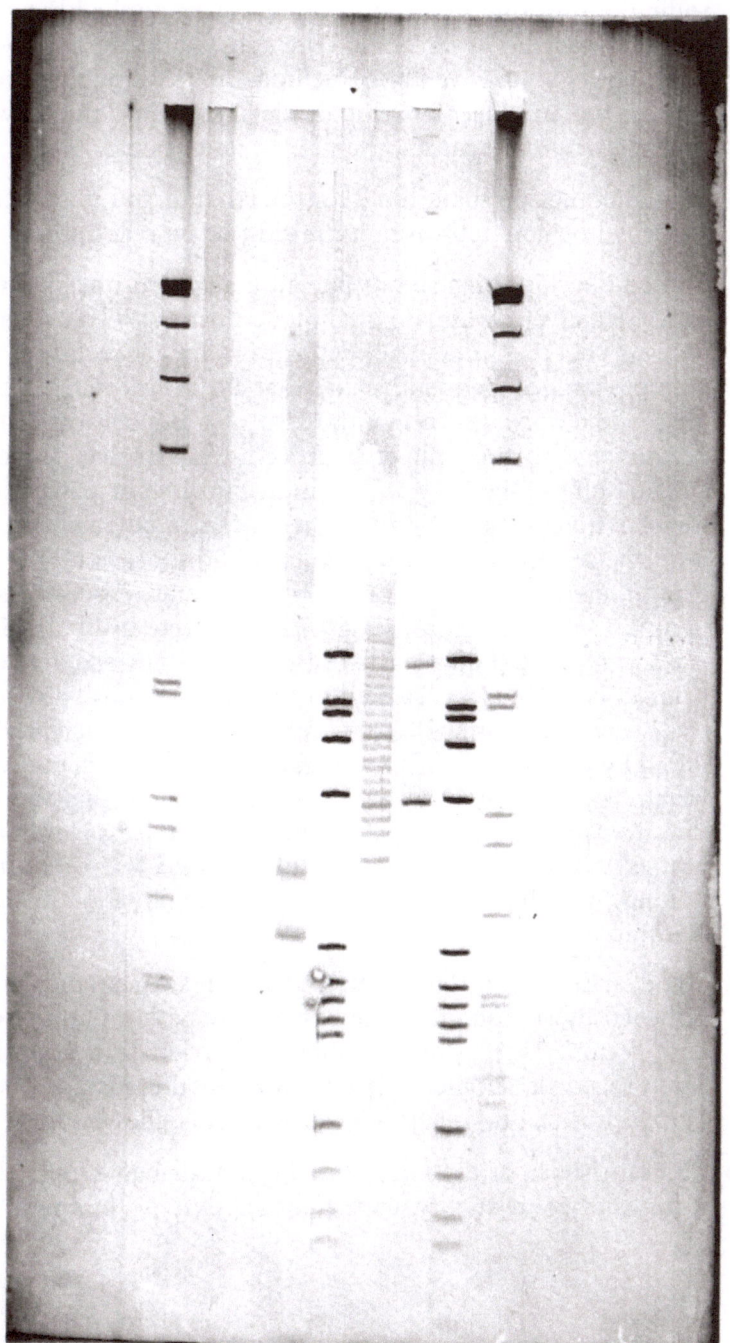

Fig. 1. A typical silver stained gel containing ladders and samples

rolling, a thin film of silicon grease can be applied between the sheet and plate on the lower end of the gel. The polyester sheet should be large enough to cover and include the gel, loading wells, and the spacers that are placed on top of the sheet. Once the polyester sheet is in place with the spacers, the gel is cast as usual.

- Methanol may be substituted for the ethanol, but it is toxic and may be absorbed by contact through the skin or by respiration.

- Cooled developer is used to permit controlled image development. A brownish silver carbonate precipitate forms as silver ions escape from the gel. This precipitate will deposit on the gel surface if the solutions are not changed frequently. Gels that are overdeveloped or that have a mirror-like appearance owing to surface deposition of silver can be destained with the following procedure: Dissolve 37 g of anhydrous cupric sulfate and 37 g of sodium chloride in 850 ml of deionized water. With constant stirring, add concentrated ammonium hydroxide until the solution becomes deep blue and there is no precipitate, then bring the volume to 1 l with deionized water. A second solution containing 436 g sodium thiosulfate pentahydrate in 1 l of deionized water is also prepared. Equal parts of these two solutions are combined just prior to use. The gel is soaked in this solution until almost completely clear. The destaining is stopped by 10 % acetic acid for 15 min, followed by a minimum of 1 h washing with several changes of deionized water. The gel may then be restained by beginning the procedure at the silver nitrate step (step 3, "Silver Staining Gels Not Bound to Membranes"). The oxidizing step is not required for restaining. This destaining procedure is also useful for the removal of silver stains from clothing.

- Some artifacts and precipitate on the surface of the gel can be removed by rubbing the dried gel gently with a damp piece of absorbent cotton. This technique should be tried in a noncritical region. The procedure should be stopped when the surface becomes tacky. However, it can be repeated after the gel is allowed to dry.

- The sensitivity of the stain may be diminished unless the deionized water is of good quality, with a conductivity of less than 1 mho.

References

Allen RC (1980) Rapid isoelectric focusing and detection of nanogram amounts of proteins from body tissues and fluids. Electrophoresis 1:32–37

Goldman D, Merril CR (1982) Silver staining of DNA in polyacrylamide gels: Linearity and effect of fragment size. Electrophoresis 3: 24–26

Kruchinina NA, Gresshoff PM (1994) Detergent affects silver sequencing. BioTechniques 17: 280–282

Merril CR (1986) Development and mechanisms of silver stains for electrophoresis. Acta Histochemica et Cytochemica 19: 655–667

Merril CR (1990) Silver staining of proteins and DNA. Nature 343: 779–780

Merril CR, Goldman D (1984) Detection of polypeptides in two-dimensional gels using silver staining. In: Celis JE, Bravo R (eds) Two dimensional gel electrophoresis of proteins. Academic Press, pp 93–109

Merril CR, Pratt M (1986) A rapid sensitive protein silver stain and assay system for proteins on membranes. Anal Biochem 156: 96–110

Merril CR, Bisher ME, Harrington M, Steven AC (1988) Coloration of silver-stained protein bands in polyacrylamide gels is caused by light scattering from silver grains of characteristic sizes. Proc Natl Acad Sci USA 85: 453–457

Chapter 6

The Analysis of Microsatellites by Silver Staining of Nondenaturing Polyacrylamide Gels

SHIRIN S. JOSEPH

Introduction

Silver staining of polyacrylamide gels is a technique which has been used in the past for detecting trace amounts of proteins and more recently to visualise nucleic acids. Here, the analysis of microsatellites using silver staining of nondenaturing polyacrylamide gels will be described in detail, while providing the reader with references for further reading on details of the use of silver staining for other analyses (Kruchinina and Gresshoff 1994; Promega Corporation 1993; Sanguinetti et al. 1994).

The silver staining technique used here is a modification of that described by Wallace et al. (1993). The theory is as follows: Silver binds to the DNA and is reduced to form a deposit of metallic silver by formaldehyde. The advantages of silver staining are numerous:

- It is less hazardous than radioactive detection methods, most reagents being harmless, requiring no special precautions for handling.

- The cost of the technique is considerably less than radioactive techniques.

- Staining can be completed within half an hour and has a minimal number of steps.

- Definition of stained DNA bands is sharper than those obtained using radioactive detection methods.

- Sensitivity is approximately 100-fold higher than that obtained by ethidium bromide staining, the limits of detection being approximately $1\,\mathrm{pg/mm^2}$.

Shirin S. Joseph, The Sanger Centre, Hinxton Hall, Hinxton, Cambridgeshire, CB10 1SA, UK
(*phone* +44-1223-494843; *fax*: +44-1223-494919; *e-mail* shirin@sanger.ac.uk)

▉ Outline

A brief outline of the procedure is as follows:

1. Determination of the optimal touchdown PCR conditions for a given microsatellite.

2. Amplification of the microsatellite under the optimized conditions.

3. Separation of the PCR products by nondenaturing polyacrylamide gel electrophoresis.

4. Visualisation of the PCR products using the silver staining procedure.

5. Recording results either by permanent storage or photography of gel.

▉ Materials

- Acrylamide:bisacrylamide solution (gas stabilised): 40 % (19:1) (National Diagnostics)
- $(NH_4)_2SO_4$: 10 %
- N,N,N',N'- tetramethylethylenediamine (TEMED)
- Sucrose loading buffer: 40 % sucrose; 0.25 % xylene cyanol; 0.25 % bromophenol blue

Store the above solutions at 4 °C.

- TBE (10×): 108 g Tris base; 55 g boric acid; 40°ml 0.5 M EDTA
- Dimethyldichlorosilane solution (BDH)
- Silver nitrate solution (0.1 %): can be re-used approximately three times without significant loss of staining efficiency
- Ethanol (10 %)/acetic acid (0.5 %)

Store the above solutions at room temperature.

Note: Silver nitrate solution should be wrapped in aluminium foil and stored in the dark, e.g. in a cupboard.

- NaOH (1.5 %)/formaldehyde (0.1 %). Make up fresh.
- Double-sided, water-cooled (20×20°cm) PAGE system (Cambridge Electrophoresis, UK)
- Spacers: 1.35 mm
- 1 kb ladder (BRL)
- PCR buffer (11.1×): 45mMTris.HCl, pH 8.8; 11 mM $(NH_4)_2SO_4$; 4.5 mM $MgCl_2$; 6.7 mM 2-mercaptoethanol; 4.4 mM EDTA, pH 8.0; 1 mM of

each of the dNTPs. Divide into 200 µl aliquots before freezing. Store at
−20 °C.
- Amplitaq DNA polymerase (Perkin Elmer)

Procedure

Amplification of Microsatellites by "Touchdown" PCR

Since silver staining is approximately tenfold more sensitive than ethidium bromide staining, it is necessary to keep the total number of cycles
to a minimum. "Touchdown" PCR involves amplification at progressively
decreasing annealing temperatures followed by a final amplification at
the lowest annealing temperature for a given number of cycles, depending on the microsatellite being amplified. The general scheme for
"touchdown" which has been used here is a modification of that initially
described by Don et al. (1991):

---

94 °C, 5 min: 1 cycle
94 °C, 30 s; 65 °C, 30 s; 72 °C, 30 s: two cycles

Repeat the latter two cycles four more times, but with variation of the
annealing temperature, such that with each subsequent set of cycles, the
annealing temperature is decreased by 2 °C, i.e. two cycles at each of the
following temperatures: 63, 61, 59 and 57 °C, amounting to a total of ten
cycles within the touchdown. This is followed by:

---

94 °C, 30 s; 55 °C, 30 s; 72 °C, 30 s; × cycles at 55 °C

Determine the final number of cycles (x) at 55 °C empirically for each
microsatellite as shown in Fig. 1.
    The above range of annealing temperatures can be increased depend-
ing on the degree of overamplification seen on initial amplification.
(Generally for primers which are 20-mers, a 65–55 °C annealing range is
sufficient for obtaining optimum amplification of a microsatellite).

Fig. 1. Determination of the optimal number of cycles required for amplification of a microsatellite. After a touchdown of ten cycles at 65 °C–57 °C, the final number of cycles at 55 °C was varied. *Lane 1* Primer only control; *lanes 2–6* 18, 16, 14, 12 and 10 cycles at 55 °C, respectively; *lane 7* 1 kb ladder

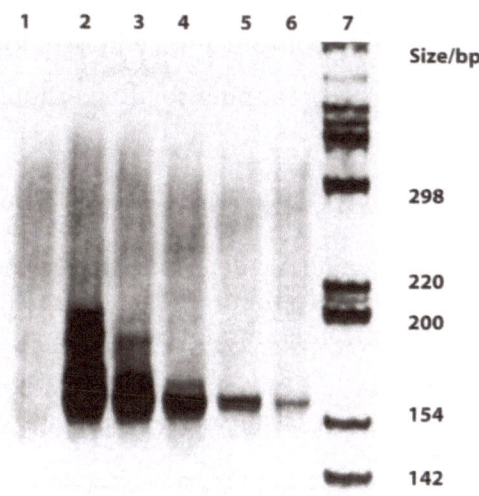

The components of PCR per 10 µl final volume are as follows:

0.9 µl 11.1× PCR buffer (1× final concentration)
0.1 unit Amplitaq DNA polymerase
1 µM of each primer
25 ng genomic DNA

Resolution of Microsatellites

Microsatellites within the 100–25 °bp product size range can be easily resolved by electrophoresis in 6–9 % nondenaturing, acrylamide gels. Variation of the acrylamide content according to the size of the micro-satellite locus amplified results in better resolution, e.g. higher percent-ages of acrylamide achieve greater levels of definition for larger microsa-tellite bands, although longer runs are necessary for suitable resolution.

Nondenaturing gels are used in preference to denaturing gels for the analysis of microsatellites, since this avoids the complications in inter-pretation often present due to sequence-dependent strand separation seen in denaturing gels. Using gels of approximately 1 mm thickness allows ease of handling during staining. Mixing of the samples with a sucrose loading buffer avoids distortion of DNA bands due to the glycerol reacting with the borate in TBE loading buffers.

Note: All silver staining solutions should be handled using gloves.

Example Analysis of a Microsatellite with Products in the 140–170°bp Size Range

1. After empirical determination of the total number of cycles required for amplification of the microsatellite, amplify the microsatellite at the optimised number of cycles.

2. Prepare a 9 % nondenaturing acrylamide gel (20×20°cm,1.35-mm-thick). Per 50 ml solution, prepare as follows:

---

5 ml 10× TBE
11.25 ml 40 % (19:1) acrylamide:bisacrylamide solution
1 ml 10 % $(NH_4)_2SO_4$
32.75 ml distilled water

Allow to set for 2 h at room temperature. Gel can be made 16 h prior to usage.

Note: It is important to ensure that the glass plates are kept scrupulously clean – immediately after use, rinse with water, wipe dry and siliconise with dimethyldichlorosilane solution (Sambrook et al. 1989).

3. Add loading buffer to PCR products (2 µl/10 µl PCR) and load 12 µl PCR product onto gel using duck-billed tips (allows better resolution; if using oil over PCR reaction, removal of oil is necessary before loading as the intake of the aqueous phase using duck-billed tips is difficult). Sample volumes should be kept small as larger volumes lead to loss of resolution.

4. Separate products at a constant voltage of 90°V for 16 h. Although shorter periods of electrophoresis at higher voltages can be used, overnight runs at lower voltages result in consistent, high quality separation of microsatellites without "smiling" of bands.

5. After the run, remove one glass plate and spacers from gel, and place gel on glass plate in a sandwich box designated for silver staining. (The glass plate allows ease of handling of the gel during addition and removal of reagents during the staining procedure).

6. Add 10 % ethanol, 0.5 % acetic acid, and leave to soak with shaking for 5 min.

7. Using the glass plate as a support, remove the gel from the sandwich box and discard the ethanol/acetic acid solution. Replace with 0.1 % silver nitrate solution. Soak gel for 15 min with gentle shaking.

8. Remove and store silver nitrate solution. Add 1.5 % NaOH/0.1 % formaldehyde and develop stain for approximately 5–10 min with shaking.

9. Once the bands are formed, the gel can be fixed and stored. Fix in 0.75 % Na_2CO_3 for 10 min and remove the solution. The gel can be stored in a sealed freezer bag at room temperature or dried down by placing it on a glass plate and covering with porous hydrophilic cellophane for approximately 24 h. Drying the gel is preferable to sealed storage as the gel is more prone to damage in freezer bags. Providing the DNA bands are well separated after electrophoresis and readily visible after staining, it is generally unnecessary to fix and store the gel, as a suitable record of the results can be obtained by photography.

10. Photograph the gel using a light box under a UVP camera (Genetic Research Instrumentation Ltd, Essex, UK) connected to a videographic printer (SONY UP-860CE) and a black and white monitor (SONY SSM-930CE).

Results

An example of the type of result obtained with silver staining is shown in Fig. 2. DNA from various male and female individuals was amplified using primers derived from a polymorphic microsatellite repeat locus on the X chromosome. Females are either heterozygous or homozygous for the alleles present at this locus, whereas males are hemizygous.

Interpretation

This is generally straightforward except for cases in which:

- Certain microsatellites show inconsistency in amplification of both alleles for a given individual.

– Conformational bands result, which are unique to the alleles amplified, but these, on segregation analysis, can be shown to cosegregate with the true alleles.

Fig. 2. Typical silver staining results obtained with an cross-specific microsatellite. *Lane a* 1 kb BRL ladder; *lanes b–m* genomic DNA from individuals in a pedigree amplified using cross-specific microsatellite; males are hemizygous, whereas females are either hetero- or homozygous. *Open circles* Female; *open squares* male

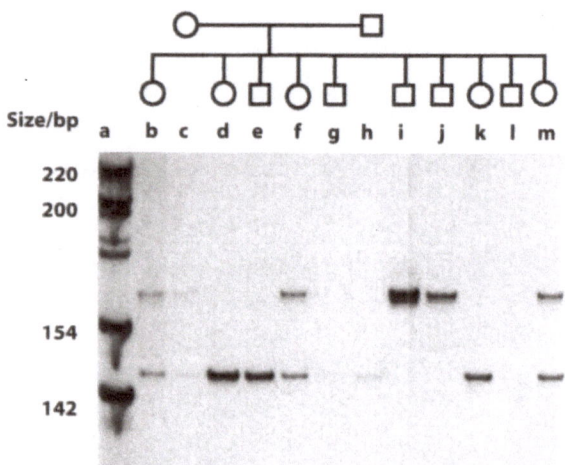

Fig. 3. Cross-specific microsatellite analysis of various male and female individuals. *Lanes 1–7* Genomic DNA from individuals in a pedigree amplified using a cross-specific microsatellite; males are hemizygous, whereas fenales are either hetero- or homozygous. Only highlighted (filled squares or circles) individuals in the pedigree have been analysed. *Lane 8* Primer only control; *lane 9* 1 kb bRL ladder. Note that although stutter bands can be seen in all cases, the bands corresponding to the true locus are of a greater intensity, and therefore easily identified. *Open* or *filled squares* Males: *open* or *filled circles* females

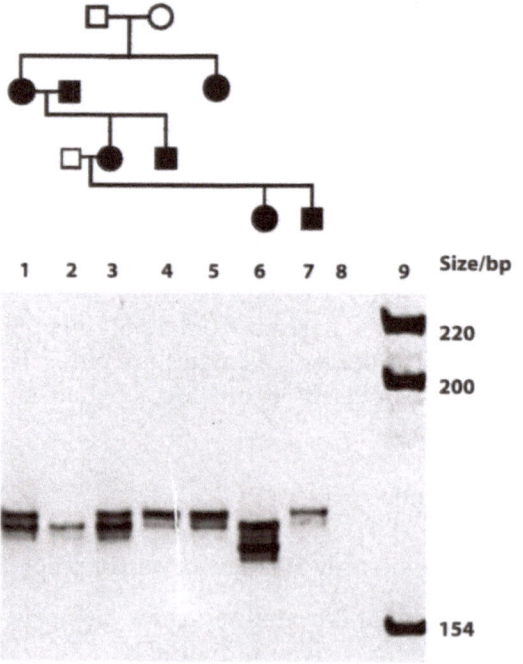

In the case of inconsistent amplification of one or other allele, this can be a feature of the design of the primer sequence, rather than the lack of stringency in the PCR conditions used. In such cases, often the simplest solution is to re-design primers or to use an adjacent microsatellite which is easier to type.

"Stutter" bands of reduced mobility (generally 2 bp smaller than the true allele and lower in intensity) can form with some microsatellites. These do not present a serious problem in interpretation due to their generally clear features of reduced size and lower intensity than the true product (Fig. 3).

References

Caetano-Anolles G and Gresshoff PM (1994) Staining with silver. Promega Notes 45:13–18

Don RH, Cox PT, Wainwright BJ, Baker K, Mattick JS (1991) 'Touchdown" PCR to circumvent spurious priming during gene amplification. Nucleic Acids Res 19:4008

Kruchinina NG, Gresshoff PM (1994) Detergent affects silver sequencing. Biotechniques 17:280–282

Promega Corporation (1993) Silver Sequence DNA Sequencing System, product Q4130. Madison, Wisconsin

Sambrook J, Fritsch EF, and Maniatis T (1989) Molecular cloning : a laboratory manual. 2nd edition, Cold Spring Harbor Press, Cold Spring Harbor, New York

Sanguinetti CJ, Dias Neto E, Simpson AJG (1994) Rapid silver staining and recovery of PCR product separated on polyacrylamide gels. Biotechniques 17:915–919

Wallace AJ, Mountford RC, Williamson P, Ramsden SC (1993) Visualisation of microsatellite repeats by silver staining. J Med Gen 30:346–347

Chapter 7

Capillary Electrophoretic Separation of DNA Fragments

John M. Butler and Dennis J. Reeder

Introduction

The desire for rapid and automated DNA analysis has resulted in a recent growth in the area of capillary electrophoretic (CE) separation of DNA fragments. The narrow glass tubes, e.g., 50 μm (internal diameter, i.d) ×27 cm, used in CE allow high electric fields due to efficient dissipation of heat generated from current flow. DNA separations an order of magnitude faster than traditional slab gel electrophoresis may be obtained with CE (Barron 1995). The advent of stable capillary coatings (Hjerten 1985) and refillable polymeric separation media (Zhu 1989) has made capillary electrophoresis a robust and reproducible technique. Detection sensitivities rivaling classical radioactive methods are now possible with laser-induced fluorescence (LIF) of double-stranded (ds) DNA labeled with intercalating dyes (Schwartz 1992; McCord 1993b).

Capillary electrophoresis has the capability of accomplishing separation tasks with greater speed and ease than other methodologies. For example, separation of megabase-size DNA in a matter of minutes was recently demonstrated on a CE system (Kim 1996), whereas traditional pulsed-field electrophoresis would take hours to days to accomplish the same degree of separation. An important benefit of CE is the ability for automation. To identify DNA fragments by CE, slab gels do not have to be prepared nor do samples have to be manually pipetted onto the gel. In commercial CE instruments, sample vials are simply loaded into an autosampler. Under computer control, the DNA sample may be loaded

John M. Butler, Dennis J. Reeder, DNA Technologies Group, Biotechnology Division, Chemical Science and Biotechnology Laboratory, Bldg. 222, National Institutes of Standards and Technology, Gaithersburg, Maryland 20899, USA
Correspondence to John M. Butler, GeneTrace Systems, Inc., 333 Ravenswood Avenue, PN088, Menlo Park, California 94025, USA
(*phone* +1-650-859-3051; *fax*: +1-650-859-2654; *e-mail* butler@genetrace.com)

Capillary Electrophoresis System

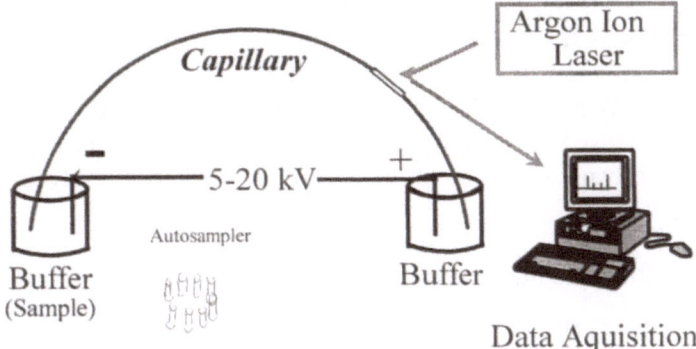

Fig. 1. The elements of a capillary electrophoresis (CE) system used for DNA fragment analysis

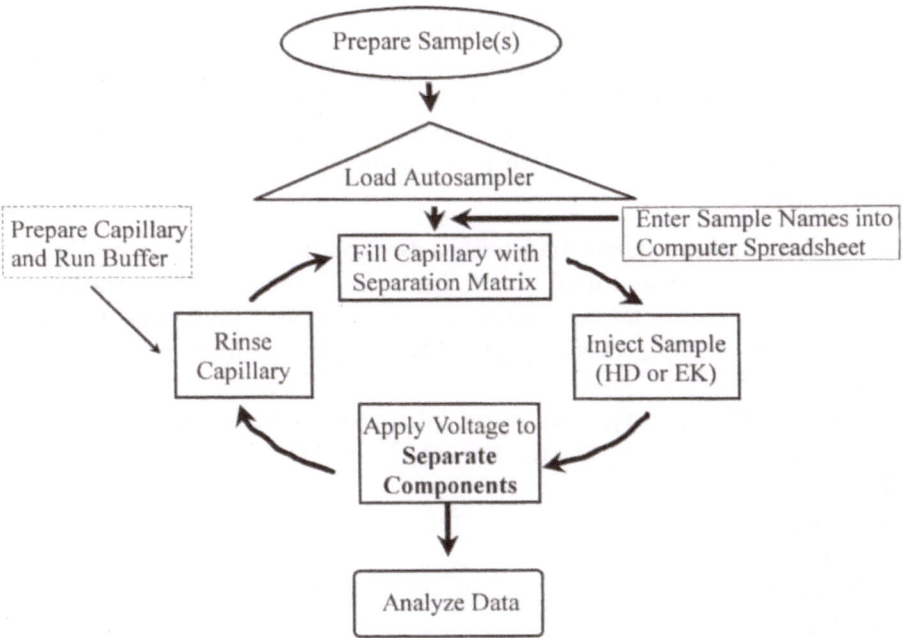

Fig. 2. Flowchart of CE sample analysis

onto the capillary via hydrodynamic pressure (HD) or electrokinetic (EK) injection (Butler 1994a). The components of a typical CE system are schematically described in Fig. 1.

In this chapter, we describe a CE method (Fig. 2) for rapidly separating DNA fragments generated from the polymerase chain reaction (PCR) (Mullis 1987) or DNA restriction endonuclease digestion. Variations of this CE method have been used to quantitate PCR products (Butler 1994a; Wilson 1995) and to size PCR-amplified short tandem repeats (STRs) used in DNA typing (Butler 1995a; Isenberg 1996). CE has distinct advantages over other methods of dsDNA analysis including:

- Ease of use

- High reproducibility

- Greater speed

- Increased sensitivity

Materials

Equipment

Note: Certain commercial equipment or materials are identified in this chapter in order to adequately specify the experimental procedure. Such identification does not imply recommendation or endorsement by the National Institute of Standards and Technology, nor does it imply that the materials or equipment identified are necessarily the best available for the purpose.

- Experiments were performed with a Beckman P/ACE 5500 capillary electrophoresis instrument with argon ion 488 nm laser module, including a computer workstation and printer (Beckman Instruments). Other CE systems (preferably with laser fluorescence) may also be used.
- DB-17 coated capillary, roll: 50 μm i.d.×10 m (Alltech; cat. no. 93450) or single capillary: 50 μm i.d.×100 cm (J and W Scientific, part number 126-1713-CE)
- Hydroxyethyl cellulose (HEC): 80–125 cP at 2 % solution (Aldrich; cat. no. 30,863-3)
- YO-PRO-1 nucleic acid intercalating dye (Molecular Probes, Inc.; cat. no. Y-3603)
- Fuming sulfuric acid (Aldrich) and soldering iron
- Cellulose acetate filter: 0.45 μm (Corning Inc., Corning, NY) or equivalent

- Methanol: HPLC grade
- DNA fragments for use as internal standards (BioVentures or Gen-Sura)
- Autosampler vials: 4 ml wide-mouth with threads (Alltech)

Buffers

Stock HEC solution (500 ml)

100 mmol/l Tris-Borate, pH 8.2
2 mmol/l EDTA
1 % (w/v) HEC

Mix and filter through a 0.45 μm pore size cellulose acetate filter.

Note: Vigorous shaking or stirring overnight is normally required to dissolve the HEC powder. Solution may be stored at room temperature.

Final run buffer

15 ml stock HEC solution
500 ng/ml YO-PRO-1 (11.9 μl of 1 mM YO-PRO-1)

Note: Solution should be covered with aluminum foil as the YO-PRO-1 is light-sensitive. New solutions should be prepared daily.

Procedure

Preparation or Replacement of CE Capillaries

1. Remove laser cartridge from the Beckman CE. The cartridge allows a coolant solution to flow around the capillary to maintain a constant temperature environment.

2. Remove the old capillary from the cartridge as shown in the capillary replacement guide provided by Beckman with the CE instrument.

3. Using the old capillary as a guide, cut the desired length of DB-17 capillary (usually 27 cm) from the capillary spool. Always leave several centimeters on either end of the capillary for final adjustments.

4. Mark the location of the detection window (using old capillary as a guide) with a black marker. The detection window is located approximately 7 cm from the outlet end. Wearing safety goggles, remove the

Making the Capillary Window

polyimide outer coating from the capillary at the black mark using a soldering iron and concentrated fuming sulfuric acid (Bocek 1991). The capillary should be placed directly underneath the soldering iron and the acid then dripped onto the tip of the hot iron. Isopropyl alcohol, in the form of alcohol hand wipes, may be used to remove the blackened polyimide coating and to clean the new window area. Approximately 5 mm of polyimide coating needs to be removed to allow passage of the laser light. Alternatively, the flame from a match may be used to remove the polyimide outer capillary coating. However, the flame method may damage the inner coating of the capillary and cause adverse effects on column performance. Precut capillaries with detection windows already prepared may also be purchased from most CE companies or column manufacturers.

Note: The fuming sulfuric acid procedure should be performed in a fume hood to avoid breathing any toxic vapors that may be produced. Diluted acid waste should be slowly neutralized with diluted sodium hydroxide solution before disposal.

5. The instructions for putting the capillary into the cartridge holder are provided in the Beckman cartridge rebuilding guide. The total length of the capillary is determined by the number of times it is wrapped around the spindle inside the cartridge. The capillary should be wrapped around the spindle once for a 37 cm length, twice for a 47 cm, or three times for a 57 cm total length. The shortest possible capillary, in which the capillary is not wrapped around the spindle, is 27 cm and has the advantage of the highest separation speed. Although many researchers use longer capillaries for DNA separations, a resolution of 3 bp has been demonstrated on a 27 cm capillary (Butler 1995a). After the capillary has been installed, carefully trim the ends of the capillary as described in the guide booklet.

6. Air bubbles may be removed by pushing water through the installed capillary with a syringe. This syringe may be prepared by gluing a glass capillary press fit union (J and W Scientific, cat. no. 705-0725) to the tip of a 5 ml plastic syringe (McCord 1993a).

7. Reinstall the cartridge in the CE instrument. Finally, lift the piercing levers with a flathead screwdriver to insure that the ends of the capillary are next to the electrodes. Improper capillary placement could result in breaking the ends.

Procedure for Preparation of CE-LIF Run Buffer

1. Weigh out 6.06 g trizma base (Tris), 3.09 g boric acid, 0.29 g ethylene-diamine tetraacetic acid (EDTA), and 5.00 g hydroxyethyl cellulose (HEC). These weights in 500 ml of solution will produce a 1 % (w/v) HEC, 100 mmol/l Tris-borate, 2 mmol/l EDTA solution. Add the HEC powder last as it takes the longest time to go into solution.

2. Stir or shake the solution overnight or until all of the HEC has dissolved.

3. Using vacuum suction, filter the HEC solution through a 500 ml 0.45 μm cellulose acetate filter.

4. The pH of the 100 mmol/l Tris-borate buffer should be approximately 8.1–8.2. The pH can be raised with 1 mol/l CsOH to pH 8.7 to benefit some separations (McCord 1993b).

Preparing the HEC stock solution

5. Place 15 ml of the above 1 % HEC solution into a 15 ml aluminum foil-covered centrifuge tube.

6. Add 11.9 μl of 1 mmol/l YO-PRO-1 stock solution (Molecular Probes nucleic acid intercalating dye) to achieve a final concentration of 500 ng/ml YO-PRO-1. Gently shake the mixture by turning the tube end-for-end. The YO-PRO-1 stock solution needs to be shield from light or kept in the dark at 4 °C or at −20 °C. It is wise to make aliquots of the stock solution for storage purposes to prevent problems with freezing and thawing.

Preparing the Run Buffer Containing the Intercalating Dye

7. Using a transfer pipette, fill three 4 ml amber vials with the run buffer. Vials should be filled to the level of the lowest thread (approximately 4 ml). Be sure to remove all bubbles from the solution surface as they may interfere with the flow of electrical current during electrophoresis. Two buffer vials will be used as the inlet and outlet vials during the separation step; the third will be used to fill the capillary with fresh separation media between each run.

Preparing Run Buffer Vials

Procedure for Sample Preparation

1. Small glass vials are commercially available and work well for holding the CE sample. A less expensive alternative is to use modified PCR tubes which are disposable. To make these CE sample tubes, carefully

remove the top portion of a 0.2 ml MicroAmp reaction tube (Perkin-Elmer, Norwalk, Connecticut) with a scapula knife or razor blade.

2. Add 24–99 μl of deionized water to the modified PCR tube depending on the desired dilution of the sample. We typically work in the 25 μl–50 μl range. It is difficult to inject reproducibly with less than 10 μl in the sample vial.

3. Add 1 μl of the PCR product and mix with the deionized water (step 2) by drawing it into and out of the pipette tip several times.

4. Place the sample tube into a 4 ml glass vial containing a spring. The spring allows the sample vial to come in contact with the electrode and inlet end of the capillary during the injection without bending the electrode or breaking the capillary.

5. Screw a gray rubber cap onto the vial to prevent evaporation. An evaporation rate of 0.05 nl/s has been reported for sample volumes of 5 μl, which corresponds to more than 4 μl per day (Watzig 1993). The rubber cap also forms a tight seal which is important during hydrodynamic injections.

6. Type the sample name in the computer spreadsheet for each autosampler position. Finally, place the prepared sample vial in the appropriate position.

Note: It is important to pay attention to the proper placement of reagents and samples in the autosampler positions. It may be difficult to sort out sample data which has been identified incorrectly.

Setting Up a Method for DNA Separation

The Beckman CE system is controlled by a programmed "method" which moves the autosampler and correlates each electropherogram to a particular sample. In a typical CE analysis of DNA, the process would proceed as illustrated in Fig. 3. When the run is complete, the process may be repeated again, with additional samples being injected from vials 12, 13, 14, and so forth. The same buffer and rinse vials are used repeatedly for typically 10–20 samples. As these steps are completely automated, a sample can be injected, separated, and detected in only a few minutes with no operator interaction.

```
P/ACE 2000 Series Version 3.0 - Beckman Instruments Inc.

Method: C:\PACE\EXAMPLE.MTD              06 Jun 95  11:52


50um x 27cm DB-17
100mM TBE, pH 8.2, 1% HEC (Aldrich), 500 ng/mL

Vial contents:
 1. outlet buffer
10. waste
11. DNA sample
31. MeOH
32. filling buffer
33. deionized water
34. inlet buffer

Display Channel A
Time:        0.00 to  15.00 Minutes
Channel A:  -0.100 to   0.500 Fluorescence
```

STEP	PROCESS	DURATION	INLET	OUTLET	CONTROL SUMMARY
1	SET DETECTOR				LIF: 488:520 nm Rate: 5 Hz Rise: 1.0 sec Gain: 100
2	SET TEMP				Temp: 25 C Wait until reached
3	RINSE	1.0 min	31	10	Forward: High Pressure
4	RINSE	2.0 min	32	10	Forward: High Pressure
5	WAIT	0.1 min	33	10	
6	INJECT	5.0 sec	11	1	Voltage: 1.00 kV
7	SEPARATE	15.0 min	34	1	Constant Voltage: 5.00 kV Current Limit: 250.0 uA Integrator On

Fig. 3. An example method, which may be set up on the Beckman P/ACE Windows software 3.0, for high resolution DNA fragment analysis (see Butler 1995b). The *numbers* under the inlet and outlet columns represent vials described under the vial contents section

Results

Optimizing a Separation Method

As there are a number of variables in CE, standards should be run on a regular basis to confirm that everything is working correctly. The resolution of DNA fragments contained in a particular standard give an indication of whether the CE system is functioning properly.

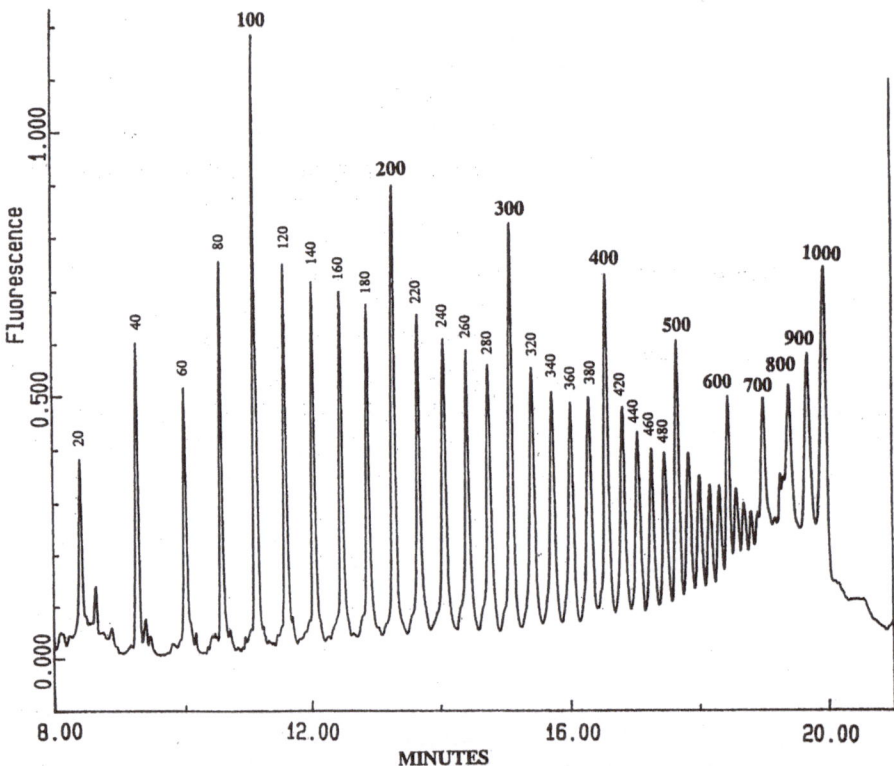

Fig. 4. GenSura 20 bp ladder with DNA fragments from 20 to 1000 bp in 20 bp increments. The declining peak signal with increasing fragment size is a result of the enzymatic reaction used to generate the ladder. The addition of a 100 bp ladder makes those peaks higher. Conditions: *capillary*: 50 μm i.d.×27 cm DB-17; *buffer*: 3 % HEC (24,000–27,000 MW; Polysciences), 100 mmol/l TBE, pH 8.2; 500 ng/ml YO-PRO-1; *temperature*: 25 °C; *injection*: 5 s at 1 kV; *separation*: 5 kV; *sample*: 160 ng/ml 20 bp ladder+40 ng/ml 100 bp ladder (GenSura)

Expected results for several standards:

- GenSura 20 bp ladder (Fig. 4): contains 50 DNA fragments from 20–1000 bp in 20 bp increments; a useful standard due to the even spacing of the peaks.
- pBR322 *Hae*III digest (Fig. 5): contains 22 restriction fragments ranging in size from 8 to 587 bp. Separation of the 123 and 124 bp fragments has been shown with intercalating dyes, such as ethidium bromide (McCord 1993a).
- φX174 *Hae*III digest (Fig. 6): contains 11 restriction fragments covering a size range of 72 bp to 1353 bp. Most PCR products fall into this size range.

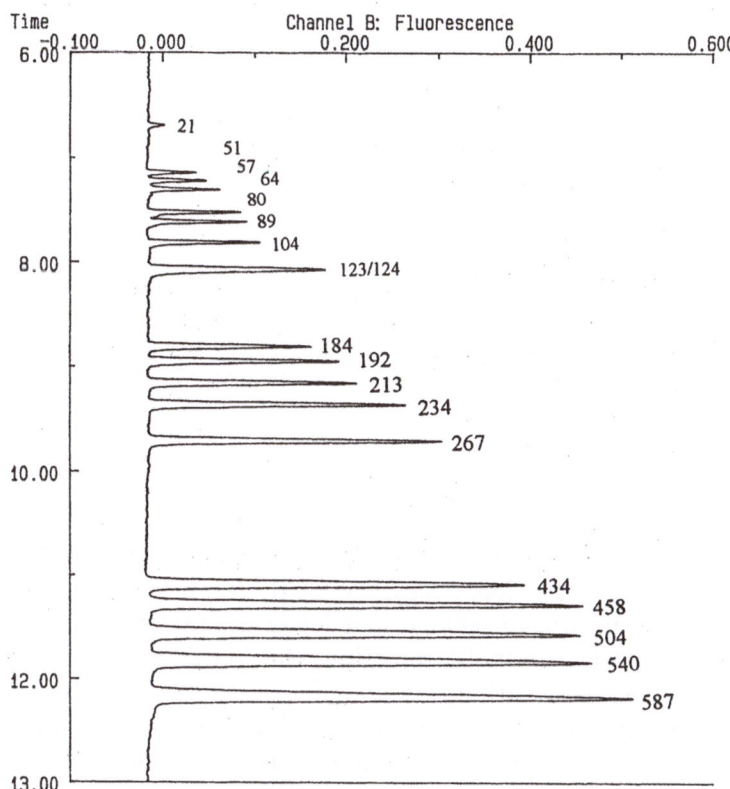

Fig. 5. The separation of pBR322 *Hae*III restriction digest. Conditions as in Fig. 4, except 1 % (w/v) HEC (Aldrich), 10^{-6} dilution of SYBR green stock solution (Molecular Probes), and hydrodynamic injection (10 s water plug, 30 s sample). *Sample*: 610 ng/ml pBR322 *Hae*III digest. *Numbers* above the peaks indicate the DNA fragment size (bp). The 123 and 124 bp fragments were not separated under these conditions nor were the fragments below 21 bp detected

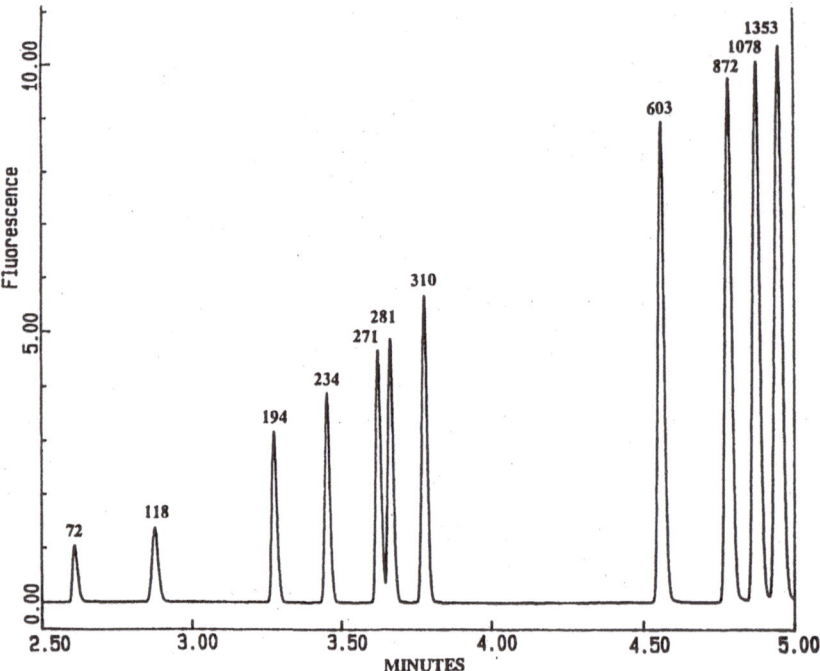

Fig. 6. The separation of φX174 *Hae*III restriction fragments with an effective length of 7 cm. Conditions as in Fig. 4, except polarity was reversed (cathode at outlet), sample vial in the outlet autosampler tray, and 1 % (w/v) HEC (Aldrich). *Sample*: 590 ng/ml φX174 *Hae*III digest. *Numbers* above the peaks indicate the DNA fragment size (bp)

Important CE Variables

CE variables that may be changed to improve a separation:

- Voltage: The lower the applied voltage, the better the resolution and the longer the time necessary for the separation.
- Injection time or voltage: The lower the injection time or voltage during an electrokinetic injection, the better the resolution, but at the expense of sensitivity.
- Temperature: The higher the capillary temperature, the lower the buffer viscosity and the faster the separation speed, but at the expense of resolution.
- Column: The shorter the capillary length, the higher the separation speed. The resolution may be improved by decreasing the capillary diameter.
- Buffer components: A higher concentration of the intercalating dye (for monomeric dyes) will improve resolution, but at the expense of

slower migration times. A longer polymer length may be used to improve resolution for larger DNA fragments, but at the expense of higher solution viscosities.

After successful separation of one of the DNA standards discussed above and after optimization of the various CE parameters for a particular application, one can proceed with sample analysis. As sample quantities are often limited, it is wise to run standards prior to the sample to ensure that all is well. The most common problem seen with the CE technique described here is loss of resolution from the failure of the column coating. When the inner capillary coating begins to break down, sample material sticks to the column wall resulting in the sudden appearance of broad peaks. The best remedy is to replace the capillary with a new one. With the CE method described in Fig. 3, a good capillary should last over 1000 runs when rinse steps are performed between each run. However, some lots of DB-17 capillary coatings have not worked as well as others, presumably due to variability in the manufacture of the capillary coating. Coated capillaries may also be obtained from Supelco or other CE manufacturers. There is currently no way to predict how well a capillary will perform until it is actually used. A poor capillary coating is immediately evident from broad peaks in a DNA standard and failure to resolve closely spaced fragments.

Importance of Standards

Increasing Separation Speed

A major advantage of CE is its speed for single samples. As CE sampling is sequential, sample throughput time will also be improved when separations are done faster. Therefore, developing rapid separation methods can be important. Typically, you should determine the minimum resolution needed for your application and then increase the voltage and/or shorten the capillary in order to increase the speed of analysis. For example, if you are only concerned with detecting two DNA fragments, which are spaced more than 100 bp apart (e.g., competitive PCR products), separation times may often be reduced to a few minutes. Several features on the Beckman CE may be employed to increase speed, often without significantly sacrificing resolution. Two of these features include using: (1) "short" capillaries with reverse injection and (2) voltage gradients.

Short Capillaries with Reverse Injection

Injecting the sample from the outlet autosampler tray results in the sample components having to travel only 7 cm to the detection region rather than 20 cm when the sample is injected from the inlet side. The result of shortening the analyte's traveling distance by one third is a threefold increase in separation speed. A sufficiently high resolution separation may be achieved in only a few centimeters using this technique (Fig. 6).

Note: The CE instrument polarity must be reversed (cathode at outlet) for this procedure.

Voltage Gradients

A single-step voltage gradient can be used to increase the peak capacity for a particular region and to reduce the overall separation time (Fig. 7). This form of voltage programming is performed by first applying a high electric field (e.g., 400 V/cm) at the beginning of a run to move the DNA quickly through the column and then, shortly before the DNA fragments would normally pass the detector window, reducing the voltage (e.g., 150 V/cm) in a single step to improve the resolution in that particular region (Butler 1994b).

Intercalating Dyes

Effects of Intercalating Dyes

The following features of an intercalating dye are important to DNA separations using CE:

- Spectral characteristics: Is the dye efficiently excited by the laser line used?

- Binding specificity : Will the dye bind all DNA fragments in a sequence-independent manner?

- Sensitivity: Is the emission signal greatly enhanced when the dye is bound to DNA verses free in solution?

The dye YO-PRO-1 meets all of these specifications and works well in CE applications involving DNA sizing or quantitation (McCord 1993b; Butler 1995b). SYBR green I, a nucleic acid staining dye also available from Molecular Probes, behaves in a similar fashion to YO-PRO-1 (Butler 1995b; Skeidsvoll 1995) and gives good results (Fig. 5). Other intercalating dyes may be used but have different characteristics.

The binding preferences of various fluorescent intercalators have been studied using the pBR322 HaeIII digest, which possesses DNA fragments of varying guanine-cytosine (GC) contents (Butler 1995b). The Molecular Probes dye TO-PRO-1 appears to have selected affinity for certain

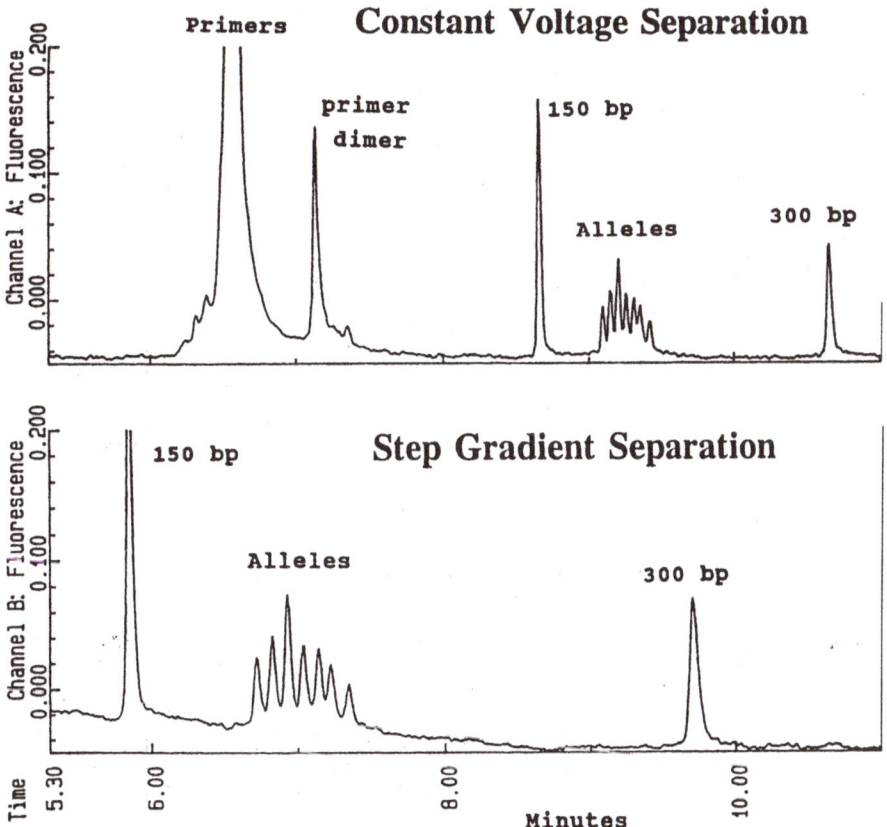

Fig. 7. A comparison of separation speed for HUMTH01 alleles between a constant voltage and a step-gradient voltage separation (Butler 1994b). The time scales are the same. In the *lower frame*, the primers and primer dimer migrate faster than 5 min and are not shown. The alleles shown are a group of PCR products ranging in size from 179 to 203 bp, each 4 bp apart. Conditions as in Fig. 6 except 37 cm capillary and 50 ng/ml YO-PRO-1. Constant voltage separation was performed at 10 kV while the step gradient was 15 kV for 5 min followed by an immediate drop to 5 kV

DNA fragments, which upon excess binding shifts the intercalator-DNA complex to slower migration times. The separation of two peaks can improve, if the larger fragment has the higher GC content (e.g., 123 bp and 124 bp fragments of pBR322 *Hae*III digest). However, comigration of two peaks may occur, if the smaller fragment in a doublet has greater affinity for the intercalator. In applications which involve sizing or quantitating DNA fragments, dyes, such as YO-PRO-1, which bind universally (based upon the total number of intercalating sites) rather than discriminately (preferring a particular sequence) are preferred.

Monointercalators in the run buffer (McCord 1993b) yield better resolution than dimeric dyes (Srinivasan 1993). Migration time reproducibility is affected by the dye concentration with higher concentrations producing better precision (Butler 1995a).

If wider bore capillaries (e.g., 100 µm i.d.) are used, the addition of >1 µM ethidium bromide, in combination with 50 ng/ml YO-PRO-1, has proven beneficial in separations involving closely spaced DNA fragments (see McCord 1993b). When using an argon ion laser (488 nm), the ethidium bromide improves the resolution but does not enhance sensitivity because the dye is not sufficiently excited at 488 nm. However, equivalent resolution may be obtained without ethidium bromide using 50 µm i.d. capillaries (Butler 1995b).

Polymer Solutions

Importance of Specifying Polymer Characteristics

Over 13 different types of polymers, including linear polyacrylamide, hydroxyethyl cellulose, and polyethylene oxide, have been used to separate DNA in a capillary format (Barron 1995). A wide variety of polymer lengths and concentrations have also been used. Polymer length, type, and concentration all have an impact on DNA separations by CE. It is important to specify the source of the HEC or other polymer used so that others may reproduce your work. For example, the HEC powder which may be purchased from Aldrich is a different length than the HEC provided by Polysciences, Inc. Various molecular weight ranges of HEC allow separation of different DNA size ranges (Isenberg 1996). In general, the best separations for DNA fragments in the 50–1000 bp range will be achieved with a longer polymer at a lower concentration than a shorter polymer at a higher concentration (Butler 1995b).

Other Details

Sample Dilution and Sample Stacking

It is important to mix the sample well during dilution. Usually drawing the solution in and out of the pipette tip several times is sufficient. In addition to increasing the volume in the sample vial, the dilution process allows better sampling due to a process known as sample stacking (Burgi 1991). The ionic difference between the sample and the buffer during electrokinetic injection is an important factor affecting how much DNA gets onto the column. PCR samples contain chloride ions and typically are dialyzed prior to CE analysis (McCord 1993a). They may be diluted

50-fold with deionized water and successfully injected (Butler 1995a). For this same reason, it is also important to dip the inlet end of the capillary into a water vial prior to electrokinetic injection to prevent contamination of the sample by the buffer ions from the run buffer filling step.

All DNA standards and samples should be kept refrigerated or frozen. Generally it is a good idea to prepare multiple aliquots of a DNA sample and only remove one at a time from the freezer. The use of gloves when handling samples can prevent contamination from DNA nucleases found on human skin.

DNA Storage and Handling

For overnight or long-term storage, coated capillaries should be filled with deionized water. In addition, both ends of the capillary should be placed in water to prevent their drying out. If the capillaries dry out, crystals may form in the buffer and clog the capillary. Plastic centrifuge tubes (0.6 ml) with holes punched in the center of the caps work well for covering the capillary ends.

Capillary Storage

Troubleshooting

- The piercing levers are the contact point between the sample vials and the CE instrument. The viscous sieving polymer can build up over time on the piercing levers causing the vials to stick and the autosampler tray to jam.
  - These problems may be avoided in many cases if one: (1) cleans the inner capillary coating between each run with methanol; (2) removes and cleans the piercing levers with water on a weekly basis

Routine Maintenance

Commonly seen problems, which can be corrected without too much difficulty, include:

- Lack of current flow, or an unstable current, during the separation step
  - This is due to capillary blockage.
    - When a new capillary is first loaded onto the CE instrument, a quick test may be performed to ensure that the capillary is not plugged or broken. After filling the capillary with run buffer, an electric potential can be applied across the capillary (e.g., 5 kV) using the front panel keys,. If the current stays at zero with the application of the voltage, then the capillary is plugged or broken. An air bubble may sometimes be the cause of no current flow.

Simply pushing water through the capillary with a syringe can remove the bubble. Trimming 1–2 mm from both ends of the capillary may remove a plugged section resulting from particulates in the buffer or sample. However, be careful not to remove too much from the inlet end, or the capillary may not come in contact with low volume samples and proper injection will not occur.

- Failure to detect peaks
  - The polarity needs to be reversed: with coated capillaries, the inlet should be cathodic (–).
  - The laser may not be turned on.
  - The fiber optic connection may need to be cleaned (usually some signal can be seen in this case, but it is lower than expected).
  - The intercalating dye was not added to the run buffer.

- High voltage discharge
  - This may be due to high humidity during the summer.
    - Humidity above 65–70 % in the laboratory can cause arcing (a phenomenon known as a corona discharge), due to water condensation on the electrodes. A dehumidifier in the laboratory may be a necessary solution in humid parts of the country, especially during the summer months.

Comments

Safety

Intercalating Dyes   Use of intercalating dyes to visualize DNA is a common practice in molecular biology laboratories. One of the more commonly used dyes is ethidium bromide, $C_{21}H_{20}BrN_3$, also known as Dormilac or homidium bromide (CAS 1239-45-8). The dark red crystals are readily soluble in water (5 g/100 ml). A dilute solution is often used to soak solid electrophoresis gels containing separated DNA fragments and then visualizing the gel under a UV light. The DNA strands containing the intercalated dye then fluoresce brightly and can be photographed or further manipulated.

No definitive studies have shown ethidium bromide to have carcinogenic or teratogenic properties in humans although it is known to be a potent mutagenic compound and is irritating to the eyes, skin, mucus membranes and upper respiratory tract. In the Ames test, 90 μg of ethidium bromide is as mutagenic as the smoke from one cigarette. Thus, the compound should be handled in the laboratory with care. Solutions should be prepared in a fume hood, protective gloves should be worn at all times, and operations capable of generating dust from the powdered form or aerosols from solutions should be avoided.

In the event of skin contact, immediately wash with soap and water and remove contaminated clothing. In case of eye contact, promptly wash with generous amounts of water and obtain medical attention. Spills should be absorbed with appropriate materials and disposed of according to local safety regulations.

For capillary electrophoresis, many new dyes, such as YO-PRO-1, have been synthesized to take advantage of their intercalating capabilities, spectral properties, and binding affinities. As with ethidium bromide, these dyes should be viewed as potentially toxic and mutagenic compounds and handled with proper precautions.

Disposal

To a solution of ethidium bromide (ca. 0.5 mg/ml) in 100 ml of water, TBE buffer or MOPS buffer, add 20 ml 5% (v/v) hypo-phosphorous acid solution and 12 ml 0.5 mol/l sodium nitrite solution. Stir the mixture briefly and after 20 h, neutralize it with sodium bicarbonate and discard (Joshua 1986; Lunn 1987). Alternatively, a commercial ethidium bromide waste reduction system is marketed by Schleicher and Schuell, (Keens, New Hampshire, cat. no. 448031).

Electrical Safety

The high voltages used in capillary electrophoresis are protected by safety interlocks in commercial CE instruments. These safety interlocks should not be defeated under any circumstances.

Other Issues

Old capillaries should be discarded in receptacles labeled for broken glass to prevent injury to those handling the regular laboratory trash. Safe laboratory practices should be followed at all times.

Additional Help

Internet Newsgroups and WWW sites

Due to the growth of information distribution through the internet, we have included sites where more information is available on CE. This is a partial list to show the kind of information that is available. Internet newsgroups that may provide additional help regarding CE work include:

- CE discussion group (cze-itp@muni.cz): Leave the subject line blank, and post "subscribe cze-itp (first name) (last name)" to listserv@ muni.cz
- sci.chem.analytical
- bionet.molbio.methds-reagnts

Helpful Web sites include:

- Pedro's biomolecular research tools: http://www.public.iastate.edu/ ~pedro/research-tools.html
- Web resources on capillary electrophoresis: http://minyos.its.rmit. edu.au/~seans/ce.html
- CESONE: Capillary Electrophoresis Society of New England: http://members.aol.com/cesone/index.html

Professional Meetings   Several conferences are held annually for discussion of capillary electrophoresis research. Two important conferences are:

- The "International Symposium on High Performance Capillary Electrophoresis and Related Microscale Techniques" (HPCE) is usually held at the end of January in California, Florida, or Europe. For more information, contact: Shirley Schlessinger, Suite 1015, 400 East Randolph Drive, Chicago, Illinois 60601 USA; +1-312-527-2011. Internet address: www.hpl.hp.com/casss/hpce97/
- The "Frederick Conference on Capillary Electrophoresis" is held at the end of October in Frederick, Maryland. For more information, contact: Margaret Fanning, SAIC Frederick, NCI-Frederick Cancer Research and Development Center, PO Box B, Frederick, Maryland 21702; phone +1-301-846-5865; fax +1-301-846-5866.

Books   A few books on CE that we personally recommend include Lander (1994) and Schwartz and Guttman (1995).

■ References

Barron AE, Blanch HW (1995) DNA separations by slab gel and capillary electrophoresis: theory and practice. Separation and Purification Methods 24:1–118

Bocek P, Chrambach A (1991) Capillary electrophoresis of DNA in agarose solutions at 40 °C. Electrophoresis 12:1059–1062

Burgi DS, Chien RL (1991) Optimization in sample stacking for high-performance capillary electrophoresis. Anal Chem 63:2042–2047

Butler JM, McCord BR, Jung JM, Wilson MR, Budowle B, Allen RO (1994a) Quantitation of polymerase chain reaction products by capillary electrophoresis using laser induced fluorescence. J Chromatogr B 658:271–281

Butler JM, McCord BR, Jung JM, Allen RO (1994b) Rapid analysis of the short tandem repeat HUMTH01 by capillary electrophoresis. BioTechniques 17:1062–1070

Butler JM, McCord BR, Jung JM, Lee JA, Budowle B, Allen RO (1995a) Application of dual internal standards for precise sizing of PCR products by capillary electrophoresis. Electrophoresis 16:974–980

Butler JM (1995b) Sizing and quantitation of polymerase chain reaction products by capillary electrophoresis for use in DNA typing. University of Virginia, PhD Dissertation

Figeys D, Arriaga E, Renborg A, Dovichi NJ (1994) Use of the fluorescent intercalating dyes POPO, YOYO, and YOYOfor ultrasensitive detection of doublestranded DNA separated by capillary electrophoresis with hydroxypropylmethyl cellulose and noncrosslinked polyacrylamide. J Chromatogr A 669:205–216

Hjerten S (1985) High-performance electrophoresis: elimination of electroendosmosis and solute adsorption. J Chromatogr 347:191–198

Joshua H (1986) Quantitative adsorption of ethidium bromide from aqueous solution by macroreticular resins. BioTechniques 4:207–208

Isenberg A, McCord BR, Koons BW, Budowle B, Allen RO (1996) DNA typing of a PCR-amplified D1S80/amelogenin multiplex using capillary electrophoresis and a mixed entangled polymer matrix. Electrophoresis 17:1505–1511

Kim Y, Morris MD (1996) Ultrafast high resolution separation of large DNA fragments by pulsed-field capillary electrophoresis. Electrophoresis 17:152–160

Lander JP (ed) (1994) Handbook of capillary electrophoresis. CRC Press, Boca Raton, Florida

Lunn G, Sansone E (1987) Ethidium bromide: destruction and decontamination of solutions. Anal Biochem 162:453–458

McCord BR, Jung JM, Holleran EA (1993a) High resolution capillary electrophoresis of forensic DNA using a nongel sieving buffer. J Liq Chromatogr 16:1963–1981

McCord BR, McClure DM, Jung JM (1993b) Capillary electrophoresis of PCRamplified DNA using fluorescence detection with an intercalating dye. J Chromatogr 652:75–82

Mullis KB, Faloona FA (1987) Specific synthesis of DNA in vitro via a polymerase-catalyzed chain reaction. Methods Enzymol 155:335–350

Schwartz H, Guttman A (1995) Separation of DNA by capillary electrophoresis. Beckman Instruments primer VII on capillary electrophoresis (Beckman part number 607397)

Schwartz HE, Ulfelder KJ (1992) Capillary electrophoresis with laserinduced fluorescence detection of PCR fragments using thiazole orange. Anal Chem 64:1737–1740

Skeidsvoll J, Ueland PM (1995) Analysis of double-stranded DNA by capillary electrophoresis with laser-induced fluorescence detection using the monomeric dye SYBR Green I. Anal Biochem 231:359–365

Srinivasan K, Girard JE, Williams PE, Roby RK, Weedn VW, Morris SC, Kline MC, Reeder DJ (1993) Electrophoretic separations of polymerase chain reactionamplified DNA fragments in DNA typing using a capillary electrophoresislaser induced fluorescence system. J Chromatogr 652:83–91

Watzig H, Dette C (1993) Precise quantitative capillary electrophoresis-methodological and instrumental aspects. J Chromatogr 636:31–38

Wilson MR, Polanskey D, Butler JM, DiZinno JA, Replogle J, Budowle B (1995) Extraction, PCR amplification, and sequencing of mitochondrial DNA from human hair shafts. BioTechniques 18:662–669

Zhu M, Hansen DL, Burd S, Gannon F (1989) Factors affecting free zone electrophoresis and isoelectric focusing in capillary electrophoresis. J Chromatogr 480:311–319

Suppliers

Capillary Electrophoresis Instrument

– P/ACE with LIF
  Beckman Instruments Inc.
  2500 Harbor Blvd., Fullerton, California 92634, USA
  Phone: 1-800-742-2345
  Fax: 1-800-643-4366
  Web site: www.beckman.com

– ABI 310 Genetic Analyzer
  PE Applied Biosystems
  850 Lincoln Drive, Foster City, California 94404, USA
  Phone: +1-415-570-6667
  Fax: +1-415-572-2743
  Web site: www.perkin-elemer.com

Software

– System Gold
  Beckman Instruments Inc.
  2500 Harbor Blvd., Fullerton, California 92634, USA
  Phone: 1-800-742-2345
  Fax: 1-800-643-4366
  Web site: www.beckman.com

– ABI Genescan
  PE Applied Biosystems
  850 Lincoln Drive, Foster City, California 94404, USA
  Phone: +1-415-570-6667
  Fax: +1-415-572-2743
  Web site: www.perkin-elemer.com

– Millenium
  Waters, Corp.
  34 Maple Street, Milford, Massachusetts 01757, USA
  Phone: 1-800-HPLC
  Fax: +1-508-478-5839
  Web site: www.waters.com

Capillary

- DB-17 coated
  J & W Scientific, Inc.
  91 Blue Ravine Rd., Folsom, California 95630-4707, USA
  Phone: 1-800-552-0413
  Fax: +1-916-985-1101
  Web site: www.jandw.com

- DB-17 coated
  Alltech Associates, Inc.
  2051 Waukegan Rd., Deerfield, Illinois 60015-1899, USA
  Phone: 1-800-255-8324
  Fax: +1-708-948-1078
  Web site: www.alltechweb.com

- Celect-N
  Supelco, Inc.
  Supelco park, Bellefonte, Pennsylvania 16823-0048, USA
  Phone: 1-800-247-6628
  Fax: +1-814-359-3044
  Web site: www.supelco.sial.com

Buffer

- Tris-borate EDTA
  Sigma-Aldrich, Corp.
  Box 14508 St. Louis, Missouri 63178, USA
  Phone: 1-800-325-3010
  Fax: 1-800-325-5052
  Web site: www.sigald.sial.com

Separation Medium

- HEC
  Aldrich Chemical Co.
  1001 West St. Paul Ave., Milwaukee, Wisconsin 53233, USA
  Phone: 1-800-558-9160
  Fax: 1-800-962-9591
  Web site: www.sigald.sial.com

- Polysciences, Inc.
  400 Valley Road, Warrington, Pennsylvania 18976, USA
  Phone: 1-800-523-2575
  Fax: 1-800-343-3291
  Web site: www.polysciences.com

Intercalating dye

- YO-Pro-1
  Molecular Probes, Inc.
  4849 Pitchford Ave., Eugene, Oregon 97402, USA
  Phone: +1-541-465-8300
  Fax: +1-541-344-6504
  Web site: www.probes.com

DNA Standards

- 200 bp ladder
  Gensura Laboratories, Inc.
  2640 Del Mar Heights Rd., Suite 219, Del Mar, California 92014, USA
  Phone: 1-800-436-7872
  Fax: +1-619-755-1758
  Web site: www.virtualad.com/gensura

- 200 bp fragment
  Bio Ventures, Inc.
  848 Scott St., Murfreesboro, Tennessee 37133, USA
  Phone: 1-800-235-8938
  Fax: +1-615-896-4837
  e-mail: bretd@telalink.net

Rapid Detection of Hepatitis C Virus in Plasma and Liver Biopsies by Capillary Electrophoresis

SAMEER A. SAKALLAH, ROBERT W. LANNING, DAVID L. COOPER

Introduction

Liver transplantation is one of the most complicated and costly medical procedures performed in hospitals and medical centers today. Candidates for liver transplantation include patients whose livers have been damaged by hepatitis C virus (HCV) infection. Since donated livers, however, sometimes carry HCV infection (Pereira et al. 1991), it would be beneficial to be able to test for HCV prior to transplantation. Because of the nature of organ transplantation, in which surgery must be performed within a few hours following organ donation, this test must be very fast but without a loss of sensitivity.

Several HCV detection methods have been developed in recent years. The most common methods are based on either ELISA (Donegan et al. 1995) or reverse transcriptase polymerase chain reaction (RT-PCR) followed by liquid hybridization (LH) to internal probes (Mateo et al. 1994; Payan et al. 1995). None of these methods, however, is suitable for testing livers prior to transplantation because they involve lengthy steps and results are usually generated only after 18–24 h. Variations of these methods remain time-consuming and unsuitable for this application. Thus, there is a need for a new method with the sensitivity of RT-PCR but which generates results in a few hours. Here, we describe a new test con-

Correspondence to Sameer A. Sakallah, Division of Molecular Diagnostics, Department of Pathology, 728 Scaife Hall, University of Pittsburgh Medical Center, Pittsburgh, Pennsylvania 15261, USA (*phone* +1-412-648-7549; *fax* +1-412-383-9594; *e-mail* ssa@med.pitt.edu)
Current address: Genetic Identification Technologies, Inc. P.O. Box 1818, Cranberry, PA 16066, USA (*phone/fax* +1-724-251-0737; *e-mail* genetic-id@hotmail.com
Robert W. Lanning, Molecular Genetics Laboratory, Carolinas Medical Center, Charlotte, North Carolina, 28232, USA
David L. Cooper, Quest Diagnostics, Inc., 33608 Ortega Highway, San Juan Capistrano, California 92690, USA

sisting of RT-PCR amplification followed by separation of the amplified fragment by capillary electrophoresis (CE) and detection with laser-induced fluorescence. The entire procedure takes 2.5 h with a sensitivity comparable to that of RT-PCR/LH methods.

Outline

Steps involved in the RT-PCR/CE test are outlined in Fig. 1. The primary objective for performing this test is to determine the HCV positivity of a donated liver in the shortest possible time. Thus, assuming the clinical specimen (liver biopsy) is received at 12:00 noon, this test will provide an answer to this question by 2:30 p.m.

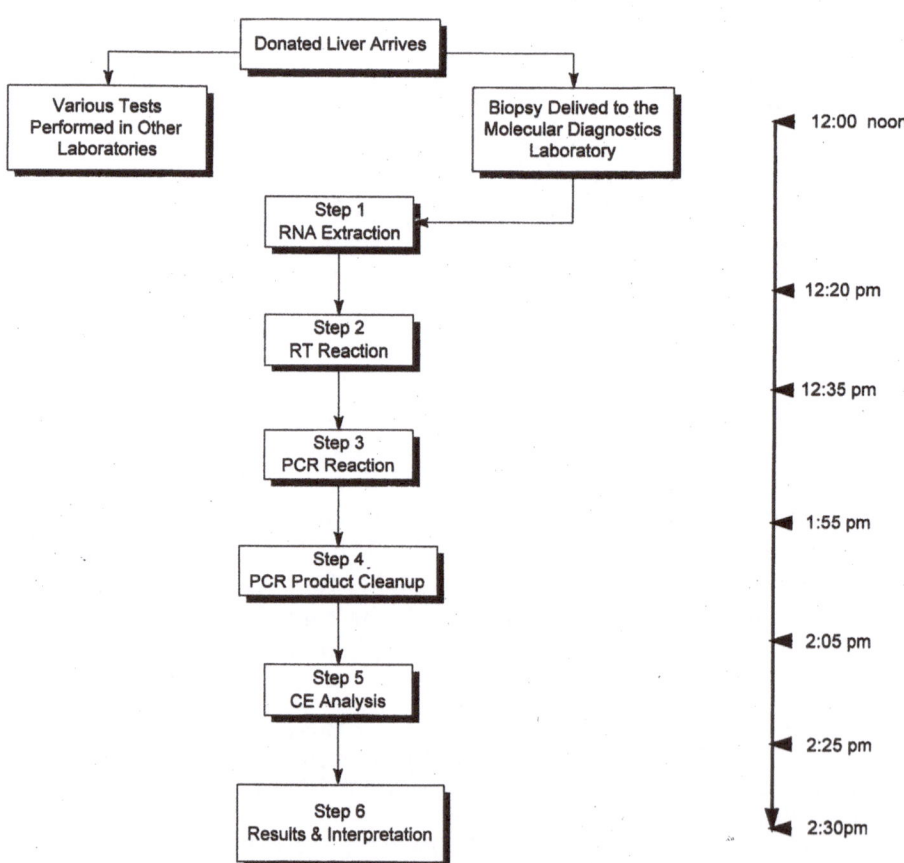

Fig. 1. Steps involved in the HCV RT PCR/CE test. See text for details

Materials

- Series 9600 thermal cycler (Perkin Elmer)
- P/ACE system 5010 capillary electrophoresis system, equipped with 488 nm argon laser-induced fluorescence (LIF) unit (Beckman Instruments)

- Primers: The PCR primer sequences used in this test are given below (Mateo et al. 1994). They amplify a 251 base pair DNA fragment derived from the mRNA of the HCV 5' untranslated region. The RT reaction utilizes random hexamers (see step 2 in "Procedure" below) to generate first strand DNA templates for subsequent PCR amplification.
  - Sense: 5'-CAC TCC CCT GTG AGG AAC TAC TGT CT
  - Antisense: 5'-TAC CAC AAG GCC TTT CGC GAC CCA ACA CTA CTC

  Both PCR primers are labeled at their 5'-ends with fluorescein.

Note: Although fluorescein may be used in the electrophoresis buffer, we found that this generates many nonspecific peaks, presumably from primer dimers and PCR unfinished products and artifacts.

- Reagent kits: RNA extraction is carried out using the RNeasy total RNA kit (Qiagen, Chatsworth, California). RT-PCR reactions are performed using the GeneAmp RNA PCR core kit (Perkin Elmer, Norwalk, Connecticut). For PCR product cleanup, QIAquick spin kit is used (Qiagen). In all cases, the manufacturer's instructions are followed without modifications. Capillary electrophoresis of the PCR fragments is carried out using Beckman's LIFluor dsDNA 1000 kit (Beckman Instruments, Columbia, Maryland), which includes a coated capillary (47-cm-long, 100 µm in diameter), and electrophoresis gel buffer (0.1 % acrylamide, >75 % polyacrylamide, >10 % Tris-(hydroxymethyl) aminomethane, >5 % boric acid).

Procedure

1. RNA extraction: Extraction of RNA from liver biopsies can be performed using any of the available standard methods. However, most of these methods are lengthy and thus not suitable for this test. We found that commercial total RNA extraction kits are more appropriate due to their speed and ease of use, and we have been using the

RNeasy total RNA kit from Qiagen following the manufacturer's recommendations.

2. Reverse transcription

2a. Prepare the RT master mix by mixing the reagents on ice in the following order:

Component	Volume, µl	Final concentration
25 mM MgCl$_2$	4	5 mM
10× PCR Buffer	2	1×
10 mM dGTP	2	1 mM
10 mM dCTP	2	1 mM
10 mM dATP	2	1 mM
10 mM dTTP	2	1 mM
RNase inhibitor, 20 U/µl	1	1 U/µl
MuLV RT, 50 U/µl	1	2.5 U/µl
Random hexamers, 50 µM	1	2.5 µM
Final volume	17	

1 × PCR buffer is 50 mM KCl, 10 mM Tris-HCl, pH 8.3.

2b. Add 3 µl of the extracted RNA template, place in a thermal cycler programmed for 10 min at 42 °C followed by 5 min at 99 °C and finally 4 °C for 5 min.

Note: A variety of reverse transcriptases from different manufacturers have been tested for their utility in this test. While all of these enzymes gave the correct product, none gave detectable product in less than 30 min. MuLV RT, however, routinely gave detectable products in 5–10 min.

Note: The amount of total RNA used as template in the RT reaction must be determined with a known positive sample. In our hands, as little as 1 ng of input RNA gave a detectable peak, although we typically use 500 ng of total RNA per reaction.

3. PCR reaction

3a. While waiting for the RT reaction, prepare the PCR reaction mix as follows:

Component	Volume (μl/reaction)	Final concentration
25 mM MgCl$_2$	4.0	2 mM
10× PCR buffer	8.0	1×
DEPC water	65.5	
AmpliTaq DNA polymerase	0.5	2.5 U
Sense primer	1.0	0.25 μM
Antisense primer	1.0	0.25 μM
Final volume	80.0	

3b. Add 80 μl of the PCR master mix to the RT reaction tube (final volume=100 μl) and start thermal cycling with an initial denaturation at 95 °C for 105 s, followed by 35 cycles of 95 °C for 15 s, and 60 °C for 30 s. Include a final extension step at 60 °C for 7 min followed by soaking at 4 °C.

4. PCR product cleanup: Cleaning up the PCR reaction products prior to analysis by capillary electrophoresis improves peak resolution and eliminates confusion about the position of the HCV peak. The easiest and fastest way to clean the PCR reactions from deoxynucleotide triphosphates, salts, and excess primers is by passing the reaction volume on a spin column. Criteria for choosing from the numerous spin columns available commercially include speed, ease of use and yield. In this laboratory, the QIAquick spin columns (Qiagen) have been used routinely according to the manufacturer's recommendations with excellent results. Samples elute in a final volume of 20 μl.

5. Detection by capillary electrophoresis.

5a. Capillary preparation: Fill the capillary with the gel buffer by applying a high pressure rinse for 4 min.

Note: Occasionally, air bubbles become trapped in the gel buffer which results in the appearance of sharp spikes in the electropherogram. This problem is easily eliminated by sonication of the gel buffer for 10–15 min just before use. The buffer temperature should be around 20 °C before separation starts.

5b. Sample preparation: Supplement the sample (20 μl) with 1 μl (1 μg) of 100 bp size marker (Life Technologies). This will serve as an internal standard for DNA fragment mobility in the capillary.

5c. Sample injection. Apply samples to the capillary gel using the pressure method for 5–30 s.

Note: Injection times of less than 5 s may fail to detect minor positive signals. Injections longer than 30 s are unnecessary and may result in peaks that are too broad.

5d. Separation: Start separation by capillary electrophoresis at 10 kilovolts (kV) for 25 min. In most cases, the HCV peak elutes at about 21 min.

Note: Fragment mobilities will shift slightly from one separation to another. Occasionally, however, there will be a more serious mobility shift of several minutes. Thus, the internal standard will eliminate any ambiguities related to mobility shifts. The HCV peak can be easily recognized by its position which is flanked by the 200 and 300 bp peaks of the size marker.

5e. Data output: The instrument can be programmed to integrate the peaks and calculate the area under each peak. While the data obtained here is not intended to be quantitative, it can provide a semi-quantitative measure of the positivity of the clinical specimen.

Results

Figure 2 shows a representative electropherogram of HCV detection from an actual clinical specimen. The HCV peak in this particular case elutes at 21.82 min, as indicated by the arrow, although in some cases this peak may elute much later (over 24 min). This effect is not unusual in capillary electrophoresis and illustrates the importance of including an internal standard or size marker. The elution pattern of the entire electropherogram may shift up or down the time scale depending on the age of the capillary or buffer temperature. Figure 3 demonstrates this shift in DNA fragment mobility from our HCV data. Please note that, regardless of the elution time, the HCV peak always represents the correct DNA fragment size of 251 bp.

Table 1 shows areas represented by peaks from Fig. 2. Please note that peaks 1–14 are not shown in Fig. 2. Although these data can provide a measure of positivity of the clinical specimen, they must not be used to quantify HCV infection.

The CE detection method described here is at least as sensitive as the standard LH method. We tested over 70 plasma specimens which were

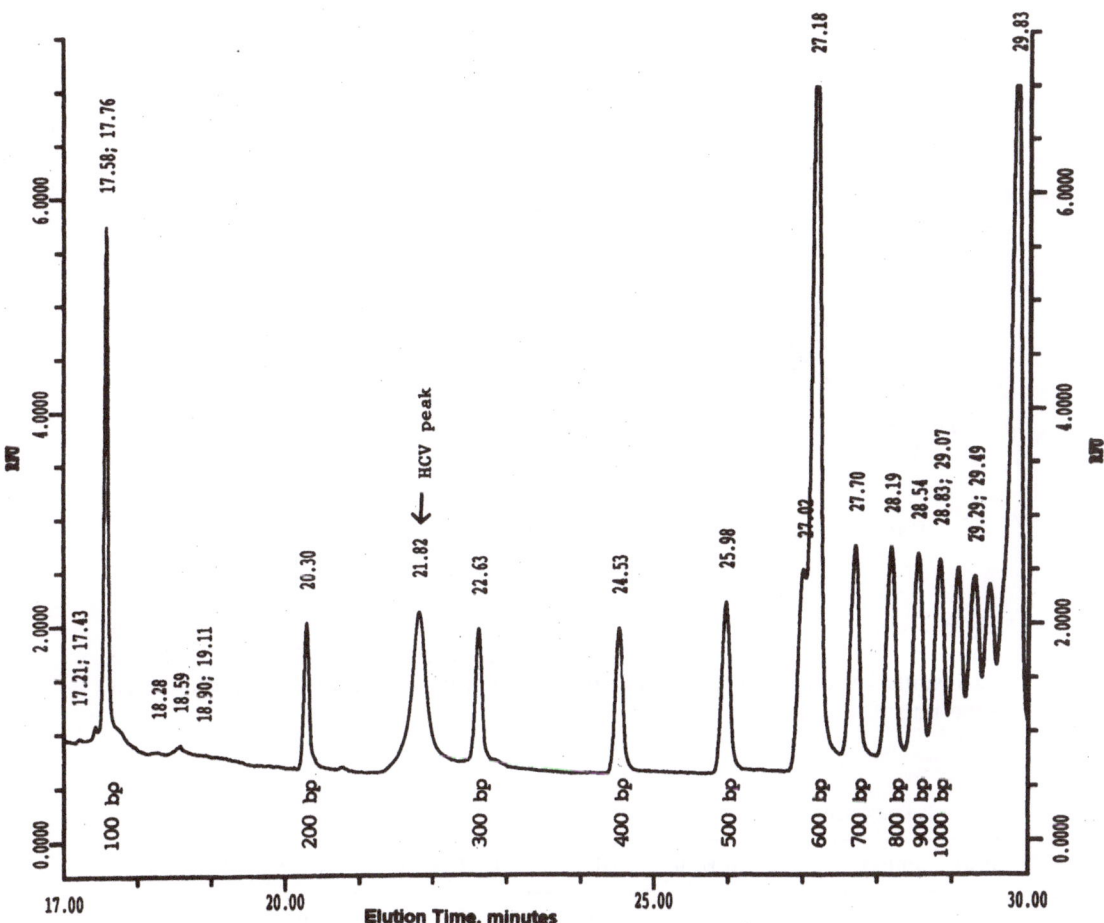

Fig. 2. Electropherogram of a HCV positive specimen. The PCR products were cleaned as described in the text, mixed with a 100 bp size marker and separated by CE. *Numbers* above the peaks are elution times in minutes and correspond to those in Table 1 (peak areas). Sizes of the marker peaks are indicated below the peaks

previously tested in our clinical laboratories for HCV using LH. The CE data were in total agreement with the LH data. We believe that the CE detection method is probably more sensitive than LH since the amount of PCR reaction products injected into the capillary gel (a few nanoliters) is several orders of magnitude smaller than that loaded on acrylamide gels following LH (several microliters).

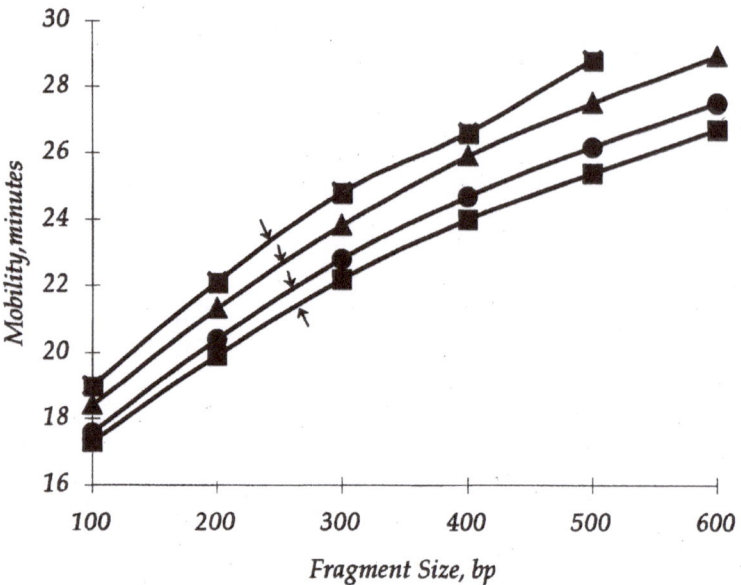

Fig. 3. Mobility shift of DNA peaks. Elution times of the 100 bp marker peaks from four separate CE experiments using HCV positive clinical specimens are plotted against marker fragment size. *Arrows* indicate elution times of the HCV peak in each experiment

Table 1. Peak areas (from Fig. 2)

Peak number	Retention time	Peak area	Peak height	Area (%)	Height (%)
1	12.050	0.1	0.0	0.0	0.0
2	12.111	0.1	0.0	0.0	0.0
3	12.191	0.2	0.0	0.0	0.0
4	12.307	0.2	0.0	0.0	0.0
5	12.387	0.6	0.0	0.0	0.1
6	12.473	1.4	0.0	0.0	0.0
7	12.518	2.8	0.0	0.1	0.1
8	12.637	0.9	0.0	0.0	0.0
9	12.692	1.1	0.0	0.0	0.0
10	12.811	2.0	0.0	0.0	0.1
11	14.771	98.4	0.6	1.3	1.2
12	15.784	369.4	2.0	4.7	4.3
13	16.633	127.1	0.5	1.7	1.0
14	16.992	320.4	0.7	4.1	1.4
15	17.211	98.6	0.7	1.3	1.5
16	17.433	93.3	0.8	1.2	1.6
17	17.576	525.4	5.5	6.7	11.4

Table 1. (Continue)

Peak number	Retention time	Peak area	Peak height	Area (%)	Height (%)
18	17.757	223.4	0.7	2.9	1.6
19	18.283	120.5	0.5	1.6	1.0
20	18.587	249.9	0.5	3.2	1.2
21	18.900	87.5	0.4	1.1	0.9
22	19.107	97.6	0.4	1.3	0.8
	20.000				
23	20.296	243.8	1.6	3.1	3.4
24	21.824[a]	565.6	1.6	7.3	3.4
25	22.629	226.0	1.4	2.9	2.9
26	24.532	177.5	1.4	2.3	2.9
27	25.977	219.0	1.6	2.8	3.3
28	27.020	162.2	1.8	2.1	3.8
29	27.177	1062.2	6.9	13.7	14.3
30	27.073	288.5	2.0	3.7	4.2
31	28.186	212.9	1.7	2.7	3.6
32	28.541	249.8	1.8	3.2	3.7
33	28.825	238.7	1.7	3.1	3.6
34	29.068	222.9	1.6	2.9	3.3
35	29.290	210.8	1.5	2.7	3.1
36	29.487	190.8	1.4	2.5	2.9
37	29.830	1074.8	6.4	13.8	13.4
Total		7766.2	47.8	100.0	100.0

Average efficiency:0.
[a] HCV peak.

References

Pereira BJ, Milford EL, Kirkman RL, Levey AS (1991) Transmission of hepatitis C virus by organ transplantation. N Engl J Med 325:454–460

Donegan E, Wright TL, Roberts J, Ascher NL, Lake JR, Neuwald P, Wilber J, Quan S, Kuramoto IK, Dinello RK, Urdea M (1995) Detection of hepatitis C after liver transplantation: four serologic tests compared. Am J Clin Path 104:673–679

Mateo R, Faruki H, Cooper DL, Demetris AJ, Ehrlich GD (1994) Detection of hepatitis C virus RNA in liver and plasma by reverse transcriptase PCR using liquid hybridization. In: Ehrlich GD, Greenberg SJ (eds) PCR-based diagnostics in infectious diseases. Blackwell Scientific Publications, Boston, pp 375–397

Payan C, Dupre T, Belec L (1995) Detection of hepatitis C virus RNA by a reliable, optimized single-step reverse transcription polymerase chain reaction. Res Virol 146:363–370

Chapter 9

Electrophoresis in the Field of Forensic Science

HILDEGARD HAAS-ROCHHOLZ

Introduction

One of the main topics in the field of forensic science is the typing of biological materials for the purpose of identification of individuals, specifically for paternity and forensic testing . This genetic characterization of biological materials is being performed increasingly at the DNA level. At present, polymorphic loci whose alleles are the result of the variable number of tandem repeats (VNTRs) (Nakamura et al. 1987) or short tandem repeats (STRs) (Edwards et al. 1991) are the most informative markers for individualizing biological material. The VNTR/STR polymorphism is based on differences in the number of core repeat units contained within alleles carried among individuals. While different protocols are performed in the analysis of these polymorphic regions, the separation of these products usually is performed by electrophoresis.

Electrophoresis is a separation technique that is based on the mobility of ions in an electric field. When negatively charged molecules (such as DNA) are loaded into a sample well at the cathode end (–) of a gel and exposed to an electric field, they will migrate towards the anode end (+). The rate of the migration depends on the charge density and/or on the shape or size of the molecule. The differences in the rate of movement of the molecules in a mixture is used to resolve many of the components in a sample. This separation has to be sufficient to resolve alleles or DNA fragments. Different electrophoretic systems are used to separate fragments containing repeat units of VNTR/STR loci, and these procedures use different gels, buffers and measurement systems. Validated protocols are necessary to be effective for allele determination. This chapter describes electrophoretic protocols used in the field of forensic science to type VNTR/STR alleles.

Hildegard Haas-Rochholz, Institute of Legal Medicine, Frankfurter Strasse 58, 35392 Giessen, Germany (*phone* +49-641-99-414-27 or 26; *fax* +49-641-99-414-19; *e-mail* Hildegard.Haas-Rochholz@forens.med.uni-giessen.de)

Analytical Methods for DNA Typing

An overview of the protocol approaches used for characterization of biological materials in the field of DNA forensic science is displayed in Fig. 1. Generally, characterization at the DNA level consists of the following steps: extraction, quantification of the DNA, enzymatic reaction (either digestion or amplification), electrophoresis, detection and interpretation of results. DNA can be extracted from different biological materials such as blood, tissue, saliva, sperm, hair and urine. For quality control in the analysis the quantity of DNA should be determined, which can be carried out by mini-gel electrophoresis (Sambrook et al. 1989), spectrophotometry (Sambrook et al. 1989), fluorometry (Lipman 1989), slot-blot hybridization with a human specific probe (Waye et al. 1989) or by quantitative PCR (Ferre et al. 1994). An appropriate quantity of DNA is subjected subsequently to an enzymatic reaction. Restriction fragment length polymorphism (RFLP) analysis via Southern transfer and hybridization with multi- or single locus probes (MLP or SLP) requires high-molecular-weight (HMW) DNA to be digested by action of a restriction enzyme. The polymerase chain reaction (PCR), which enables amplification of target sequences, is used to investigate smaller fragments of DNA such as amplified repeat polymorphisms (AmpFLPs) which include as VNTR regions, e.g. D1S80 (Kasai et al. 1990), YNZ22 (Horn et al. 1989) and SE33 (Polymeropoulos et al. 1989) and as STR loci, e.g. HumTH01 (Polymeropoulos et al. 1991), HumVWFA31 (Kimpton et al. 1992), Hum-FESFPS (Polymeropoulos et al. 1991), HumCD4 (Edwards et al. 1991), HumF13B (Nishimura et al. 1992), HumF13A01 (Polymeropoulos et al. 1991) (see Table 1). VNTRs and STRs (a subset of VNTRs) are comprised of tandem repetitive sequences. Depending upon the size of the repeat units and their number, PCR fragments of different sizes are produced. The PCR itself can be performed by using singleplex (one locus amplified at a time) or multiplex (two or more loci amplified simultaneously) amplification systems (Lygo et al. 1994; Kimpton et al. 1994). Another enzymatic approach to analyze the extracted and/or amplified DNA is sequencing. In forensics, the sequencing reaction generally is used to investigate sequence polymorphisms of mitochondrial DNA or at times to determine the sequence of alleles from VNTR or STR regions.

To resolve the various restriction digested or amplified fragments a wide range of electrophoretic systems are employed. These include slab gel electrophoresis using an agarose medium, or native or denaturing polyacrylamide gel electrophoresis (PAGE) and capillary electrophoresis with different sieving media (Fig. 1).

204   HILDEGARD HAAS-ROCHHOLZ

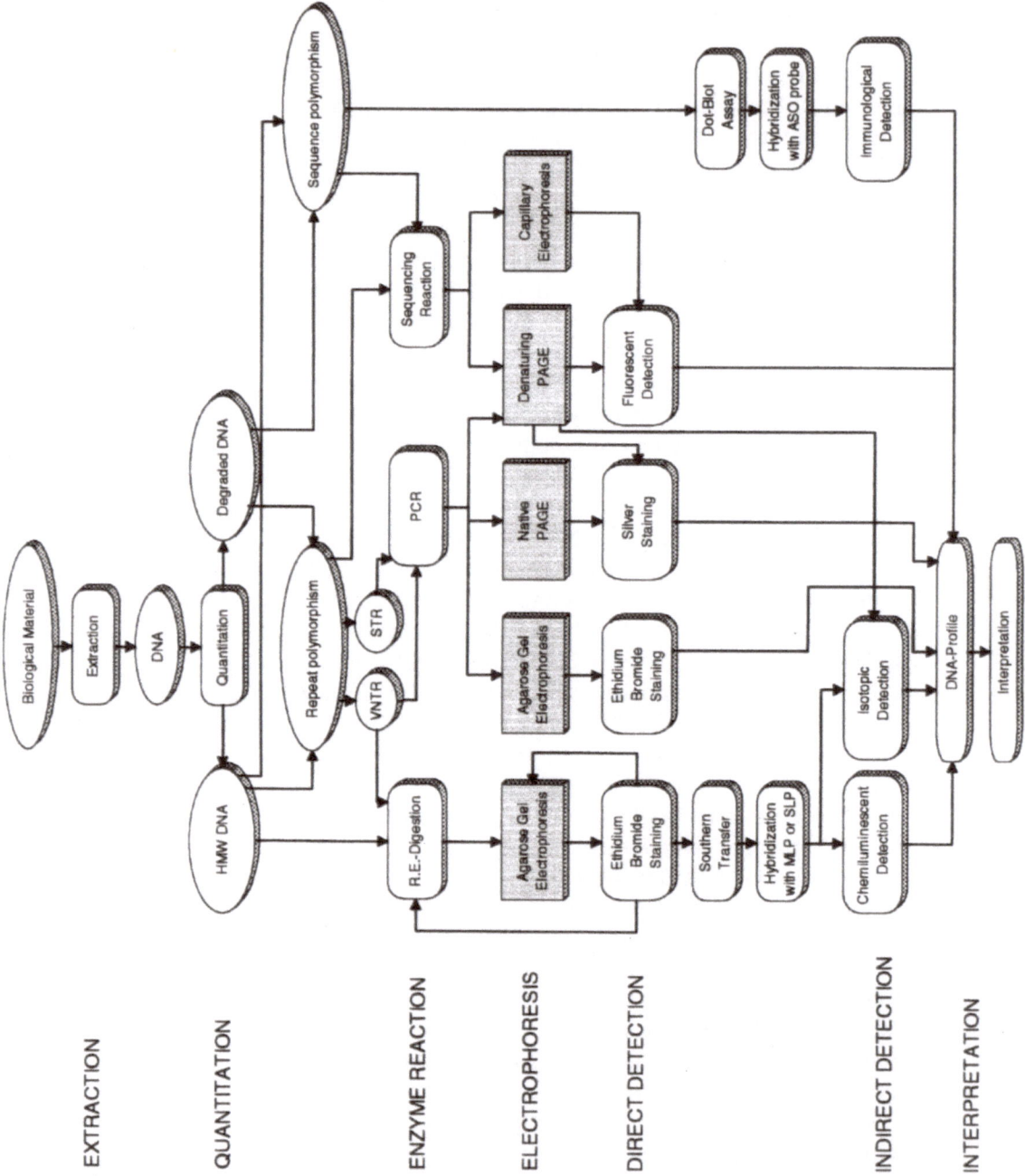

Fig. 1. Overview of the protocols used for characterization of biological materials in the field of forensic science

Table 1. Proposed gel concentrations for separation of VNTRs/STRs by native discontinuous PAGE[a]

VNTR/STR locus	Chromo-somal location	Size and sequence of repeat unit	PCR product size (bp)	%T	%C$_{PDA}$	Reference
HumAmelX	Xp22.1–p22.3	Single copy	106	9	5	Nakahori et al. (1991)
HumAmelY	Yp11	Single copy	112	9	5	Nakahori et al. (1991)
HumCD4	12p	5 bp (TTTTC)*	86–126	9	5	Edwards et al. (1991)
HumLPL	8p22	4 bp (TTTA)*	108–136	9	5	Zuliani et al. (1990)
HumVWFA31	12p12–pter	4 bp (TCTA)*	134–166	8	5	Kimpton et al. (1992)
HumF13B	1q31–q32.1	4 bp (AAAT)*	169–189	8	5	Nishimura et al. (1992)
HumFIBRA	4q28	4 bp (TCTT)*	177–231	7	5	Mills et al. (1992)
HumTH01	11q15–15.5	4 bp (AATG)*	179–207	7	5	Polymero-poulos et al. (1991)
D21S11	21	4 bp (TCTR)*	205–249	6.5	5	Sharma and Litt (1992)
HumFES	15q25–qter	4 bp (AAAT)*	222–250	6.5	5	Polymero-poulos et al. (1991)
HumTPOX	2p13	4 bp (AATG)*	224–252	6.5	5	Anker et al. (1992)
HumHPRTB	Xq26	4 bp (AGAT)*	259–303	8	2	Edwards et al. (1992)
D18S51	18q21.3	4 bp (AAAG)*	275–323	8	2	Strauss et al. (1993)
HumF13A01	6p24–p25	4 bp (AAAG)*	281–331	8	2	Polymero-poulos et al. (1991)
HumCSF1PO	5q33.5–34	4 bp (AGAT)*	295–327	8	2	Hampe et al. (1989)
D1S80	1p	16 bp*	340–780	7.5	2	Kasai et al. (1990)
HumACTBP2	2p24–p23	14–16bp*	580–910	7	3.3	Boerwinkle et al. (1989)

AmpFLPs marked with an asterisk (*) are loci with microvariations in between the repeat unit.
[a] Examples of VNTR or STR regions with different size ranges, and proposed gel concentrations for separation by native discontinuous PAGE. The final mM formate concentration for all recommended %T and %C$_{PDA}$ concentrations is 60 mM.

Direct or indirect detection systems are used to visualize the electrophoretically separated DNA fragments. Ethidium bromide stained DNA fragments, visualized under UV light, is often used after agarose gel electrophoresis. Silver stain detection, fluorescent labelling followed by laser excitation or ^{32}P labelling observed by autoradiography is usually performed after PAGE.

One exception in which electrophoresis is not used is the dot-blot hybridization assay using allele-specific oligonucleotide probes to detect sequence polymorphisms. HLA-DQA1 and the polymarker systems are the most characterized PCR-based systems using a reverse dot-blot format for analysis of amplified DNA products (Saiki et al. 1989).

The last step in genetic characterization at the DNA level is the interpretation of DNA profiles. Generally such a profile consists of one or two bands or dots, designated as alleles. For the size polymorphisms, alleles generally represent RFLP or PCR fragments with various numbers of core repeat units. A comparison of DNA profiles from different samples is performed. If the profiles are similar then evidence is given with a statistical evaluation. Therefore the gel itself, a photograph, a film or data on file are interpreted. With RFLP analysis, fragment sizes are estimated based on their mobilities relative to those of fragments in base pairs (bp) of known sizes, called analytical markers, using for example the local Southern method (Southern 1975). When applying PAGE for the analysis of Amp–FLPs, the alleles of the sample are designated by either fragment length in bps or by number of repeats contained within the fragment. For AmpFLP-typing an allelic ladder is necessary as a reference. It contains a mixture of several or all common alleles at a locus, and it serves as a size standard allowing rapid and precise allele designation (Puers et al. 1994).

Electrophoretic Systems for DNA Typing

Allelic products of the various VNTR/STR loci differ from one another in size. In order to determine the alleles represented in a DNA sample, the products produced by the different enzymatic reactions must be subjected to electrophoresis. The different electrophoretic systems that can be applied are listed in Fig. 1.

Agarose gel electrophoresis is often used to separate DNA molecules that are generally 1000–20000 nucleotides in size. Such DNA fragments are VNTR alleles, produced by digestion with a restriction enzyme, such as *Hin*fI (cuts at sequence GANTC) or *Hae*III (cuts at sequence GLCC).

This RFLP analysis is valid and reliable for forensic and paternity testing, but has certain limitations, which include a requirement of a sufficient quantity of HMW DNA (usually at least 50 ng) (Budowle and Baechtel 1990). Additionally, it is laborious as well as time consuming, involving several manipulations such as hybridization with MLPs or SLPs and chemilumescent or isotopic detection methods. Furthermore, the RFLP technique cannot resolve unequivocally the alleles of most VNTR loci. Agarose gel electrophoresis can also be used to separate VNTR alleles after the PCR, e.g. for the loci D1S80 and YNZ22, but its resolution capacity is not as high as that of polyacrylamide gels (Budowle et al. 1991). Due to these limitations for typing PCR products, agarose gel electrophoresis will not be described further.

Another electrophoretic approach that should be noted in the analysis of repeat or sequence polymorphism is capillary electrophoresis (CE) combined with laser induced fluorescence. For CE, different media can be used for analysis (Butler et al. 1994, 1995; Liu et al. 1995; Wang et al. 1995). These include agarose, polyacrylamide and hydroxyethylcellulose. Because of limited space for discussion the reader should consult the literature for a basic description of CE.

Polyacrylamide gel electrophoresis (PAGE) is the easiest, most common and least expensive approach for VNTR/STR typing after PCR. Amplifications of STRs by PCR yield DNA fragments that generally can differ by as little as 2–6 bp, depending on the length of the core repeat unit. However in some STR systems, alleles which differ by only 1 bp have been observed (Puers et al. 1994; Urquhart et al. 1994). With PAGE it is possible to separate easily fragments that range from 6 to 2000 bp (Allen et al. 1989; Sambrook et al. 1989). Native (nondenaturing) and denaturing polyacrylamide gels with continuous or discontinuous buffer systems are applicable for the analysis of DNA fragments. The most common methods for typing employ PAGE followed by silver stain detection or automated fluorescent detection. Although the used automated fluorescent detection is becoming more prevalent in the analysis of PCR fragments, the manual approach including gel preparation, the electrophoretic run and silver stain detection of the separated fragments is used in many laboratories. Silver staining of gels still represents an easy, sensitive and relatively inexpensive method. Therefore, the manual detection silver staining method will be described in detail. It should be noted that the electrophoretic chemistry applies equally well to silver staining and fluorescent detection approaches.

In this chapter native, discontinuous PAGE in a horizontal mode and denaturing, continuous PAGE in a vertical mode are described. Because at times the PAGE gel should be capable of resolving as little as 1 bp differences in a single analysis the gel quality and chemistry are important. This chapter will serve to explain some factors that affect the quality of electrophoretic separations.

Table 2. Excerpt from the Material Safety Data Sheets (MSDS) for chemicals

Reagent	Precautionary labelling			Laboratory protective equipment
Acetic acid – glacial	Health	2	Moderate	Goggles and shield; lab coat and apron; vent hood; proper gloves; class B extinguisher
	Flammability	2	Moderate	
	Reactivity	2	Moderate	
	Contact	3	Severe (corrosive)	
Acrylamide	Health	3	Severe (cancer causing)	Goggles; lab coat; vent hood; proper gloves
	Flammability	1		
	Reactivity	3	Slight	
	Contact	3	Severe (explosive) Severe (life)	
Ammonium persulfate	Health	1	Slight	Safety glasses; lab coat
	Flammability	0	None	
	Reactivity	3	Severe (oxidizer)	
	Contact	2	Moderate	
Binding solution	Insufficient data available			
Bisacrylamide	See acrylamide			
Boric acid	Health	2	Moderate	Safety glasses; lab coat; vent hood; proper gloves
	Flammability	0	None	
	Reactivity	0	None	
	Contact	2	Moderate	
Bromphenol blue	Health	1	Slight	Goggles; lab coat; vent hood; proper gloves
	Flammability	0	None	
	Reactivity	20	None	
	Contact	1	Slight	
EDTA	Health	1	Slight	Saftey glasses; lab coat
	Flammability	1	Slight	
	Reactivity	0	None	
	Contact	1	Slight	
Formaldehyde	Health	3	Severe (cancer causing)	Goggles and shield; lab coat and apron; vent hood; proper gloves; class B extinguisher
	Flammability	2		
	Reactivity	2	Moderate	
	Contact	3	Moderate Severe (corrosive)	

Table 2. (Continue)

Reagent	Precautionary labelling			Laboratory protective equipment
Formamide	Health	3	Severe (life)	Goggles and shield; lab coat and apron; vent hood; proper gloves
	Flammability	1	Slight	
	Reactivity	2	Moderate	
	Contact	2	Moderate	
Formic acid	Health	2	Moderate	Goggles and shield; lab coat and apron; vent hood; proper gloves; class B extinguisher
	Flammability	2	Moderate	
	Reactivity	1	Slight	
	Contact	3	Severe (corrosive)	
Nitric acid	Health	3	Severe (poison)	Goggles and shield; lab coat and apron; vent hood; proper gloves
	Flammability	0	None	
	Reactivity	3	Severe (oxidizer)	
	Contact	4	Extreme (corrosive)	
Piperazine diacrylamide	See acrylamide			
Silver nitrate	Health	3	Severe (poison)	Goggles; lab coat; vent hood; proper gloves
	Flammability	0	None	
	Reactivity	3	Severe (oxidizer)	
	Contact	3	Severe (corrosive)	
TEMED	Health	2	Moderate	Goggles; lab coat; vent hood; proper gloves; class B extinguisher
	Flammability	4	Extreme	
	Reactivity	1	(flammable)	
	Contact	2	Slight Moderate	
TRIS (buffer)	Health	1	Slight	Safety glasses; lab coat
	Flammability	1	Slight	
	Reactivity	1	Slight	
	Contact	1	Slight	
Urea	Health	0	None	Safety glasses; lab coat
	Flammability	1	Slight	
	Reactivity	0	None	
	Contact	1	Slight	
Xylene cyanol FF	Insufficient data			

Precautions for Chemical Handling

A number of chemicals used in the following protocols are hazardous. All manufacturers of hazardous materials are required by law to supply the user with pertinent information on any hazards associated with their chemicals. This information is supplied in the form of Material Safety Data Sheets (MSDS) (Table 2). The MSDS is available from gopher:// gopher.chem.utah.edu:70/11/MSDS/). This information contains the chemical name, CAS no., health hazard data including first aid treatment, physical data, fire and explosion hazard data, reactivity data, spill or leak procedures, and any special precautions needed when handling the chemical. In Table 2 all chemicals used in these protocols are listed with their hazard ratings (0–4; 0=no hazard; 4=extreme hazard), and laboratory protective equipment.

Disposal of Chemicals

Any uncontaminated nonhazardous solution can be discarded in the trash, not in the sink, and the bottles rinsed well. Any media that becomes contaminated should be promptly autoclaved before discarding it. Other biological waste should be discarded in biohazard containers which will be autoclaved prior to disposal. Organic reagents, e.g. phenol, should be used in a fume hood and all organic waste should be disposed of in a labeled container, not in the trash or sink.

9.1
Horizontal, Native and Discontinuous PAGE

Generally, the relevant design requirements for horizontal PAGE include temperature control, humidity control, and the nature of the gel buffer and electrophoretic buffer (Chrambach et al. 1985). When using a horizontal discontinuous PAGE system the DNA molecules can be separated in the double stranded, native conformation (or denatured if desired, see Buscemi et al. 1995; Neuhuber et al. 1996).

Discontinuous buffer systems in electrophoresis are characterized by the presence of one or more moving boundaries. They consist of a leading ion (with relatively high net mobility) and a trailing ion (with relatively low net mobility). The counter ion of leading and trailing ions usually is identical in both and thus designated as the common ion of the system. Examples of discontinuous buffer systems are Tris formate/Tris-borate, Tris sulfate/Tris-borate and Tris glycine/Tris-borate, in which Tris is the counter ion, formate, sulfate and glycine are the leading ions and borate is the trailing ion. In the discontinuous PAGE system presented here, Tris formate is used in combination with Tris-borate. Allen et al. (1989) first applied this buffer system in horizontal PAGE to separate VNTR/STR alleles. The gel matrix used was polyacrylamide with the cross-linker piperazine diacrylamide (PDA). Advantages of choosing PDA over the more traditional cross linker bis-acrylamide (Bis) are the increased gel strength with low total monomer concentration (%T) gels and reduced background in silver staining. Some possible variations of this discontinuous system by changing the %T, the cross linker concentration (%C), the separation distance and the concentration of the leading ion formate (mMF) for effecting separation of DNA fragments are described by Haas et al. (1994). Advantages of using the horizontal, native PAGE include the easy preparation of the gel by the flap technique, simple electrophoretic setup, use of only a small volume of electrode buffer, relatively fast separations and a highly sensitive silver staining procedure. The most desirable feature of this system is that it can be carried out in most laboratories.

Materials

- Electrophoretic apparatus with a cooling plate: Several commercial electrophoresis apparatuses are available, e.g. Pharmacia Multiphor II horizontal electrophoresis system (cat. no. 18-1018-06); Sigma electrophoresis unit horizontal, isothermal controlled (cat. no. E5515; operates on 220 V AC). The apparatus used here was the BioPhoresis horizontal cell (BioRad cat. no.170-2900) with ceramic cooling platform (125×360 mm), location grid on the cooling platform surface, built-in condensation control, unique electrofocusing electrode holder, direct connect electrofocusing electrodes; platinum wire electrodes, leveling feet, safety interlock
- Benchtop cryostat (horizontal PAGE; model UWK): Necessary for maintaining the temperature of the cooling plate during electrophoresis at 15 °C (BioRad cat. no. 339-0001)
- Power supply: provides constant voltage to dual output jacks under a variety of operating conditions. The voltage range is 5–1000 V, producing up to 250 W of power. The current range is 0–500 mA. Model 1000/500 from BioRad (cat. no. 165-4710) was used.

Equipment

Gel and Electrolyte Buffer

- Tris-formate buffer solution: 120 mM formic acid titrated with Tris to pH 9.0, 1 l. Dissolve 901.75 g Tris base (e.g. Sigma cat. no. T 8524) in about 700 ml double distilled water (ddH$_2$O). While stirring, add 5.0 ml concentrated formic acid (e.g. J.T. Baker Chemical Company cat. no.0129-1). Verify that the pH is 9.0. Bring the final volume to 1 l with ddH$_2$O. Recheck that the pH is 9.0.
- Borate-Tris solution: 280 mM boric acid titrated with Tris to pH 9.0, 1 l. Dissolve 125.9 g Tris base in about 700 ml ddH$_2$O with stirring. Add 17.31 g boric acid (e.g. Sigma cat. no.B6768). Bring to a final volume of 1 l with ddH$_2$O. Check that the pH is 9.0. This is the stock borate-Tris buffer. For use in the electrophoresis tank, the stock must be diluted 1:10 with ddH$_2$O to obtain 28 mM borate-Tris.

The electrode buffer also contains bromphenol blue (BPB) at a final concentration of 0.01 % w/v. BPB serves as a tracking dye to mark the borate/formate moving boundary.

- Bromphenol blue solution (1 %): Weigh out 1 g bromphenol blue and add approximately 70 ml ddH$_2$O while stirring. When the solid is diluted, bring the final volume to 100 ml with ddH$_2$O. For 1 l 280 mM boric acid, use 50 µl 1 % bromphenol blue solution.

Gel Preparation

The supplies necessary for the preparation of a PAGgel in the dimensions 240×120×0.5 mm and a final gel concentration of 7 %T, 5 % cross linking with PDA (C$_{PDA}$) containing 60 mM formate are:

- Gel preparation plates: Float glass plates with dimensions 250×130×5 mm with polished sides
- Dymo tape: used as a spacer (Esselte cat. no.5247.09)
- Gel Bond PAG film: 124×258 mm (FMC Bioproducts cat. no. 894727). A transparent, flexible polyester film designed to support polyacrylamide gels and to facilitate handling of the ultrathin layer gel after electrophoresis. The acrylamide monomers covalently attach to the hydrophilic coating on one side of the mylar film during the polymerization process.

Note: Wear gloves when handling.

- 14 % Total acrylamide/5 % piperazine diacrylamide (PDA). Weigh out 13.3 g electrophoretic grade acrylamide (e.g. GIBCO BRL cat. no. 5512 UC) and 0.7 g electrophoretic grade PDA (BioRad Laboratories cat. no. 161-0202) and add both solids to approx. 70 ml ddH$_2$O while stirring. When the solids have dissolved, bring the final volume to 100 ml with ddH$_2$O. Store the solution at 4 °C.

Note: Always take the utmost care when weighing out and dissolving the acrylamide and cross-linker powders (Bis or PDA). Both are neurotoxins and can enter the body by inhalation or by absorption through the skin. Wear gloves, face mask, and use a fume cupboard.

- TEMED: N,N,N',N'-tetramethylethylenediamine (Sigma cat. no. T 9281). Store at 4 °C.
- Ammonium persulfate (10 %): Weigh out 1 g ammonium persulfate (Sigma cat. no.A9164) and dissolve in ddH$_2$O to a final volume of 10 ml. Store at 4 °C. Prepare fresh every 7 days.

The gel is cast by the flap technique (Budowle et al. 1991). Gels with different dimensions can be used; if this is the case, the glass plate dimensions and gel volume should to be adjusted. Different gel concentrations are proposed for the separation of VNTR/STR alleles of the loci listed in Table 1. The final concentration of C$_{PDA}$ in the gel serves as guideline for choosing the stock solution.

- Blotting pads are used as electrode plugs (wicks, 120×12 mm): The use of agarose plugs instead of blotting paper is acceptable to house the electrode buffer (Wiegand et al. 1993).

Sample Preparation

- Allelic ladder: Each locus has a characteristic allelic ladder. Promega offers GenePrint STR products containing specific primer pairs and ladder plus other components sufficient to perform 100 reactions. The reader should consult the literature for a basic description of the multigenerational amplification of a reference ladder for alleles (see Baechtel et al. 1993; Puers et al. 1993).
  - HumLPL (Promega cat. no. DC4071)
  - HumVWFA31 (Promega cat. no. DC4031)
  - HumF13B (Promega cat. no. DC4001)
  - HumTH01 (Promega cat. no. DC1191)
  - HumFES (Promega cat. no. DC4021)
  - HumTPOX (Promega cat. no. DC4051)
  - HumHPRTB (Promega cat. no. DC4061)
  - HumF13A01 (Promega cat. no. DC4010)
  - HumCSF1PO (Promega cat. no. DC4011)
  - D1S80 (Perkin Elmer cat. no. N808-0064)
- Visual markers
  - 1 kb DNA ladder (Gibco BRL cat. no. 15615-016) (250 μg): 12,216 bp, 11,198 bp, 10,180 bp, 9162 bp, 8144 bp, 7126 bp, 6108 bp, 5090 bp, 4072 bp, 3054 bp, 2036 bp, 1636 bp, 1018 bp, 517 bp, 506 bp, 396 bp, 344 bp, 298 bp, 220 bp, 201 bp, 154 bp bp, 134 bp, 75 bp

- 100 bp DNA ladder (Gibco BRL cat. no. 15628-019) (50 µg): 2072 bp, 1500 bp, 1400 bp, 1300 bp, 1200 bp, 1100 bp, 1000 bp, 900 bp, 800 bp, 700 bp, 600 bp, 500 bp, 400 bp, 300 bp, 200 bp, 100 bp
  - pGEM marker (Promega cat. no. G174) (1 mg): 2645 bp, 1605 bp, 1198 bp, 676 bp, 517 bp, 460 bp, 396 bp, 350 bp, 222 bp, 179 bp, 126 bp, 75 bp, 65 bp, 51 bp, 36 bp
- Sample application pieces: 200 pieces (Pharmacia cat. no. 80-1129-46)
- K562-DNA: K562 genomic DNA in TE buffer (30 µg) (Promega cat. no. DD2011)

Silver Stain
- 1 % Nitric acid (HNO_3), 10 l: Add 100 ml concentrated HNO_3 (69–71 %) to 9 l ddH_2O. Mix and bring to a final volume of 10 l with ddH_2O. Store at room temperature.

Note: Nitric acid is a strong oxidizer, is corrosive, and will cause severe burns. Wear gloves and safety glasses when preparing this solution.

- 0.2 % Silver nitrate ($AgNO_3$) (500 ml): Dissolve 1 g $AgNO_3$ (Sigma cat. no. S0139) in 500 ml ddH_2O. Prepare at least 15 min before time of intended use.

Note: Silver nitrate is toxic and light sensitive. Wear gloves when handling. The solution should be stored in the dark.

- Sodium carbonate formaldehyde developing solution: 280 mM sodium carbonate/0.5 % (v/v) formaldehyde, 10 l: Add 296.8 g anhydrous sodium carbonate (Sigma cat. no. S6139) to 9 l ddH_2O. Stir until dissolved and bring to a final volume of 10 l with ddH_2O. Add 5 ml of 37 % formaldehyde (Sigma cat. no. F1635) to the sodium carbonate solution and mix thoroughly. This solution is stable for 1 week. Store at room temperature.
- Acetic acid stop solution (10 % acetic acid, 10 l): Combine 1 l glacial acetic acid (CH_3COOH, Baker Chemicals analysis grade; cat. no. 9508-05, 99.8 %) with 9 l ddH_2O. Mix thoroughly. Store at room temperature.
- Glycerol wash: 5 % (v/v) glycerol, 5 l: Add 250 ml glycerin (ACS grade, 99.5 %, Fisher Scientific cat. no. G33-4) to 4 l ddH_2O. Mix thoroughly and bring to a final volume of 5 l with ddH_2O. Store at room temperature.

▓ Procedure

Preparation of Ultrathin Native Polyacrylamide Gels: Horizontal Electrophoretic Separation

1. Clean the two glass plates with isopropanol.

2. To make a spacer and frame, fix two layers of Dymo tape (Esselte, 9 mm × 3 m, cat. no. 524706) around all four edges of one glass plate. The total height of the two layer spacer is approximately 0.5 mm.

3. Place the glass plate at a slight angle by laying the glass plate with the Dymo tape (face up) horizontally on the benchtop and put something (e.g. a pipette tip) under the top edge of the long side. Trim the Gel Bond PAG film to match the size of the other the glass plate (top plate).

4. Check the hydrophobicity of the Gel Bond PAG film by spraying some ddH$_2$O on the edge of the film. If the water beads, it is the hydrophobic side.

5. Spray some ddH$_2$O on the glass plate without Dymo tape (i.e. the top plate) and put the Gel Bond PAG film on it with the hydrophobic side facing the top plate.

6. Place a paper towel on the Gel Bond PAG film.

7. To fix Gel Bond PAG film on the glass plate use a roller and squeeze out the excess water by rolling over the film.

Preparation of the Glass Plates

Note: Conduct steps 1–3 listed below in a hood, and wear gloves and safety glasses. When mixing the ingredients avoid producing air bubbles.

1. To prepare the gel solution (total volume 24 ml) combine the following ingredients in a 50 ml falcon tube: 12 ml acrylamide stock solution (14 %T; 5 %C$_{PDA}$); 12 ml Tris-formate stock solution (0.75 M Tris; 0.12 M formate, pH 9.0). Mix carefully.

2. Add 24 µl of TEMED and 240 µl of 10 % APS. Mix carefully. The APS forms a free radical to enable gel polymerization, while TEMED is the catalyst for the polymerization. After adding these, the gel solution must be poured immediately.

Preparation of a Gel Solution with a Final Concentration of 7 %T, 5 %C$_{PDA}$ and 60 mM Formate

Note: Work on a lab bench protected with plastic-backed material since some acrylamide solution may spill while casting the gel.

Casting the Gel

1. Pour the prepared gel solution on the glass plate with Dymo tape, starting at one corner of the lower edge of the Dymo tape frame. Pour the gel solution gently over the length of the gel plate adjacent to the Dymo tape.

2. Take the top plate with Gel Bond PAG film and with the hydrophilic side directed towards the gel solution, align the lower edges to the plates. Carefully lower or "flap" the glass plate with the Gel Bond PAG film over the gel solution. If bubbles are formed, gently tap the plates until all bubbles have risen towards the edge of the Dymo tape.

3. Remove the pipette tip.

4. Lay the plate flat and check (with the level) that the plates are generally in a horizontal position.

5. Let the gel polymerize for at least 30 min.

Note: If the gel is to be stored, cover the glass plates with wet paper towels (ddH$_2$O) and wrap with plastic. Store the polymerized gel in the refrigerator. The gel is useable for up to 1 week.

Electrophoretic Setup

A scheme of the electrophoretic setup is shown in Fig. 2. The polyacrylamide was bound to a Gel Bond PAG film (mylar film) before electrophoresis. The mylar sheet is sufficiently thin to enable heat transfer between the gel and cooling plate. The sample application pieces (tabs) are placed and aligned on the gel and the PCR samples are loaded directly on the

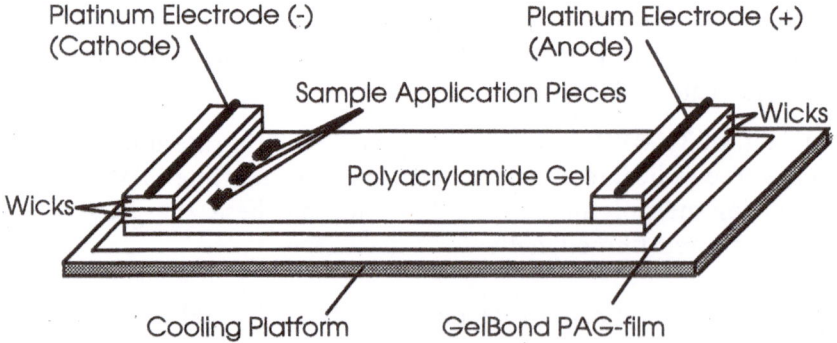

Fig. 2. The electrophoretic setup for horizontal PAGE

tabs. Special sample preparation is unnecessary. Different sample volumes can be applied by choosing tabs of different sizes and thickness. The electrolyte buffer is identical for the anode and cathode (Tris-borate), and contains BPB as a visualizing marker. The electrolyte buffer contained in the blotting pads (wicks) is sufficient for the electrophoretic run. During electrophoresis, the temperature is controlled to provide efficient dissipation of the Joule heat. The temperature of the cooling plate is maintained by the cooling unit. The status of the electrophoretic run can be monitored by migration of the BPB marker, visible under this pH as a blue band. For 240 mm separation distance, the gel run lasts 3–3.5 h.

Starting the Gel

1. Start the benchtop cryostat in order to maintain the cooling plate of the electrophoretic chamber at 15 °C.

2. To remove the gel, separate the glass plates by using a spatula.

3. Separate the Gel Bond film (and gel) from the top plate.

4. To ensure good contact between the cooling plate and Gel Bond PAG film remove excess polymerized acrylamide by cleaning the hydrophobic side (back) of the Gel Bond PAG film with isopropanol.

5. Spray some ddH$_2$O on the cooling plate and place the Gel Bond PAG film, gel side up onto the cooling plate. Avoid bubbles between the film and the cooling plate.

6. Put sample application pieces (tabs) onto the gel 20 mm apart from the cathodal wick (see Fig. 2). For 9 µl PCR volume, cut the original tab size (10×5 mm) in half, so that the tab size is 5×5 mm, for 4 µl volume cut it again, so that the tab has a dimension of 2.5×5 mm. When putting the tabs on the gel make sure that the fibers of the tabs are in line with the direction of ion flow.

Preparation of the Gel for Electrophoresis

Note: The distance between each tab should be a minimum of 2 mm in order to prevent cross-contamination.

1. Pipette the PCR sample onto each tab. Take care that the whole application tab is in contact with the gel.

2. Lane number 1 should contain a visual marker, e.g. 50 ng of the 1 kb or 123 bp marker (Gibco BRL). From a 50 µl PCR reaction, 9 µl of the

Sample Loading

sample are used. K562 cell line DNA is used as a PCR control and should be loaded into lane 2. The allele constitution of K562-DNA should be known for each VNTR or STR system. The allelic reference ladder (volume according to the instructions of the supplier) should be loaded adjacent to the K562 sample (lane 3) and is loaded on every third lane, flanking two samples of unknown allele profiles.

Note: Larger or smaller volumes may be used along with appropriate tab sizes.

Blotting Pad (Wick) Loading

1. Soak four strips of blotting pads (120×12 mm) in the electrolyte buffer (Tris-borate buffer containing BPB).

2. Remove excess buffer by blotting the wicks gently on paper towels.

3. Align two of each wick on the ends of the gel as shown in Fig. 2.

Note: Take care that at least half of a blotting pad is in contact with the gel end.

Electrophoretic Run

1. Position the electrodes in the middle of the wicks and close the chamber.

2. Connect the leads from the electrophoresis chamber to the power supply.

Note: Make sure that the gel is oriented correctly, sample application pieces at negative end (cathode), because the negatively charged DNA will migrate to the positive end (anode).

3. Start the run by setting the limits at 600 V, 25 mA and 15 Watts.

4. Monitor the length of the run by observing the BPB marker. This will appear as a moving yellow line.

Note: Always switch off the power before opening the electrophoresis chamber.

5. Run the gel until the BPB marker has migrated to the anodal wick.

6. Turn off the power.

7. Remove the wicks from the gel.

Detection of the DNA Fragments by Silver Staining

Amplified products can be detected using a silver staining procedure described by Budowle et al. 1991. This is a fast and sensitive method allowing rapid evaluation of amplified VNTR/STR fragments.

1. Place the gel (gel side up) into a 250×25 mm plastic dissecting tray, cover the gel with 1 % HNO_3 and place the tray on an orbital shaker. Shake the gel until the BPB dye line turns blue (about 5 min.).

   Silver Stain Detection

2. Decant the 1 % HNO_3 and rinse the gel briefly with ddH_2O.

3. Cover the gel with 100 ml 0.2 % (w/v) $AgNO_3$; place the tray on the orbital shaker for 20 min. If necessary (although not recommended), the gel can be left at this stage overnight.

4. Decant the $AgNO_3$ into a special silver recovery vessel and rinse the gel with water.

5. The next step is the visualization of the DNA products. Treatment is composed of repetitive rinses and decantations of the development solution, which consists of 0.28 M sodium carbonate with formaldehyde (250 µl of 37 % formaldehyde per 500 ml sodium carbonate). The first rinse will turn milky brown very quickly and should be decanted immediately. Wash the gel with the development solution until the DNA products become suitable visible.

6. Quickly rinse the gel with ddH_2O.

7. Decant the water and cover the gel with 10 % acetic acid and place it on an orbital shaker with gentle agitation for 5 min.

8. Decant the acetic acid and wash the gel with ddH_2O on an orbital shaker with gentle agitation for 10 min.

9. Decant the ddH_2O and cover the gel with 5 % glycerol and place on an orbital shaker for 5 min with gentle agitation.

Gel Drying and Storage

After drying, the gel can be stored as a permanent record.

1. Remove the gel from the glycerol and air-dry by hanging it in a hood over night. Alternatively the gel can be dried at 60 °C for 30 min.

2. Place another support film or plastic wrap on top of the dried gel to protect it.

3. Tape the edges of the gel.

Results

Examples of STR products from locus HumTH01 and from locus HumVWFA31 after electrophoresis with different concentrations of %T and %C$_{PDA}$ and silver stain detection are shown in Fig. 3a,b, respectively. In both parts of the figure, lanes 2, 5 and 8 show the allelic ladder. For HumTH01 (Fig. 3a) the gel contains alleles ranging from 5 to 11 and for HumVWFA31 (Fig. 3b) alleles ranging from 13 to 21. The alleles are designated operationally based on the number of repeat sequences contained within the fragment. Lanes 1, 3, 4, 6, 7 and 9 show the PCR products of seven different individuals. The interpretation of the DNA profile (allelic designation) of each individual sample is accomplished by comparison with the allelic ladder. For example, the allelic status of the individual in lane 3 (Fig. 3a) can be designated as 6/8. For HumVWFA 31 (Fig. 3b) the allelic status of the individual in lane 3 can be designated as 14/17.

One advantage of using polyacrylamide as gel medium is the ability to change the pore size by varying the amount of %T and %C$_{PDA}$. Table 1 shows different gel concentrations for separation of various VNTR/STR systems that have different fragment sizes. The loci in this table are listed with their size and sequence of the repeat unit. They are listed in increasing PCR product size. For the analysis of VNTR regions such as D1S80, which contains a repeat sequence 16 bps in length, the PCR products range generally between 340 and 780 bp and require a cross-linker concentration %C$_{PDA}$ to be 2 %. At a low %C the pore size is large and larger fragments can migrate through the gel. For STRs (HumTH01 contains a repeat sequence size of 4 bp, thus is smaller in size) the %C$_{PDA}$ is at a maximum of 5 % and the %T is increased.

Some authors describe artefacts associated with horizontal discontinuous PAGE. These additional double allele bands are observed in STR systems such as HumFESFPS, HumF13A01 and, to a lesser extent, HumVWFA31. Anomalous migration of PCR products using nondenaturing PAGE is described by Eng et al. (1994) and Kimpton et al. (1995). The reason for such anomalies are not known, but it is likely to be related to the product sequence and may be due to heteroduplex formation or mispriming events during PCR.

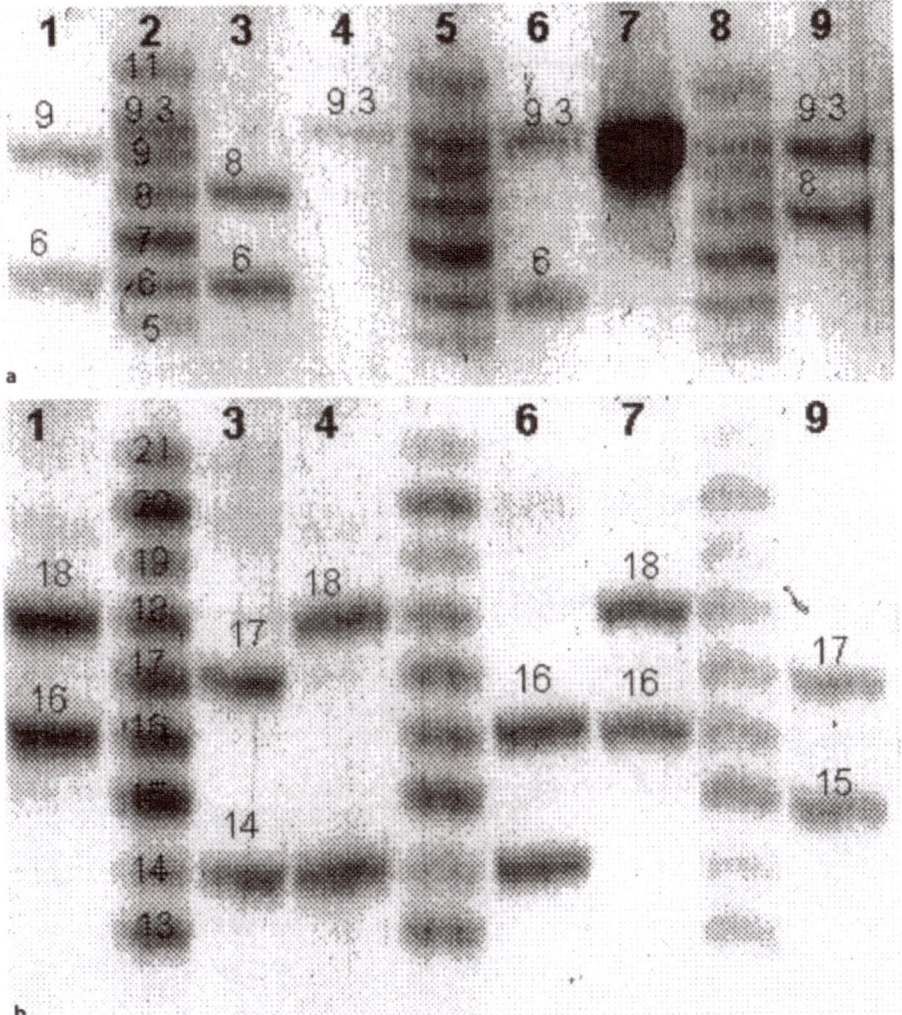

Fig. 3. a Horizontal, native, discontinuous PAGE (7 %T, 5 %C$_{PDA}$, 60 mM formate) after silver staining detection with fragments of the locus HumTH01. *Lanes 2, 5* and *8* show the allelic ladder for the STR locus HumTH01. *Lanes 1, 3, 4, 6, 7* and *9* show DNA profiles (including the allelic status in numbers) of seven different individuals. **b** Horizontal, native, discontinuous PAGE (8 %T, 5 %C$_{PDA}$, 60 mM formate) after silver staining detection with fragments of the locus HumVWFA31. *Lanes 2, 5* and *8* show the allelic ladder for the STR locus HumTH01. *Lanes 1, 3, 4, 6, 7* and *9* show DNA profiles (including the allelic status in numbers) of seven different individuals

9.2
Vertical, Denaturing and Continuous PAGE

Polyacrylamide slab gels that are employed in a vertical mode are usually enclosed between two glass plates. The gel is poured between the two glass plates and the sample wells reside at the top, such that the DNA molecules migrate down through the gel. The electrophoresis system has two buffer reservoirs, top and bottom, cathode and anode respectively, which are filled after the plates and gel complex have been clamped in position. Figure 4 shows the general electrophoretic setup.

For gel preparation, the dimensions of the glass plates and the desired gel thickness must be known. These dimensions depend on the desired separation requirements and the size of the apparatus that is available. In the experiment discussed here, a vertical apparatus (Gibco BRL, model SA 32) was used.

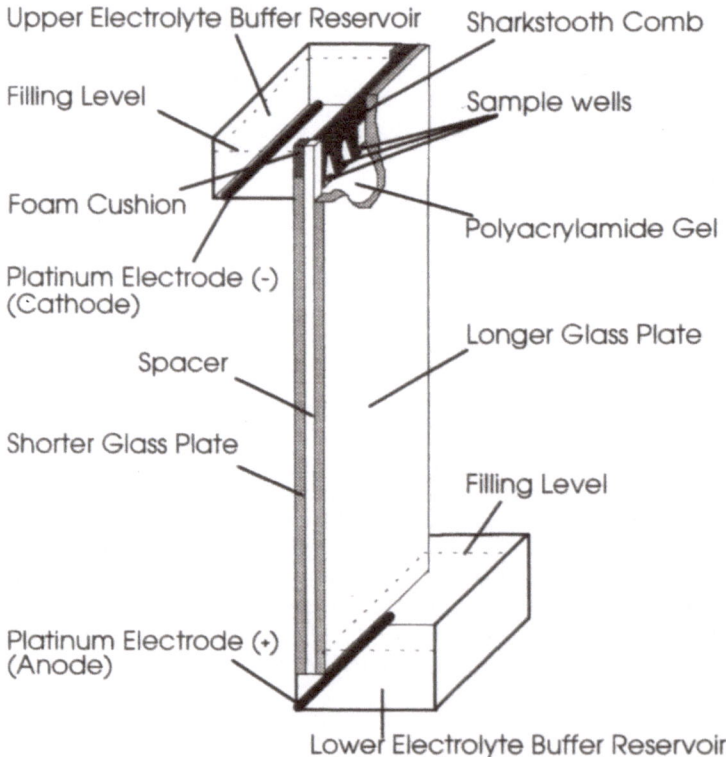

Fig. 4. The electrophoretic setup for vertical PAGE

The gel solution is composed of acrylamide and bis-acrylamide, with a final concentration of 4% (up to 6%) T and 5 %C_{Bis}. Urea is the denaturing agent in the gel. It modifies the molecule to single-stranded DNA to achieve better resolution. Furthermore hairpin loops and other secondary structures are less likely to form in the DNA molecules being electrophoresed. Usually urea is used at a final concentration of 7 M. The gel buffer, in a continuous buffer system, is identical to the electrolyte buffer. A buffer with a high buffer capacity, usually TBE (Tris, boric acid, EDTA), is used. For a vertical system the buffer volume is always higher (up to 350 ml for the upper and lower buffer reservoir) than in a horizontal system. Since denaturing PAGE in the vertical mode usually is not cooled the gel itself can reach a temperature of 45–80 °C. The DNA fragments can be detected by silver staining or laser excitation/fluorescence.

Materials

Equipment

- Electrophoretic apparatus and power supply: Several commercial vertical electrophoresis apparatus are available (e.g. BioRad Sequi-Gen II System cat. no. 165-3600; Sigma-Aldrich nucleic acid sequencing unit cat. no. Z35, 185-7). In this protocol the vertical apparatus used was from Gibco BRL (SA 32 cat. no. 31096027). The dimensions of the two glass plates (Gibco BRL cat. no. 1093KG, sold as a pair) are 368×196×5 mm and 338×196×5 mm for apparatus SA 32. The power supply should be able to deliver up to 500 mA at 1500 V. A BioRad Power Pac 3000 (cat. no. 165-5056) was used.
- Vacuum filter: Filter holder with receiver (Nalgene cat. no. 300-4000)
- Spacer set: 0.4 mm plastic spacers (Gibco-BRL cat no. 21093-042) including foam cushion
- Sharkstooth comb: 24 lanes, 25 pt, 0.4 mm thick (Gibco BRL cat. no. 21045-018)

Gel and Electrolyte Buffer

The gel and electrolyte buffer for denaturing PAGE is TBE buffer, with a final concentration of 89 mM Tris, 89 mM boric acid and 2 mM EDTA (pH 8.3) (1× TBE buffer).

- 10× TBE (0.89 M Tris; 0.89 M boric acid, 20 mM EDTA): Dissolve 216 g Tris base (e.g. Sigma Trizma base cat. no. T8524 molecular biology grade), and 110 g boric acid (e.g. Sigma cat. no. B-6768) in approximately 1600 ml ddH_2O. Add 80 ml 0.5 M EDTA. Check pH, it should be 8.0–8.3. If it is greater than 8.3 adjust it to 8.3 with boric acid. Adjust final volume to 2 l. Store at room temperature.

Gel Preparation

- Gel Slick: for treatment of the longer glass plate (AT Biochem cat. no. 219-GSP)
- Binding solution: Methacryloxypropyltrimethoxysilane (bind silane, Sigma Chemical Company cat. no. M 6514). Used in vertical electrophoresis for treatment of the shorter glass plate; prepare fresh before use. Add 3 µl bind silane to 1 ml 0.5 % acetic acid in 95 % ethanol in a 1.5 ml microcentrifuge tube.

Note: Bind silane is toxic and should be used in a chemical fume hood.

- 40 % acrylamide:bis stock solution (BioRad cat. no. 161-0146)
- Urea (Sigma cat. no. U5378): used as denaturing agent
- TBE (10×): 107.8 g Tris base (Sigma cat. no. T8524); 7.44 g EDTA (Na$_2$DTA·2H$_2$O; Sigma cat. no. E5134); 55 g boric acid. Dissolve the Tris base and EDTA in 800 ml deionized H$_2$O. Slowly add the boric acid and monitor the pH until it reaches 8.3. Bring the volume to 1 l with deionized H$_2$O.
- Ammonium persulfate (10 %): Weigh out 1 g ammonium persulfate (Sigma cat. no. A9164) and dissolve in ddH$_2$O to a final volume of 10 ml. Store at 4 °C. Prepare fresh every 7 days.
- TEMED: (Sigma cat. no. T 9281) Store at 4 °C.

Sample Preparation

- Loading solution (2×): 10 mM NaOH; 95 % formamide; 0.05 % BPB; 0.05 % xylene cyanol FF (XC)
- Microcapillary flat tips: Sample loading for sharkstooth combs in combination with finnpipette (Sigma cat. no. T0906)
- K562 DNA: K562 genomic DNA in TE buffer (30 µg) (Promega cat. no. DD2011)

Silver Stain   See Sect. 9.1.

- Automatic processor compatible (APC) film: 30×40 cm, 25 sheets (Promega cat. no. DQ4411)

Procedure

Preparation of Ultrathin Denaturing Polyacrylamide Gel

The gel is prepared between two glass plates of different size. The shorter one may have "ears" or "notches" at either side. The shorter glass plate is be treated with bind silane and the longer glass plate is treated with Gel Slick. The spacers determine the thickness of the gel (spacer thickness here is 0.4 mm). Gel thickness influences the sharpness of the bands and temperature dissipation. One end of the spacers has a foam block attached. A sharkstooth comb provides a space to apply the sample to the gel. This type of comb does not form physical separation between the wells in the gel, but instead forms spaces along the top edge of the gel in which the samples can be loaded. When using a sharkstooth comb the lanes are adjacent, without small gaps between them. This enables band comparison in the poorer resolving portion of the gel, but it is more difficult to load samples using the sharkstooth comb.

When using the glass plates from Gibco BRL and the SA 32 apparatus the gel dimension is 338×172×0.4 mm and the gel volume is 25.625 ml. The final concentration of a continuous, denaturing polyacrylamide gel system is 6 %T, 5 %C_{Bis} and 7 M urea.

1. Prior to use, the glass plates must be cleaned thoroughly. Place plates in an upright position. Wash each plate with 0.4 M NaOH, rinse with ddH_2O and rinse with 100 % ethanol. Wipe dry the plates with a lint-free laboratory tissue. The inner surfaces of the two plates must be relatively dust-free. Wash the comb with ddH_2O and let it dry.

Preparation of the Glass Plates

2. Apply 3 ml of Gel Slick onto the middle of the longer glass plate. Spread it over the entire surface by using a dry paper towel, moving the towel in a circular motion. Allow to dry for 5 min. Remove excess Gel Slick with a paper towel saturated with ddH_2O. Dry the glass plate with a dry paper towel.

3. To treat the shorter glass plate with bind silane, prepare freshly a solution of 3 µl bind silane solution in 1 ml of 0.5 % acetic acid in 95 % ethanol in a 1.5 ml microcentrifuge tube. Saturate a Kimwipe tissue with the whole volume of the solution and wipe it on the shorter glass plate. Allow to dry for 5 min. Remove excess bind silane solution by wiping the plate four times with a Kimwipe tissue saturated with 95 % ethanol.

4. Check the hydrophobicity of the longer glass plate by spraying some ddH_2O on the edge of the plate. If the water beads, it is the hydrophobic side.

5. Place the long plate, hydrophobic side face up, on a flat surface with the long axis of the plate parallel to the edge of the laboratory bench. Place the two 368×12×0.4 mm plastic spacers on either side of the longer glass plate with the foam cushions at the left and right side facing upward. Align the glass plates at what will be the bottom of the

plates, where there are no foam cushions. Make sure that there is no gap between the foam cushion and the upper glass plate. Then press the spacer tightly against the top of the short plate. Place three binder clips on each side of the gel cassette.

6. Place the gel cassette on a level surface.

Preparation of a Gel Solution with a Final Concentration of 6 %T, 5 %C$_{Bis}$, 7 M urea and 0.5× TBE

Note: Carry out steps 1–3 in a fume hood and wear gloves and safety glasses. When mixing the ingredients avoid producing air bubbles.

1. Combine the gel ingredients (total gel volume 25.625 ml) in a 50 ml capacity glass beaker or 50 ml Falcon tube:

15.75 g urea
18.125 ml ddH$_2$O
1.875 ml 10× TBE
5.625 ml acrylamide stock solution: 40 % acrylamide : bis (19:1)

---

2. Dissolve the urea by heating the mixture on a magnetic stirrer at 50 °C.

3. Vacuum filter the acrylamide solution in a filter cup with receiver (Nalgene).

4. Add 25 µl of TEMED and 250 µl of 10 % APS. TEMED is a catalyst for gel polymerization. Thus, the gel solution has to be poured immediately.

Casting the Gel

Work on a lab bench protected with plastic-backed material since some acrylamide solution may spill when casting the gel.

1. Draw the gel solution into a 50 ml syringe, taking care to introduce no air bubbles, and expel the solution into the cassette at the corner between the foam cushion and the short glass plate. With the palm of your hand, tap continuously on the cassette assembly to prevent bubble formation. Continue to inject the gel solution until the front of the solution reaches the bottom of the cassette. The length of time for applying the gel solution should be no more than 60 s.

2. Before the acrylamide polymerizes insert the flat edge of the sharkstooth comb between the glass plates to a depth of 2–3 mm below the top of the short plate. Permit the gel to polymerize for a minimum of 90 min. at room temperature. Cool the gel at 4 °C at least 30 min prior to use.

Note: If the gel is to be stored, cover the glass plates with wet paper towels (ddH$_2$O) and wrap with plastic. Store the polymerized gel in the refrigerator. The gel is useable for up to 1 week.

Electrophoretic Setup

A scheme of the electrophoretic setup is shown in Fig. 4. The polyacrylamide gel is held between two glass plates during electrophoresis. The electrolyte buffer is identical for the anode and cathode reservoirs (1× TBE). If necessary prerunning the gel can remove charged components in the gel. The PCR samples, loaded into slots, require special sample preparation. The progress of the analytical electrophoretic run can be monitored by migration of the BPB marker. At this pH the BPB marker is a blue band and it is loaded with the sample in each lane. For a 170 mm distance the gel run lasts approximately 1 h.

Starting the Gel

1. Install the gel cassette in the electrophoretic apparatus with the long plate on the outside. Tighten the gel clamps firmly and make certain that the drain clamp is closed. Add 250 ml of 1× TBE buffer to the bottom tank of the electrophoretic apparatus. Add the same volume to the top tank.

2. Remove any air bubbles that may have been trapped at the bottom of the gel by using a plastic syringe with a 1.5 inch needle affixed. With this needle flush out the wells by repetitively forcing the buffer into the wells.

3. Reinsert the comb between the glass plates with the teeth toward the gel. Insert the comb until the teeth just make contact with the surface of the gel. Leave the comb inserted.

4. Close the tank cover and attach the electrode wires to the electrophoresis apparatus and to the power supply.

5. Start the prerun with the following maximum settings: 1000 V; 500 mA and 50 Watts. The minimum prerun time should be 30 min.

Prerunning

1. Prepare samples for loading by placing 3 μl aliquots of 2× loading solution into a tube for each sample. To each tube add 5 μl amplified

Sample Preparation and Loading

DNA and mix thoroughly. In this 6 % gel the BPB migrates at the same rate as a 105 bp fragment.

2. Just prior to loading denature the samples at 95 °C in a thermocycler for 5 min or in a water bath, then immediately chill them on ice (i.e. snap cooling).

3. Use special pipette tips (microcapillary flat tips, Sigma cat. no. T0906) to load the total sample volume (8 µl) into the wells. The sample volume will be loaded underneath the TBE buffer already in the wells. Slowly pipette the samples into the wells so as to form a smooth layer under the buffer.

4. Lane 1 should contain a visual marker (e.g. 50 ng 1 kb or 50 ng 123 bp ladder). K562 is used as PCR control and should be loaded into lane 2. The allelic reference ladder (volume according to the instructions of the supplier) should be loaded adjacent to K562 (lane 3) and is loaded on every third lane, flanking two samples of unknown allelic profile.

Electrophoretic Run

1. Start the run with the following maximum settings: 1000 V, 500 mA and 50 Watts. The running time usually takes 60–90 min.

2. Allow the electrophoretic separation to proceed until the BPB marker line has reached the top of the buffer in the lower tank. Stop the run by switching off the power supply. Drain the buffer from the top tank and remove the gel from the electrophoresis unit. Open the gel cassette using a thin metal spatula and remove the longer glass plate. Handle the gel only by the edges of the shorter glass plate. The gel (fixed on the shorter glass plate) is now ready for staining.

Silver Stain Detection

For the silver staining procedure see "Detection of DNA Fragments by Silver Staining."

Gel Drying and Photography

Since the gel is attached to a glass plate, it is necessary to photograph the gel to maintain a permanent record.

1. Place the dry, silver-stained gel attached to the shorter plate (gel side up) on a white fluorescent light box.

2. In the darkroom under a safelight, position the APC film, emulsion side down, over the gel to be copied.

Note: The emulsion side of the film can be identified as the glossy white surface; the non-emulsion side has a gray tint.

3. Turn on the white light box and expose the film for 10–30 s, depending on the gel background level and the intensity of the white light.

4. Develop the film as recommended by the manufacturer.

Note: The image produced on APC film is a mirror image of the gel.

Photography of the Gel

Results

The difference in size between alleles of STR loci generally ranges from 2 to 5 bp, but for some STR loci, alleles with only 1 bp differences in size may be observed. Therefore, depending upon the STR locus, resolution approaching that of sequencing gels (denaturing PAGE) may be necessary.

Figure 5 shows the separation of DNA fragments after multiplex PCR of the loci HumVWFA31, HumF13B and HumTH01. The gel concentration (6 %T, 5 % C_{Bis}, 7 M urea and 0.5× TBE) shows a good separation for DNA fragments for the loci (100–600 bp) (see Table 1 for product size). Lanes 1, 4, 7, 10, 13 and 16 show the allelic ladder for the STR loci HumVWFA31, HumF13B, HumTH01 from top to bottom show DNA profiles of different individuals.

Starting from the cathode the gel contains the alleles of the locus HumVWFA31 in the range of 21 to 13, followed by the F13B with alleles ranging from 10 to 6 and HumTH01 with alleles from 11 to 5. Lanes 2, 3, 5, 8, 9, 11, 12, 14 and 15 show the PCR products of different individuals. Lane 6 shows a mixture of DNA two different individuals in one lane. The interpretation of the DNA profile (allele designation) of each individual is carried out by comparing it with the allelic ladder. For example, the allelic status of the individual in lane 8 can be designated by 16/17 for HumVWFA31, 6/8 for HumF13B and 7/9 for HumTH01. As seen in Fig. 5 denaturing PAGE sometimes reveals two fragments at an allele. This is caused by differential strand migration for products of the same length but different sequence (Frank et al. 1979). The degree of differential

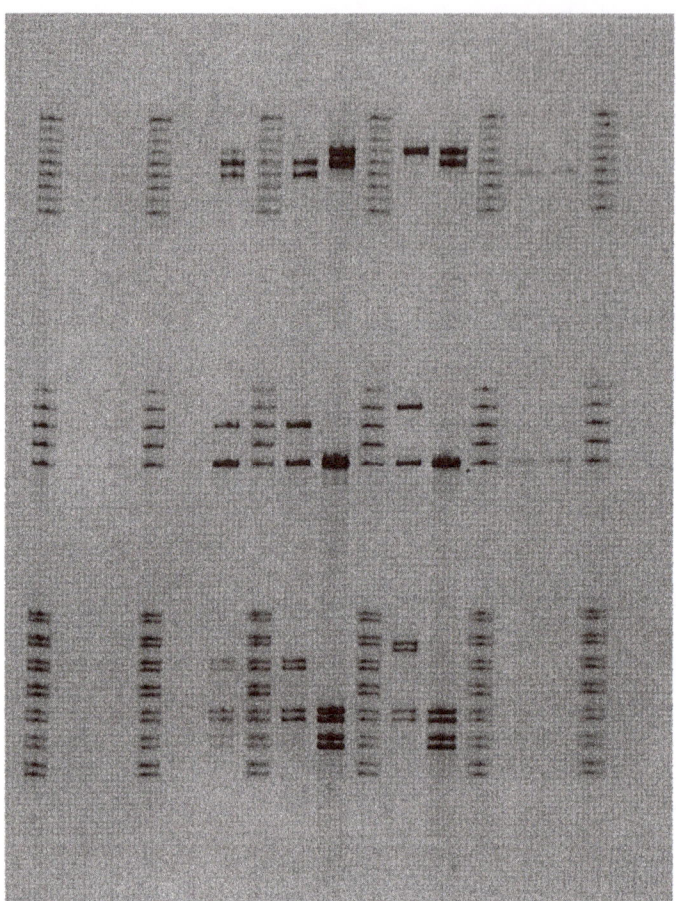

Fig. 5. Vertical, denaturing, continuous PAGE (6 %T, 5 %C$_{Bis}$, 7 M urea, 0.5× TBE) after silver staining detection with fragments of the loci HumVWFA31, HumF13B, HumTH01. *Lanes 1, 4, 7, 10, 13* and *16* show the allelic ladder for the STR loci HumVWFA31, HumF13B, HumTH01 from top to bottom. *Lanes 2, 3, 5, 8, 9, 11, 12, 14* and *15* show DNA profiles of different individuals. *Lane 6* shows a mixture of DNA two different individuals in one lane

migration of the opposing strands depends on the specific amplified sequences being separated. Thus it varies with the locus being amplified, the distance the fragments have been subjected to electrophoresis, the concentration of the gel matrix and, to some degree, on the selection of primers for amplification. Some loci, such as HumFIBRA and Hum-FESFPS, show significant differences, while others, e.g. HumCSF1PO and HumTPOX, reveal comigration of the two fragments (i.e., single strands) at a single allele. Employing one radioactive or fluorescently labelled primer and one unlabelled primer allows the detection of only one of the two complementary strands (data not shown).

The electrophoretic mobility of some STR loci in nondenaturing and denaturing polyacrylamide gels with different %T and %C values has been evaluated by several authors (Pestoni et al. 1995; Gill et al. 1995). For some STR loci anomalous mobility has been found in the ACTBP (SE33) locus (Lareau et al. 1999a) and has been shown to be an issue for forensic standardization of this system (Gill et al. 1994; Möller and Brinkmann 1994). Like SE33, HumF13A01 is another AT-rich STR, suggesting that changes in DNA conformation that can produce noticeable changes in mobility are more likely to occur in AT-rich STRs. This is an issue in native, but not generally in denaturing, systems.

Comments

For the separation of AmpFLPs two protocols are presented to type VNTR/STRs with a size range of 85 to 1000 bp. Both protocols can be carried out in laboratories that have a horizontal or a vertical electrophoretic system.

To be successful and able to interpret a DNA profile it is desirable to separate DNA molecules whose lengths differ by four nucleotides such as STR loci. Some STR loci, e.g. HumTH01, HumVWFA31 and HumFESFPS, show microvariation and carry alleles that may differ by only a single nucleotide. Therefore, depending upon the STR, resolution approaching that of sequencing gels (denaturing PAGE) may be necessary.

Both systems have their advantages and limitations. The horizontal system presented here is a discontinuous PAGE system that is based on the use of the buffers Tris-borate/Tris formate. The gel contains the cross-linker PDA which results in gels with low background after silver staining. The discontinuous system separates the DNA fragments by maintaining the double-stranded structure. The gel run itself takes 3–4 h depending on how long the gel polymerized. This systems enables several possibilities for adjusting the pore size by varying the %T and %C_{PDA} concentrations (Haas et al. 1994). Native gels remain permanently attached to the Gel Bond film through electrophoresis and the subsequent staining procedure. Gels adhered to a solid support retain their original dimensions (except thickness) after drying.

This discontinuous PAGE system is suitable for the analysis of VNTR and STR loci in both horizontal and vertical directions (Hochmeister et al. 1991; Sajantila et al. 1992). The limitations of using horizontal PAGE are anomalous migration of PCR products and relatively poor allele resolution.

The vertical PAGE system presented here is a continuous system. In such a system the same buffer is used for the gel and the electrolyte buffer. The vertical position of the gel itself dictates the migration direction. In the denaturing gel, the natural conformation is eliminated by the action of different reagents (urea, NaOH, formamide) and other reactions (heating). Therefore in a denaturing gel the DNA separates into single strands. Under denaturing conditions two fragments are produced for each allele. So if there are two alleles four fragments could be visible. All described gel systems can be combined with the silver stain detection method.

The methods described here are easy to perform and are very sensitive. In the field of field of forensic science sensitivity is one of the main considerations for typing a DNA profile. Further aspects to consider are precision and reproducibility of the results. To present evidence in court these features must be demonstrated by each laboratory. For every laboratory, quality control is necessary and may be enhanced by participation in inter-laboratory exchanges. Good results should be practical with minimum equipment, so that many laboratories should be able to carry out analyses at the DNA level. One main part of these analyses still is electrophoresis.

In some cases it is necessary to obtain results immediately. Electrophoresis produces results in less than 4 h, so that a complete DNA profile may be obtainable in one day.

Electrophoresis itself is a method that presents many possibilities for adaptation to the migration of various DNA molecules. These adaptations generally concern altering the composition of the gel itself, the buffer and the electrophoretic parameters. With respect to these alterations, electrophoresis is a facile method which can be modified readily to meet many separation requirements by making slight changes in the protocol. One example in which only small changes are made is adjustment of the protocol for denaturing, vertical PAGE to be suitable for fluorescence detection. Here the gel preparation differs in only a few aspects: No treatment with special chemicals such as binding solution or Gel Slick is necessary. Therefore electrophoresis represents a method, which is easily performed in routine forensic work and yet adaptable to further developments.

References

Alford RL, Hammond HA, Coto I, Caskey CT (1994) Rapid and efficient resolution of parentage by amplification of short tandem repeats. Am J Hum Genet 55:190–195

Allen RC, Graves G, Budowle B (1989) Polymerase chain reaction amplification products separated by rehydratable polyacrylamide gels and stained with silver. BioTechniques 7:736–744

Anker R, Steinbrueck T, Donnis-Keller H (1992) Tetranucleotid repeat polymorphism at the human thyroid peroxidase (hTPO) locus. Hum Mol Genet 1:137

Bassam BJ, Caetano-Anolles G, Gresshoff PM (1991) Fast and sensitive silver staining of DNA in polyacrylamide gels. Anal Biochem 196:80–83

Boerwinkle W, Fourest F, Chan L (1989) Rapid typing of tandemly repeated hypervariable loci by the polymerase chain reaction: application to the apolipoprotein B 3' hypervariable region. Proc Natl Acad Sci USA 86:212–216

Budowle B, Baechtel S (1990) Modifications to improve the effectiveness of restriction fragment length polymorphism typing. Appl Theor Electrophoresis 1:181–187

Budowle B, Chakraborty R, Giusti AM, Eisenberg AJ, Allen R (1991) Analysis of the VNTR locus D1S80 by the PCR followed by high resolution PAGE. Am J Hum Genet 48:137–144

Budowle B, Sajantila A, Hochmeister M, Comey C (1994) The application of PCR to Forensic Science. In: Mullis KB, Ferre F, Gibbs RA (eds) The polymerase chain reaction. Birkhäuser, Boston, pp 244–256

Buscemi L, Cururachi N, Mencarelli R, Tagliabracci A, Wiegand P, Ferrara SD (1995) PCR analysis of the short tandem repeat (STR) system HUMVWA31. Allele and genotype frequencies in an Italian population sample. Int J Legal Med 107:171–173

Butler JM, McCord BR, Jung JM, Allen RO (1994) Rapid analysis of the short tandem repeat HUMTH01 by capillary electrophoresis. Biotechniques 17:1062–1068

Butler JM, McCord BR, Jung JM, Lee JA, Budowle B, Allen RO (1995) Application of dual internal standards for precise sizing of polymerase chain reaction products using capillary electrophoresis. Electrophoresis 16:974–980

Chrambach A (1985) The practice of quantitative gel electrophoresis. VCH, Weinheim

DNA Commission of the International Society of Forensic Haemogenetics (1994) DNA recommendations – 1994 report concerning further recommendations of the DNA Commission of the ISFH regarding PCR-based polymorphisms in STR (short tandem repeat systems). Int J Leg Med 107:159–160

Edwards MC, Clemens PR, Tristan M, Pizzuti A, Gibbs RA (1991) Pentanucleotide repeat length polymorphism at the human CD4 locus. Nucleic Acids Res 19:4791

Edwards MC, Civitello A, Hammond HA, Caskey T (1992) DNA-typing and genetic mapping with trimeric and tetrameric tandem repeat. Am J Hum Genet 49:746–756

Eng B, Ainsworth P, Waye JS (1994) Anomalous migration of PCR products using nondenaturing polyacrylamide gel electrophoresis: the amelogenin sex-typing system. J Forensic Sci 39:1356–1359

Ferre F, Marchese A, Pezzoli P, Griffin S, Buxton E, Boyer V (1994) Quantitative PCR: an overview. In: Mullis KB, Ferre F, Gibbs RA (eds) The polymerase chain reaction. Birkhäuser, Boston, pp 67–88

Frank R, Koster H. (1979) DNA chain length markers and the influence of base composition on electrophoretic mobility of oligodeoxyribonucleotides in polyacrylamide gels. Nucleic Acids Res 6:2069–2087

Gill P, Evett I (1995) Population genetics of short tandem repeat (STR) loci. Genetica 96:69–87

Gill P, Kimpton C, D'Aloja E, Andersen JF, Bar W, Brinkmann B, Holgersson S, Johnsson V, Kloosterman AD, Lareu MV (1994) Report of the European DNA profiling group (EDNAP) – towards standardisation of short tandem repeat (STR) loci. Forensic Sci Int 65:51–59

Haas H, Weiler G (1994) Horizontal polyacrylamide gel electrophoresis for the separation of DNA fragments. Electrophoresis 15:153–158

Hampe A, Shamoon BM, Gobet M, Sherr CJ, Galibert F (1989) Nucleotide sequence and structural organization of the human FMS proto-oncogene. Oncogene Res 4:9–17

Heller MJ (1994) Fluorescent detection methods for PCR analysis. In: Mullis KB, Ferre F, Gibbs RA (eds) The polymerase chain reaction. Birkhäuser, Boston, pp 134–141

Kasai K, Nakamura Y, White R (1990) Amplification of a variable number of tandem repeats (VNTR) locus (pMCT118) by the polymerase chain reaction (PCR) and its application to forensic science. J Forensic Sci 35:1196–1200

Kimpton C, Walton A, Gill P (1992) A further tetranucleotide repeat polymorphism in the vWF gene. Hum Mol Genet 1:28

Kimpton C, Gill P, D'Aloja E, Andersen JF, Bar W, Holgersson S, Jacobsen S, Johnsson V, Kloosterman AD, Lareu MV (1995) Report on the second EDNAP collaborative STR exercise. European DNA Profiling Group. Forensic Sci Int 71:137–152

Kline MC, Redman J, Schumm J, Reeder DJ (1993) A comparison of methods for electrophoretic separation for short tandem repeats. Promega Symp 223

Knott TJ, Wallis SC, Pease RJ, Powell LM, Scott J (1986) A hypervariable region 3' to the human apolipoprotein B gene. Nucleic Acids Res 14:9215–9216

Lipman JM (1989) Fluorophotometric quantitation of DNA in articular cartilage utilizing Hoechst 33258. Anal Biochem 176:128–131

Liu MS, Rampal S, Evangelista RA, Lee, Chen FT (1995) Detection of amplified Y chromosome-specific sequence by capillary electrophoresis with laser-induced fluorescence. Fertil Steril 64:447–451

Lygo JE, Johnson PE, Holadaway DJ, Woodroffe S, Whitaker JP, Clayton TM, Kimpton CP, Gill P (1994) The validation of short tandem repeat (STR) loci for use in forensic casework. Int J Legal Med 107:77–89

McCord B, McClure DL, Jung J (1993) Capillary electrophoresis of polymerase chain reaction amplified DNA using fluorescence detection with an intercalating dye. J Chomatogr 652:75–82

Mills KA, Even D, Murray JC (1992) Tetranucleotide repeat polymorphism at the human alpha fibrinogen locus (FGA). Hum Mol Genet 1:779

Mullis K, Fallona F, Scharf S, Snilks R, Horn G, Erlich H (1986) Specific amplification of DNA in vitro: the polymerase chain reaction. Cold Spring Harbor Symp Quant Biol 51:260

Nakahori Y, Hamano K, Iwaya M, Nakagome Y (1991) Sex identification by polymerase chain reaction using X–Y homologous primers. Am J Med Genet 39 472–473

Nakamura Y, Leppert M, O'Connell P, Wolff R, Holm T, Culver M, Martin C (1987) Variable number of tandem repeat markers for human gene mapping. Science 235:1616–1622

Neuhuber F., Radacher M, Sorgo G (1996) Analysis of STR–PCR products with high-resolution denaturing PAGE. Int J Legal Med 108:225–226

Nishimura DY, Murray JC (1992) A tetranucleotide repeat for the F13B locus. Nucleic Acids Res 20:1167

Odelberg SJ, Plaetke R, Eldridge JR, Ballard L, O'Connell P, Nakamura Y, Leppert M, Lalouel JM, White R (1989) Characterization of eight VNTR loci by agarose gel electrophoresis. Genomics 5:915–924

Orban L, Chrambach A (1991) Discontinuous buffer system for polyacrylamide and agarose gel electrophoresis of DNA fragments. Electrophoresis 12, 233–240

Polymeropoulos MH, Xiao H, Rath DS, Merril CR (1991) Tetranucleotide repeat polymporphism at the human tyrosine hydrolase gene (TH01). Nucleic Acids Res 19:3753

Polymeropoulos MH, Rath DS, Xiao H, Merril CR (1991) Tetranucleotide repeat polymporphism at the human c-fes/fps protooncogene (FES). Nucleic Acids Res 19:4018

Polymeropoulos MH, Rath DS, Xiao H, Merril CR (1991) Tetranucleotide repeat polymporphism at the human betaactin related pseudogene H-beta-Ac-psi-2 (ACTBP2). Nucleic Acids Res 20:1432

Puers C, Hammond HA, Caskey CT, Lins AM, Sprecher CJ, Brinkmann B, Schumm JW (1994) Allelic ladder characterization of the short tandem repeat polymorphism located in the 5'flanking region to the human coagulation facxtor XIII a subunit gene. Genomics 23:260 – 264

Robertson J, Ziegle J, Kronick M, Madden C, Budowle B (1991) Genetic typing using automated electrophoresis and fluorescence detection. Experientia Suppl 58:391–398

Sajantila A, Budowle B, Ström M, Johnsson V, Lukka M, Peltonen L, Ehnholm C (1992) PCR amplification of alleles at the D1S80 locus: comparison of a Finnish and a North American Caucasian population sample, and forensic case-work evaluation. Am J Hum Genet 50:816–825

Sajantila A, Lukka M (1993) Improved separation of PCR amplified VNTR alleles by vertical polyacrylamide gel electrophoresis. Int J Legal Med 105:355–359

Sambrook J, Fritsch EF,Maniatis T (1989) Molecular cloning: a laboratory manual, 2nd edn. Cold Spring Harbor Laboratory Press, New York

Schneider PM, Fimmers R, Woodroffe S, Werrett DJ, Bär W, Brinkmann B, Eriksen B, Jones et al. (1991) Report of an European Collaborative Exercise comparing DNA typing results using a single locus VNTR probe. Forensic Sci Int 49:1–15

Sharma V, Litt M (1992) Tetranucleotide repeat polymorphism at the D21S11 locus. Hum Mol Genet 1:67

Southern E (1975) Detection of specific sequences among DNA fragments separated by gel electrophoresis. J Mol Biol 98:503–517

Storts, DR, Wu LC, Mendoza L, Oler JK (1993) Silver staining: a new approach to non-radioactive DNA sequencing. J NIH Res 5:79 – 80

Straub RE, Speer MC, Luo Y, Rojas K, Overhauser J, Ott J, Gilliam TC (1993) Microsatellite genetic linkage map of human chromosome 18. Genomics 15:48–56

Urquhart A, Kimpton CP, Downes TJ, Gill P. (1994) Variation in short tandem repeat sequences – a survey of twelve microsatellite loci for use as forensic identification markers. Int J Leg Med 107:13–20

Vesterberg O (1993) A short history of electrophoretic methods. Electrophoresis 14:1243–1249

Walsh PS, Metzgar DA, Higuchi R (1991) Chelex 100 as a medium for the simple extraction of DNA for PCR-based typing from forensic material. Biotechniques 1:91-98

Walsh PS, Varario J, Reynolds R (1992) A rapid chemiluminescent method for quantitation of human DNA. Nucleic Acids Res 20:5061–5065

Wang Y, Ju J, Carpenter BA, Atherton JM, Sensabaugh GT, Mathies RA (1995) Rapid sizing of short tandem repeat alleles using capillary array electrophoresis and energy–transfer fluorescent primers. Anal Chem 67:1197–1203

Waye JS; Willard HF (1989) Concerted evolution of alpha satellite DNA: evidence for species specificity and a general lack of sequence conservation among alphoid sequences of higher primates. Chromosoma 98:273 – 279

Wiegand P, Budowle B, Rand S, Brinkmann B (1993) Forensic validation of the STR systems SE33 and TC11. Int J Leg Med 105:315–320

Zuliani G, Hobbs HH (1990) Tetranucleotide repeat polymorphism in the LPL gene. Nucleic Acids Res 18:4958

■ Suppliers

Amersham International
Amersham Place, Little Chalfont, Buckinghamshire, HP7 9NA, England
Phone: +44-1494-544000
Fax: +44-1494-542266
Web site: www.amersham.co.uk

AT Biochem
30 Spring Mill Dr., Malvern, Pennsylvania 19355, USA
Phone: +1-610-889-9300; USA toll-free number: 1-800-282-4626
Fax: +1-610-889-9304

Bio-Rad Laboratories
Life Sciences Group, 1000 Alfred Nobel Dr., Hercules, California 94547, USA
Phone: +1-510-741-1000; USA toll-free number: 1-800-424-6723
Fax: +1-510-741-1060
Web site: bio-rad@inlink.com

Fisher Scientific International Inc.
Liberty Lane, Hampton, New Hampshire 03842, USA
Phone: +1-603-929-2650
Fax: +1-603-926-0222
Web site: www.Fisher1.com

FMC Corporation
200 E. Randolph Dr., Chicago, Illinois 60601, USA
Phone: +1-312-861-6000
Fax: +1-312-861-6176
Web site: www.fmc.com

Life Technologies, Inc.
P.O. Box 6009, Gaithersburg, Maryland 20884-9980, USA
Phone: +1-813-345-9371; USA toll-free number: 1-800-828-6686
Fax: 1-800-331-2286
Web site: www.lifetech.com

Nalgene
75 Panorama Creek Drive, Rochester, New York 14602-0365, USA
Phone: +1-716-264-3898
Fax: +1-716-264-3706

Perkin Elmer Corporation
761 Main Ave., Norwalk, Connecticut 06859, USA
Phone: USA toll-free number: 1-800-345-5224
Fax: +1-415-572-2743
Web site: www.perkin-elmer.com

Pharmacia Biotech Inc.
800 Centennial Ave. 1327, Piscataway, New Jersey 08855-1327, USA
Phone: USA toll-free number: +1-800-526-3593
Fax: 1-800-329-3593
Web site: www.biotech.pharmacia.se

Promega Corporation
2800 Woods Hollow Rd., Madison, Wisconsin 53711, USA
Phone: +1-608-274-4330; USA toll-free number: 1-800-356-9526
Fax: +1-800-356-1970
e-mail: info@promega.com
Web site: www.promega.com

Sigma Chemical Company
6050 Spruce St., St. Louis, Misouri 63103, USA
Phone: +1-314-771-5750; USA toll-free number: 1-800-521-8956
Fax: 1-800-325-5052
e-mail: sigma@sial.com
Web site: http://www.sigma.sial.com

VWR Scientific
1310 Goshen Pkwy., Chester, Pennsylvania 19380, USA
Phone: +1-610-431-1700; USA toll-free number: 1-800-932-5000
Fax: +1-610-936-1761
e-mail: solutions@vwrsp.com
Web site: www.vwrsp.com

The Electrophoretic Mobility Shift Assay (EMSA) for Detection and Analysis of Protein-DNA Interactions

MICHAEL G. FRIED AND MARK M. GARNER

Introduction

The interaction of proteins with specific DNA sites is a principal step in the control of many cellular processes. Over the past 20 years, our knowledge of protein-DNA interactions has grown geometrically. This growth has been facilitated by the availability of fast, reproducible assays for binding. Most popular of these is the electrophoresis mobility-shift assay (EMSA), also known as the "gel shift" or "gel mobility shift" assay, first described by Garner and Revzin, and by Fried and Crothers (Garner and Revzin 1981; Fried and Crothers 1981). A particular strength of this method is its ability to detect the simultaneous binding of several proteins to a single DNA species, or one (or more) protein(s) to multiple DNA species (Fried and Crothers 1981; Garner and Revzin 1986; Fried 1989), because DNA-complexes containing multiple proteins play central roles in many important cellular transactions (e.g., DNA replication, DNA repair, transcription and recombination) and because measurement of these interactions allows the assessment of cooperative interactions between proteins within an assembly (Hudson and Fried 1990; Vossen et al. 1996). With minor variations, the same assay techniques can be used to follow the rates of assembly and decay of protein-nucleic acid complexes containing one or several proteins (reviewed in Gerstle and Fried 1993). Assays can be carried out with cellular or nuclear extracts, allowing activity-monitoring during protein isolation and purification (Varshavsky 1987), or with a purified protein and an ensemble of nucleic acids; the second variation forms the basis of widely used binding site selection strategies (Boffini and Prentki 1991; Pierrou et al. 1995). The

Correspondence to Michael G. Fried, Dept. of Biological Chemistry, Pennsylvania State University Medical College, P.O. Box 850, Hershey, Pennsylvania 17033, USA (*phone* +1-717-531-8585; *fax* +1-717-531-7072; *e-mail* mfried@bcmic.hmc.psu.edu) Mark M. Garner, FMC BioProducts, 191 Thomaston St., Rockland Maine 04841, USA

availability of synthetic oligoribonucleotides and oligodeoxyribonucleotides facilitates the acquisition of nucleic acid binding sites for most proteins and allows their systematic variation, making oligonucleotide-based binding assays routine. Finally, the EMSA is one of the few easily accessible methods that give information about the conformations of the DNA and protein in the complex (Wu and Crothers 1984; Crothers et al. 1991; Strahs and Brenowitz 1994), in addition to simply detecting the presence or absence of binding.

As a rule, protein-nucleic acid interactions are highly sensitive to the concentrations of components of the buffer in which they take place (see below). Buffer compositions that produce the greatest binding affinities and/or specificities for one protein-DNA system may differ significantly from the optima for another. Thus, no single set of binding conditions will be applicable to all interactions, and optimization of binding variables (pH, salt concentration, temperature) should be performed for each protein/nucleic acid pair. An added complication is that the degree of retardation of a DNA fragment by a bound protein is sensitive to the conformations of free and bound DNA as well as that of the bound protein (Crothers et al. 1991), and is therefore idiosyncratic for every interacting system, gel concentration and type, running buffer, etc. These system-specific effects limit the usefulness of "cookbook" protocols, except as guides to initial conditions to try when first examining a system.

As is true for most popular methods, the EMSA technique has been reviewed a number of times from both theoretical and experimental perspectives by investigators studying the binding of purified prokaryotic repressor/activator protein as well as those studying eukaryotic nuclear transcription factors in crude nuclear extracts (Varshavsky 1987; Chodosh 1988; Fried 1989; Carey 1991). Since this chapter is intended to supplement, rather than replace, information contained in these earlier reviews, we will give only a general outline of binding techniques, but will focus on those experimental parameters that require strict control for precise results, and describe new results that shape our understanding of processes that take place during EMSAs. We hope that this information will help readers to set-up new EMSA studies, and to make ongoing work more successful.

Relevant Characteristics of Protein-DNA Complexes

The sequence-specific complex formed between a DNA binding protein and its cognate DNA site is stabilized by highly stereospecific interactions between the two macromolecules. Thermodynamic analyses of binding reactions have shown that the stabilizing interactions are the result of both the protein and DNA molecule exchanging interactions with solvent (ionic contacts and van der Waals interactions) with the solvent milieu for interactions with each other. Useful discussions of thermodynamic features of protein-DNA complexes can be found in Record and Mossing (1987); Ha et al. (1989); Record et al. (1991); Fried and Stickle (1993); Overman and Lohman (1994); Record and Spolar (1990); Spolar and Record (1994); Parsegian et al. (1995); Lesser et al. (1990). The practical result for the experimentalist is that the degree of binding depends in a sensitive way on the concentrations of many solution components, especially ionic species. Reproducible results will only be achieved under carefully controlled experimental conditions, so assays should be designed with this sensitivity in mind.

Factors That Affect the Equilibrium Stabilities of Protein-DNA Complexes

In addition to their sequence-specific binding activity, most DNA binding proteins possess a substantial affinity for noncognate sequences. Often, this sequence-nonspecific interaction is qualitatively distinct from the specific interaction (features that may differ for specific and nonspecific binding modes include binding site size, number of ionic and nonpolar contacts, and degree of cooperativity of protein-protein interactions; Saxe and Revzin 1979; Fried and Crothers 1984; Hudson et al. 1990; Fried and Stickle 1993). Although nonspecific binding is likely to play a role in the quantitative regulation of specific site occupancy in vivo (von Hippel et al. 1974; Stickle et al. 1994), it can cause difficulties for the experimentalist who wishes to focus on sequence-specific binding in vitro. Nonspecific sites exist both in *cis*, i.e., the other DNA base pairs on the same fragment as the specific site, and in *trans*, on a separate DNA molecule from the target site. *Cis*-type nonspecific binding can occur even with a purified DNA binding protein, if the protein has low intrinsic sequence specificity, or if the protein is present in excess over DNA, and when it occurs, it can interfere with the detection of sequence-specific binding. A similar interference may also be produced by other DNA binding activities that may be present in the protein sample. Both effects

Specific vs Non-specific Binding

can be minimized by judicious use of competing nonspecific DNA as a sink for nonspecific binding proteins (i.e., provision of excess nonspecific site in *trans* – see below), and/or by use of short "target" or "probe" DNAs that contain few DNA sites other than the specific target.

Salt Concentration Effects

Electrostatic interactions between negatively charged DNA phosphates and positively charged protein groups contribute importantly to the stability (and occasionally the specificity) of protein-DNA interactions. On binding, the protein displaces some of the solvent cations that are closely associated with the DNA due to its very high surface charge density. In addition, ions may be displaced from, or taken up by, the protein as a consequence of DNA binding (Record et al. 1976, 1978, 1991; Ha et al. 1992; Fried and Stickle 1993). As a result of these transactions, the equilibrium affinities of most proteins for DNA are strongly dependent on salt concentration. The exact degree of dependence cannot be predicted a priori, and, indeed, may sometimes be different for the same protein binding to two different sites. In a number of cases, nonspecific binding is more sensitive to changes in salt concentration than is specific binding, which is stabilized by forces in addition to electrostatic interactions. It may be beneficial to perform the binding experiments at a high enough salt concentration to minimize nonspecific binding, but one that is not so high as to significantly reduce or eliminate specific binding.

For some systems, the use of glutamate instead of chloride as the dominant solvent anion can dramatically enhance the stability of protein-DNA complexes (Leirmo et al. 1987). This "glutamate effect" appears to increase both nonspecific and specific binding, although not necessarily to the same degree. It is not clear how universal this phenomenon is (see Hudson et al. 1993), but the substitution of glutamate for other anions may be beneficial for the study of weak interactions, provided that the enhanced nonspecific binding does not overwhelm detection of specific complexes.

Hydration Effects

Studies on a number of systems have indicated that, in addition to salt concentration, the total water activity in a DNA binding solution is an important variable in determining the extent of binding. Addition of moderate concentrations of small, neutral osmolytes (osmotic stress) has been shown to dramatically enhance the stability of specific protein-DNA complexes in a number of systems (Garner and Rau 1995; Sidorova and Rau 1996: Lundbäck and Härd 1996; Vossen et al. 1997). Like the salt concentration (or "polyelectrolyte") effect, hydration effects arise from the replacement of solvation interactions by protein-DNA interactions:

increasing osmolyte concentration lowers the water activity, which favors release of water molecules from the interface and therefore favors macromolecular association. In most systems examined to date, osmotic stress preferentially increases specific binding. Thus, in addition to providing detailed information about the interface between the protein and DNA site, osmotic stress can be a useful tool to preferentially enhance specific complex formation without increasing competing nonspecific binding.

Macromolecular Crowding

It has been shown experimentally that the presence of even low-to-moderate concentrations of neutral polymers dramatically enhances the binding of a protein to DNA. This effect, termed macromolecular crowding, is thought to be due to the simple fact that the background neutral polymers occupy a significant fraction of the volume of the solution which would otherwise be available to the DNA and protein species of interest, i.e., the volume occupied by the background neutral polymers is "excluded" to the protein and DNA. The theory and demonstration of crowding phenomena have mostly been worked out by Allen Minton and Steven Zimmerman and their groups (for useful reviews see Minton 1983; Zimmerman and Minton 1993). Far from being merely an esoteric physicochemical phenomenon, macromolecular crowding is a manifestation of the fact that the intracellular concentrations of protein and nucleic acid are far higher than any normally encountered in vitro. Indeed, solutions with high concentrations of neutral polymer are probably much more accurate representations of the intracellular milieu than simple salt solutions. For example, the concentration of hemoglobin inside an erythrocyte is normally somewhat higher than that found in a hemoglobin crystal. These very high macromolecular concentrations must have profound consequences for macromolecular associations in vivo (Zimmerman and Trach 1991; Garner and Burg 1994; Garner 1997). Macromolecular crowding has been postulated to be the sole transducing factor involved in the regulation of erythrocyte volume (Colclasure and Parker 1991, 1992; Minton et al. 1992).

Kinetic Considerations

Since the equilibrium constant specifies the ratio of association and dissociation rates (Eq. 1), the kinetic features of a protein-DNA interaction will, in general, be sensitive to any factor that perturbs its equilibrium stability. Notable among these are ionic and water activities, as discussed above. Several additional factors contribute importantly to the observed rates of formation and decay of protein-DNA complexes, and can determine whether a particular interaction will be detected by standard mobility shift techniques.

$$\text{Protein} + \text{DNA} \overset{k_a}{\underset{k_d}{\leftrightarrow}} \text{Protein} \cdot \text{DNA} \qquad K_{eq} = k_a/k_d \qquad (1)$$

Attainment of Equilibrium   Consider first the simple binding reaction shown above. A system of this kind approaches equilibrium exponentially, with a relaxation time τ dependent on both the equilibrium free protein and DNA concentrations ($[\text{Protein}]_e$, $[\text{DNA}]_e$), as well as the forward and reverse rate constants k_a and k_d (Eisenberg and Crothers 1979):

$$\frac{1}{\tau} = k_a \left([\text{Protein}]_e + [\text{DNA}]_e\right) + k_d \qquad (2)$$

Since the rate constants are unique properties of the reacting system, and since the protein and DNA concentrations used in mobility shift assays can lie within a range that spans at least nine orders of magnitude ($10^{-12}\text{M} \leq [\text{protein, DNA}] \leq 10^{-3}\text{M}$), the time required to reach binding equilibrium is expected to differ dramatically for different protein-DNA interactions, and within one system, for different initial protein and DNA concentrations. Thus, in equilibrium binding assays, it is essential to verify the attainment of equilibrium. This is easily accomplished by preparing two identical gels, applying an aliquot of each sample to the first gel after equilibration interval t, and an aliquot of each sample to the second gel after an equilibration time much greater than t. The absence of detectable change in the mole fractions of complex and free DNA with prolonged equilibration is an indication that the interval t may be sufficient for the attainment of equilibrium under those reaction conditions.

DNA is not always a passive reaction component. Often binding reactions are carried out in the presence of a molar excess of protein binding sites, situated either in *cis* or in *trans* with respect to the target site of interest. This circumstance is deliberately produced when competing DNA is used as a sink for binding proteins in studies of dissociation

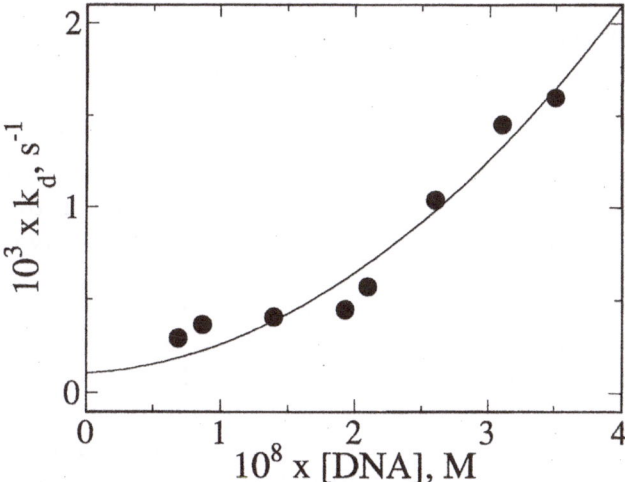

Fig. 1. Kinetics of CAP dissociation from a 219 bp lactose promoter fragment, as a function of DNA concentration. CAP (0.7 nM) and ^{32}P-labeled lactose promoter fragment (0.58 nM) were equilibrated at 21±1 °C, in buffer containing 20 mM Tris (pH 7.5 at 21 °C), 10 mM sodium acetate, 0.5 mM EDTA, 5 % glycerol, 20 μM cAMP, 0.1 mg/ml bovine serum albumin. Dissociation was initiated by addition of unlabeled lactose promoter fragment to attain the indicated concentrations. Reactions were monitored by EMSA, using 10 % polyacrylamide gels. (Figure redrawn from Fried and Liu 1994)

kinetics (Fried and Crothers 1981; Gerstle and Fried 1993) and in equilibrium assays when competing DNA is used to reduce nonspecific binding to a labeled probe (see above). However, at least two well-studied gene regulatory proteins (the *E. coli* CAP protein and the *E. coli lac* repressor) possess DNA concentration-dependent as well as DNA concentration-independent mechanisms of dissociation from cognate DNA sites (Fried and Crothers 1984; Brown and Crothers 1989; Fickert and Muller-Hill 1992; Vossen and Fried 1997; Fried and Bromberg 1997), and other proteins may be found to possess such mechanisms when investigated. For such proteins, any unbound DNA in the system will accelerate the dissociation of preformed complexes (Fig. 1). While this will clearly affect the result of kinetic studies, it is not yet widely recognized that DNA-catalyzed dissociation can affect the outcome of an equilibrium EMSA experiment. As discussed below, samples are typically far from equilibrium during the electrophoretic step of an EMSA. During that period, DNA catalyzed reactions can reduce the fraction of an initially formed complex that remains to be detected at the end of the electrophoretic run. As a consequence, the use of competing DNA as a sink for unwanted DNA binding activities must be balanced against its effects on the detection and quantitation of the complex(es) of interest. Other strategies to reduce dissociation during the gel run are discussed below.

Stability of Protein-DNA Complexes During Electrophoresis

In an EMSA, the purpose of electrophoresis is to separate reactants from products. However, this separation places the protein-DNA system far from equilibrium, driving binding reactions in the direction of dissociation. It follows that the greatest EMSA sensitivity is obtained under conditions that minimize the rates of relaxation processes that occur during the period of disequilibrium. The disequilibrium period consists of two phases; an initial interval of free electrophoresis as components migrate from the sample solution into the gel (sometimes called the "dead time" of the technique), and a subsequent interval of gel electrophoresis, necessary for the resolution of reactants and products.

During free electrophoresis, relaxation rates can be can be quite rapid, so it is useful to minimize the length of this interval. As shown in Fig. 2, the electrophoretic dead time depends strongly on the loading voltage, but only weakly on the molecular weight of the DNA. Thus, the applica-

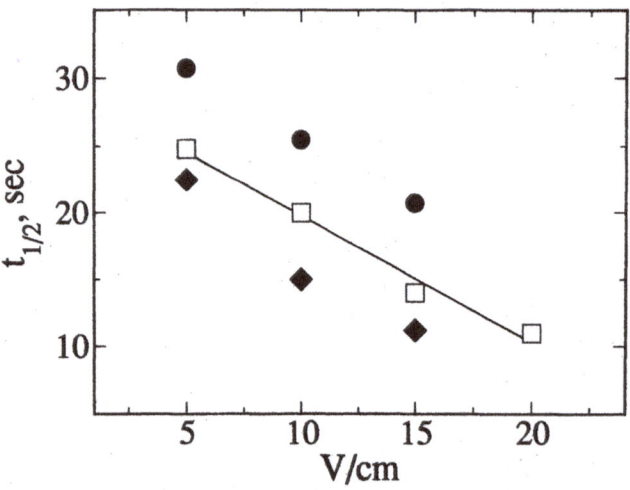

Fig. 2. The relationship between loading voltage and $t_{1/2}$ for migration of ^{32}P-labeled duplex DNA fragments into polyacrylamide gels. Samples (10 μl) were applied to 5 % polyacrylamide gels equilibrated with 45 mM Tris-borate (pH 8.4), 2.5 mM EDTA buffer. The loading buffer was gel buffer supplemented with 5 % glycerol. At intervals following the start of electrophoresis, the sample wells were flushed with running buffer to remove unincorporated DNA. Incorporated DNA was quantitated by scintillation counting of gel slices. DNA fragment sizes: *filled triangles* 82 bp; *open squares* 214 bp; *filled circles* 2686 bp. (Redrawn from Fried 1989)

tion of high voltages at the beginning of electrophoresis is one means of minimizing the dead time (Fried 1989). When this is done, care should be taken to avoid the generation of excessive heat within the gel, since protein-DNA complexes generally become less stable with increasing temperature. In addition, the minimization of sample volumes and the choice of sample-well geometries that permit samples to be applied to the gel as narrow bands can decrease the distance, and hence duration, of free electrophoresis.

The composition of the gel buffer can strongly affect the lifetimes of protein-DNA complexes during electrophoresis. Some components, (for example, salts) influence most, if not all binding interactions. Other components (e.g., inducers, co-repressors) may be specific to a particular protein-DNA interaction. Because many buffer compositions are compatible with high resolution gel electrophoresis, the gel running buffers used in the EMSA can be optimized for the interactions under study.[1] Running buffers with low salt concentrations (<20 mM) can stabilize interactions that are mediated in part by ionic contacts. For very dilute buffers of this kind, circulation between cathodic and anodic reservoirs is sometimes necessary to maintain constant pH (Fried and Crothers 1981; Fried 1989). For interactions of adequate stability, buffers of moderate ionic strength (~150 mM) are often used for their relevance to physiological conditions. Only rarely are buffers of high ionic strength (>350 mM) used, since high salt concentrations generally destabilize protein-DNA complexes (see above), reduce the electrophoretic mobilities of DNA and protein-DNA complexes, and can cause undesirable heat generation during electrophoresis. Gel buffers for EMSA can contain components not found in standard electrophoresis buffers. Examples of additives that are helpful in some circumstances include reducing agents (e.g., 2-mercaptoethanol, dithiothreitol, thioglycolic acid), low molecular weight osmolytes (e.g., ethylene glycol, glycerol, glycine), allosteric affectors, e.g., cAMP for reactions including the cAMP receptor protein (Fried and Crothers 1984), and enzymatic substrates, e.g., ribonucleoside triphosphates for reactions containing RNA polymerase (Straney and Crothers 1985). For incorporation into the gel, many low

---

[1] Often it is feasible to use an electrophoresis running buffer that is identical, or at least closely similar, to the buffer in which the binding reaction is carried out (Fried 1989). This allows the investigator to minimize the potential effects of transferring reaction components from free solution into the gel. Use of identical sample and gel running buffers is not always necessary, however. If care is taken to ensure that the transfer of reaction components from binding buffer to gel running buffer does not cause undesirable artifacts, the use of standard electrophoresis buffers is justified.

molecular weight solutes can be added directly to acrylamide or agarose solutions prior to polymerization. Solutes that react with components of the polymerization mixture (e.g., thiols, amides) may be introduced after polymerization, either by diffusion (for neutral components) or by electrophoresis (charged components; Fried 1989).

Even under sub-optimal optimized buffer conditions, protein-DNA complexes often persist through gel runs that are many times longer than their free-solution lifetimes (Fried and Crothers 1981; Vossen and Fried 1997; Fried and Bromberg 1997; Fig. 3A). This observation prompted the suggestion that the gel itself might contribute to the stability of complexes during electrophoresis (Fried and Crothers 1981). Computer simulations of transport coupled to reversible binding provide theoretical support for this idea (Cann 1989, 1990, 1993). Mechanisms that have been proposed to account for this stabilization include gel-mediated osmotic stress, direct interactions with the gel matrix, and volume exclusion (reviewed in Fried 1989). Recent experiments indicate that osmotic stress and gel-matrix interactions do not contribute significantly to the

Fig. 3A,B. The lifetimes of some protein-DNA complexes are greater within gels than they are in free solution. **A** Dissociation of *lac* repressor-DNA complexes in free solution, detected by EMSA. Complexes of *lac* repressor with a ^{32}P-labeled 203 bp *lac* operator DNA fragment were formed at a protein:DNA ratio of 4.9 in a solution containing 50 mM NaCl, 10 mM Tris (pH 7.4 at 21 °C), 1 mM EDTA, 50 µg/ml bovine serum albumin, 5 % glycerol, and 3.3 nM *lac* operator DNA fragment. After equilibration, dissociation was initiated by addition of unlabeled lac operator fragment to give a final concentration of 66 nM. Aliquots were withdrawn and applied to a running 5 % polyacrylamide gel at 0, 1, 2, 4, 8, 16, 22, and 32 min (*lanes a–h*, respectively). Electrophoresis was at 8 V/cm for 2 h. The bottom band in each lane consists of free DNA. Each band above the free DNA contains one equivalent of repressor (tetramer) more than its predecessor. Bands of equivalent stoichiometry differ in position because the samples were applied sequentially to a running gel. (Figure reproduced from Fried 1989, with permission.) **B** Dissociation rates of migrating CAP-DNA and *lac* repressor-DNA complexes depend on gel concentration. Mixtures containing 1:1 CAP or 1:1 *lac* repressor complexes with lactose promoter-operator restriction fragments were applied to acrylamide or agarose gels, and the dissociation rates determined from the amounts of complex remaining after electrophoresis for different periods of time (Fried and Bromberg 1997; Vossen and Fried 1997). In all cases the gel buffers contained 20 mM Tris-acetate (pH 8.0 at 21 °C), 0.5 mM EDTA. For analysis of CAP-DNA interactions, the gel buffer contained, in addition, 20 µM cAMP. *Filled squares* Rate constants for CAP-DNA dissociation within polyacrylamide gels (redrawn from Fried and Liu 1994); *filled circles* rate constants for *lac* repressor-DNA dissociation within polyacrylamide gels; *open squares* rate constants for lac repressor-DNA dissociation within agarose gels (redrawn from (Vossen and Fried 1997). *Error bars* represent 95 % confidence intervals

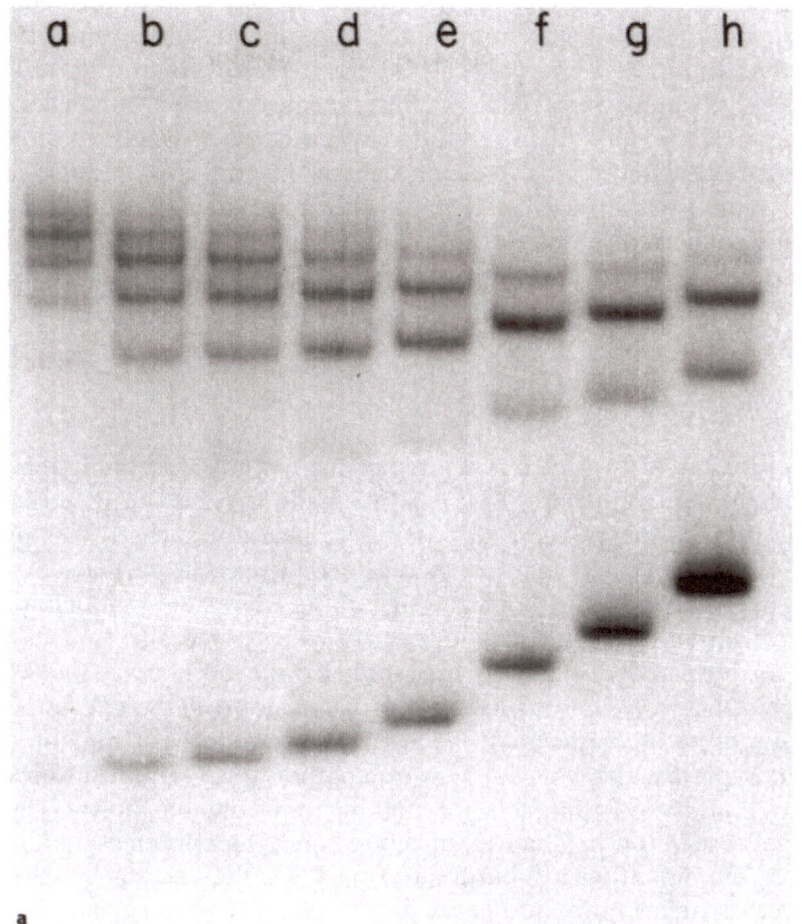

a

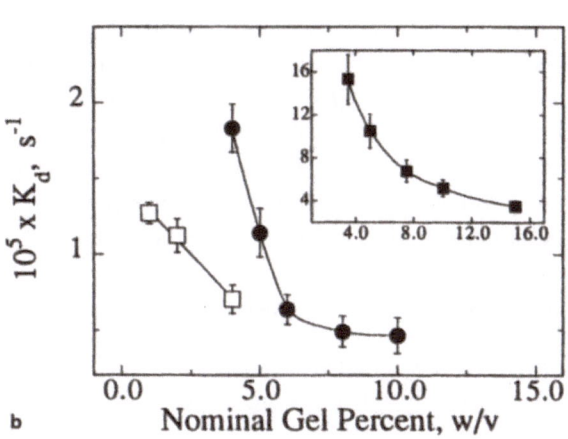

b

stabilities of DNA complexes formed with *E. coli* CAP or *lac* repressor proteins, but confirm the importance of excluded volume mechanisms (Fried and Liu 1994; Fried and Bromberg 1997; Vossen and Fried 1997). Since the volume accessible to macromolecules of given dimensions is a property of the gel system, excluded volume effects ought to affect all protein-DNA interactions to some degree. In addition, because accessible volume within a gel decreases with increasing gel concentration, these results indicate that the stabilities of protein-DNA complexes should be greatest in the most concentrated gels that is consistent with electrophoretic resolution[2] (Fig. 3B).

Basic Experimental Outline

Gel   Except for cases in which one is attempting to study very high molecular weight protein complexes, EMSA is normally carried out in polyacrylamide gels, although it is also possible to perform such studies in gels formed from specialty agaroses (see below). The exact gel concentration to be used depends mostly on the size of the DNA fragment being used. For restriction fragments on the order of 200 bp, typically 5–7.5 %T gels with an intermediate level of cross-linker (30:1–50:1, or 2–3.3 %C) are almost universal. When using short oligonucleotides, the gel concentration might be increased to 10 %T, with the same proportions of cross-linker. Typically, slab gels 5–15 cm in length and 0.5–1 mm in thickness are used. Detailed protocols for making reproducible polyacrylamide slab gels are found in Chap 1. Since one is relying on detecting a difference in the mobilities of bound and free DNA, the exact gel concentration, cross-linker ratio, and exact dimensions are not crucial, provided sufficient resolution is achieved.

Gel running-buffer systems for protein-DNA complexes are almost always the low ionic strength, continuous buffers commonly used for electrophoresis of native DNA. A 0.25–1× TBE buffer (1× TBE=89 mM Tris base, 89 mM boric acid, 2.5 mM Na_2EDTA) is almost universal. However, one should be aware that the use of borate is known to cause artifactual splitting of bands in native protein electrophoresis, so one should be somewhat dubious of multiple species which are only observed in TBE buffer.

---

[2] These results are also useful because they indicate that the type of gel used in the assay (agarose, polyacrylamide) can be varied to optimize electrophoretic resolution and/or convenience. Note, however, that relatively concentrated agarose gels are required in order to obtain stabilities comparable to those observed in polyacrylamide.

Because the EMSA techniques relies on the fact that both the protein and DNA in the complex are native while they are migrating through the gel, it is important that the gel not overheat during electrophoresis, which will cause one or both components to denature. Since most commercially available slab gel equipment makes little provision for temperature control, the field strength must be limited to prevent appreciable Joule heating of the gel during electrophoresis. Facile heat dissipation is one of the major reasons why capillary electrophoresis can be performed at very high field strengths (200–300 V/cm) on native complexes without denaturing the components (Xian et al. 1997).

It is difficult to describe a basic experiment measuring the interaction of a protein with its specific DNA, since the major experimental parameters which control the specificity and stability of such complexes need to be determined for each system (see above). However, several general guidelines can be offered: **Sample**

- It is usually helpful to carry out the reaction at the highest salt concentration which allows the detection of specific binding, since the higher salt concentration will often suppress competing nonspecific binding.

- If a nonlabeled competitor DNA is used, it is best if it is a simple sequence or synthetic polynucleotide, rather than random sequence DNA (e.g. calf thymus or salmon sperm DNA), since the specific sequence to which the protein binds will occur with random frequency in such DNA. Given the vast excess of competitor DNA typically used, even a relatively rare target sequence may "titrate out" the specific binding interaction. However, it is also important that the competitor DNA has the same overall conformation (usually double stranded B-form DNA) as the target sequence. It is for these reasons that poly d(I-C)·d(I-C) is often used as a competitor DNA (Singh et al. 1986).

- The optimal ratios of protein to specific and nonspecific DNAs depends on the ratio of K_S/K_{NS}, the fractional activity of the specific protein, and the absolute concentrations of all components. Since, in practice, few if any of these parameters are known, it is usually necessary to titrate both specific and nonspecific DNAs with the protein. If the DNA concentration is great enough, even a moderate excess of protein over specific binding site can cause nonspecific binding elsewhere on the fragment (if there is sufficient flanking DNA). This nonspecific binding can obscure the specific interaction, without necessarily interfering with the specific interaction itself. This is especially true for proteins whose nonspecific binding is highly cooperative (see, for example Fig. 3 of Garner and Revzin 1982). A shorter DNA fragment will have fewer sites for nonspecific binding which can interfere with the detection of specific binding. There are definite advantages to using a relatively short oligo-or polynucleotide, provided that it is long enough to include the entire specific binding site. Alternatively, if the ratio of specific to nonspecific binding affinities (K_S/K_{NS}) is large, nonspecific competitor DNA can be used as a sink for nonspecific binding activities.

- Often, divalent cations are not required for protein binding to DNA, although they may be necessary co-factors for enzymatic activity (e.g., DNA cleavage, polymerization). For example, in the absence of Mg^{2+} the restriction enzyme *Eco*RI binds to its specific sequence very tightly, but DNA cleavage catalyzed by the enzyme does not occur until the divalent cation is added (Sidorova and Rau 1996). Thus, omission of divalent cations from binding buffer may allow one to detect transient catalytic complexes, in addition to protecting DNA from degradation by contaminating nucleases.

It is important that the DNA fragments or oligonucleotides used in the reaction be scrupulously clean. Failure to detect specific complex formation is sometimes due to the presence of contaminants, which either compete with the protein of interest for binding to the target DNA (e.g. polymerases used in PCR amplification) or bind to the protein and inhibit DNA binding (e.g. polysulfates found in some grades of agarose, which co-elute with DNA fragments from agarose gels). Restriction digests should be phenol-chloroform extracted and ethanol precipitated, or purified using a commercial DNA purification kit prior to use. This is also true for synthetic polynucleotides used as competitors, since these are often contaminated with the enzymes used in their synthesis. PCR products should be purified to remove free primers, unincorporated nucleotides, and the enzyme(s) added for synthesis. Synthetic oligonucleotides should be purified to remove contaminating synthesis products and, if gel electrophoresis is used, scrupulously cleaned-up after elution from either agarose or polyacrylamide gel.

Detection and Quantitation of Gel-Resolved Complexes

Visualization of gel-resolved protein-DNA complexes is most often accomplished by staining the DNA with a fluorescent dye[3], followed by photography using ultraviolet illumination (Garner and Revzin 1981;

---

[3] Ethidium bromide is most widely used, although newly developed dyes such as SYBR-Green offer advantages in terms of sensitivity (Sidorova and Rau 1996)

Fried 1989). Post-electrophoresis staining is preferable, since the dye may influence the stability of protein-DNA complexes before and during the electrophoretic run. Fluorescent protein-DNA complexes may be recorded by photography, and quantitated by densitometry of the negative, or recorded digitally with a video densitometer, and integrated using commercially available software packages. While standard photographic methods yield a useful record of the gel pattern, acquisition of valid quantitative data from a photographic negative requires special care to ensure that the recorded optical densities are proportional to the amount of DNA present in each electrophoretic species (for a discussion of these issues, see Fried 1989). This caveat also applies to data acquired by video devices.

Alternatively, the DNA may be visualized by autoradiography if it has been isotopically labeled (Fried and Crothers 1981). Autoradiography is sensitive and offers high spatial resolution, but, as with photography, care must be taken to ensure that the exposure and development conditions yield images that lie within the dynamic range of the film. During the past decade, data capture methods that use two-dimensional ß-particle detectors, or phosphorescent imaging plates, have become widespread (see, for example, Garner and Rau 1995). These technologies offer advantages in sensitivity and linearity of response over typical X-ray films. In all cases, we recommend the use of DNAs of known specific activity as exposure standards in lanes adjacent to those used to resolve complexes to be quantitated. The preparation of a calibration curve based on that data allows vagaries of the visualization method to be taken into account.

The protein component of gel-resolved complexes is more rarely visualized, although useful information can be acquired when it is. Protein detection is often accomplished by post-electrophoresis gel-staining.[4] Greater sensitivity and improved quantitation is afforded by use of radioisotope-labeled proteins, with detection by autoradiography, scintillation counting, or related techniques. Double-label experiments carried out with radioisotopic proteins and DNAs can yield direct measures of protein:DNA stoichiometries of resolved complexes (Fried and Crothers 1983). A variation of this approach involves the use of an isotopically labeled low-molecular weight ligand, which is bound tightly by the protein. If the ligand:protein stoichiometry is reliably known, and if

---

[4] Useful stains for this purpose include coomassie blue R-250 (Fried 1989), protein-specific fluorescent dyes (e.g., SYPRO red) and radioactive ligands bound by one or more proteins in the system (Hudson et al. 1990).

its dissociation during electrophoresis is sufficiently slow, the quantity of ligand present in resolved bands at the end of the gel run will be proportional to the amount of protein present in the bands (Garner and Revzin 1982). Finally, the detection and identification of specific proteins can be carried out with antibodies against those proteins, either by western blotting (Demczuk et al. 1993), or by in situ antibody staining. These antibody methods are also useful alternatives to "supershift" techniques, in which the binding of an antibody changes the gel mobilities of complexes containing its cognate protein epitope (Kristie and Roizman 1986; Lane et al. 1992).

Analysis of Multiple Interactions

A particular strength of the EMSA method is its ability to detect the simultaneous binding of several proteins to a single DNA species, or one (or more) protein(s) to multiple DNA species (Fried and Crothers 1981; Garner and Revzin 1986). Such interactions are of special interest, because DNA-complexes containing large numbers of proteins play central roles in important cellular transactions including DNA replication, DNA repair, transcription and recombination. Although mobility shift assays are most frequently used for qualitative purposes, they can, under appropriate conditions, provide quantitative data for the determination of binding affinities and the evaluation of cooperative interactions. Here we provide a brief theoretical background to the analysis of multiple interactions, as it applies to the mobility shift assay, and a pair of examples demonstrating applications of this method. The thermodynamic approach described below is quite general and can be easily extended to other experimental systems (see Fried and Crothers 1981; Ackers et al. 1982; Brenowitz et al. 1991).

Prelude   At equilibrium, the molecules in an interacting protein-DNA system exist in a series of binding states, each differing in stoichiometry and/or distribution of protein(s) among the available DNA sites. The probability with which a given state (j) of the system occurs is proportional to $\exp(-\Delta G_j/RT)$, in which ΔG_j is the free energy change that accompanies the formation of state j from a reference state (Hill 1985; Wyman and Gill 1990). If we take the state in which the DNA molecule has no protein bound as the reference state, the term $\exp(-\Delta G_j/RT)$ for the binding of a protein to a particular DNA site, with association constant K_j, is equal to $K_j[P]$. The partition function, Q is just the sum of terms for all states of the system:

$$Q = \sum_j \exp(-\Delta G_j/RT) \tag{3}$$

The probability P_j with which state j occurs is equal to the term in the partition function representing the jth state, divided by Q. Since the samples employed in the mobility shift assay typically contain large numbers of molecules, the formation probability P_j will be very closely approximated by the relative frequency with which state j occurs. Thus, for samples at equilibrium, the mole fraction of the DNA that is present in a given binding state is a good measure of that state's formation probability[5], normalized electrophoretic band intensities provide useful measures of the formation probabilities of resolved states of the system.

Often it is useful to learn whether the interaction of one DNA-binding protein with its cognate site influences the binding of a second protein to an adjacent site. An example of such heterotypic association is the formation of complexes containing the E. *coli* CAP and lactose repressor proteins, bound to their highest affinity binding sites in the lac promoter C1 and O1, with association constants K_{C1} and K_{O1} respectively (Fig. 4A) (Hudson and Fried 1990; Vossen et al. 1996). The partition function[6] for this system is:

The binding of two different proteins to discrete sites on the same DNA molecule.

$$Q = 1 + K_{C1}[C] + K_{O1}[R] + K_{C1}K_{O1}[C][R]\omega_{C1O1} \tag{4}$$

Here, the term 1 corresponds to the free DNA reference state (the formation $\Delta G=0$, so $\exp(-\Delta G/RT)=1$); $K_{C1}[C]$ represents the state in which only CAP is bound, $K_{O1}[R]$ represents the state in which only repressor is bound, while $K_{C1}K_{O1}[C][R]\omega_{C1O1}$ represents the state in which both CAP and lac repressor occupy their cognate sites. The cooperativity term, ω_{C1O1}, is defined by $\omega_{C1O1}=\exp(-\Delta G_{int}/RT)$ in which ΔG_{int} is the free energy of protein-protein interaction. If $\Delta G_{int} > 0$, then ω_{C1O1} lies between 0 and 1, and binding is competitive. If $\Delta G_{int} < 0$, then $\omega_{C1O1} > 1$, and binding is synergistic. If $\Delta G_{int}=0$, then $\omega_{C1O1}=1$, and the two proteins bind independently. The probabilities for the allowed states of the system are $P_{free}=1/Q$, $P_C=K_{C1}[C]/Q$, $P_R=K_{O1}[R]/Q$, and $P_{CR}=K_{C1}K_{O1}[C][R]\omega_{C1O1}/Q$. Because CAP protein and lac repressor bind sites C1 and O1 with high specificity (Hudson and Fried 1990), each gel band is dominated by a

---

[5] The requirements are that the relevant complexes are cleanly resolved and that the dissociation of noncovalent complexes during electrophoresis is negligible.

[6] Note that if occupancy of secondary sites is allowed, the partition function contains more terms than that shown here, but this does not change the expression for the ratio of state probabilities given in Eq 4.

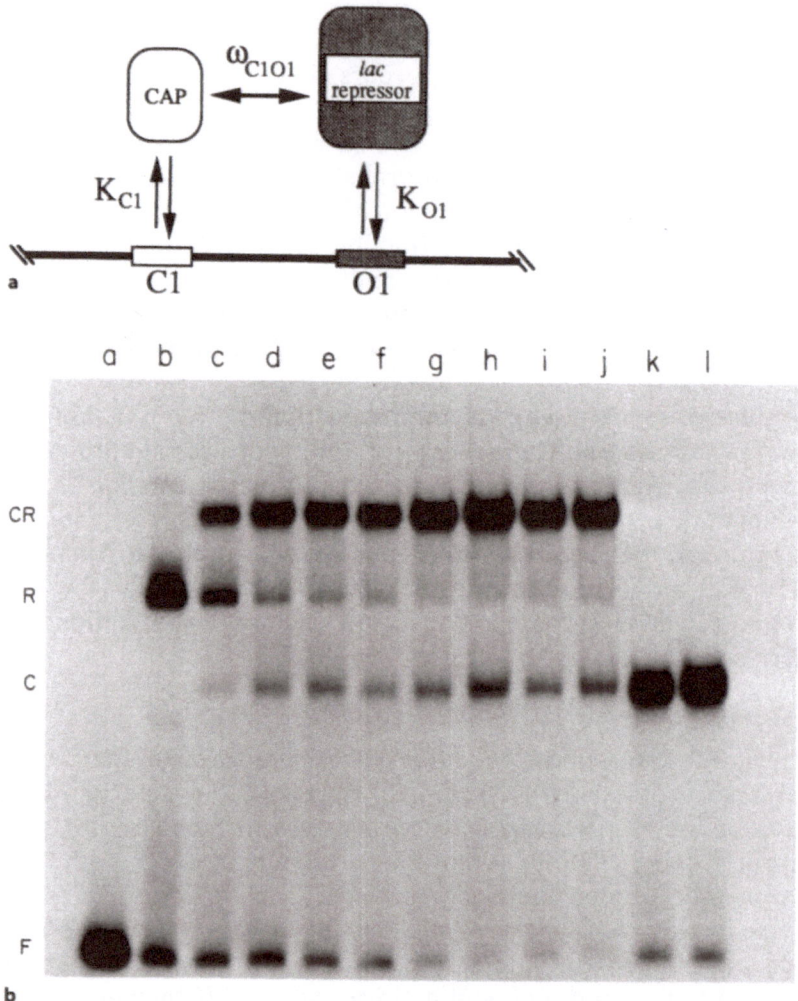

Fig. 4A,B. The binding of *E. coli* CAP and *lac* repressor to their highest affinity sites in the lactose promoter. **A** Partial map of binding sites C1 and O1 within the *lac* promoter, with CAP and *lac* repressor interactions indicated schematically. **B** EMSA of CAP and *lac* repressor binding to a *lac* promoter fragment as depicted in the schema shown in A. Addition of CAP to a solution containing a 214 bp *lac* promoter fragment (*F*) and the 1:1 repressor-DNA complex (*R*) (samples *c–j*), results in the formation of the ternary complex (*CR*) plus a small amount of 1:1 CAP-DNA complex (*C*). All reactions were carried out in 10 mM Tris (pH 8.0 at 21 °C), 1 mM EDTA, 20 μM cAMP, 100 μg/ml bovine serum albumin, 5 % glycerol. All samples contained 2.7 nM DNA fragment. Samples *b–j* contained 2.0 nM *lac* repressor, and samples *c–j* contained, in addition, 1.5, 3.1, 4.6, 6.2, 7.7, 9.3, 10.8, and 12.4 nM CAP, respectively. Samples *k* and *l* contained 2.7 nM DNA and 6.2 and 12.4 nM CAP, respectively. (Data from Hudson and Fried 1990, with permission)

single binding state. The state probabilities (measured by band intensities) can be combined to yield the ratio

$$\frac{I_{free}}{I_C} \frac{I_{CR}}{I_R} = \frac{P_{free}}{P_C} \frac{P_{CR}}{P_R} = \omega_{C101} \tag{5}$$

Analysis of band intensities for experiments like that shown in Fig. 4B gave ω_{C101}=11.8±3.4, clear evidence of a cooperative (synergistic) interaction between CAP and lac repressor (Vossen et al. 1996).

Many DNA-binding proteins alter the conformation of DNA molecules with which they interact. If the conformational change affects the local twist or writhe of the DNA duplex, the free energy difference of binding (and hence the observed association constant, K_{obs}) will depend on the initial topological state of the DNA. An experiment that measures protein distribution between topologically distinct DNA populations can quickly reveal a topological binding preference. Such an experiment is shown in Fig. 5. Here, wheat germ RNA polymerase II was used to titrate pBR322 plasmid DNA present in a mixture of topological forms. Four unbound DNA species form prominent electrophoretic bands. In order of decreasing migration, these are: supercoiled monomer, relaxed circular monomer, supercoiled tandem dimer, and relaxed tandem dimer. If all sites on a DNA molecule are occupied independently (i.e., binding is not cooperative), the polynomial describing RNA polymerase II (P) binding to DNA species α can be written as

Protein Distribution Between Topologically Distinct DNA Populations

$$F(\alpha)=1+K_{\alpha,1}[P]+K_{\alpha,1}K_{\alpha,2}\,[P]^2+K_{\alpha,1}K_{\alpha,2}K_{\alpha,3}[P]^3+... \tag{6}$$

in which α denotes the DNA topological species and the numerical subscript denotes the 1st, 2nd, 3rd, etc., binding event. In the limit of low RNA polymerase concentration, this simplifies to $F(\alpha)=1+K_{\alpha,1}[P]$. For the binding of RNA polymerase to the mixture of DNA species, the partition function is just the product of the polynomials for the four DNA topological species present (represented by α–δ):

$$Q = \prod_{\text{all species } \nu} F(\nu) = (1 + K_{\alpha,1}[P](1 + K_{\beta,1}[P](1 + K_{\gamma,1}[P](1 + K_{\delta,1}[P]) \tag{7}$$

For two DNAs at identical concentrations, the ratio of binding affinities $K_{\alpha,1}/K_{\beta,1}$ is given by

$$\frac{P_{\alpha,1}}{P_{\beta,1}} = \frac{K_{\alpha,1}(K_\beta[P] + K_\gamma[P] + K_\delta[P] + \text{terms in } [P]^2 \text{ and } [P]^3)}{K_{\beta,1}(K_\alpha[P] + K_\gamma[P] + K_\delta[P] + \text{terms in } [P]^2 \text{ and } [P]^3)} \tag{8}$$

This expression simplifies in the limit of low protein concentration to give:

$$\lim_{[P] \to 0} \frac{P_{\alpha,1}}{P_{\beta,1}} = \frac{K_{\alpha,1}}{K_{\beta,1}} \tag{9}$$

In general, DNAs α and β will not be present at equal concentrations and will not contain equal numbers of binding sites. To avoid bias in the estimation of $K_{\alpha,1}/K_{\beta,1}$ it is necessary to take into account any differences in the concentrations of binding sites present on DNAs α and β. This is done by dividing each probability term by the product of the mole fraction of the DNA form to which it refers and the number of binding sites (n_s) present in that DNA form.[7]

$$\lim_{[P] \to 0} \frac{P_{\alpha,1}}{P_{\beta,1}} \cdot \frac{f(\beta) \, n_s(\beta)}{f(\alpha) \, n_s(\alpha)} = \frac{K_{\alpha,1}}{K_{\beta,1}} \tag{10}$$

Shown in Fig. 5B is a graph giving the dependence of $I_{\alpha,1} \cdot f(\beta) \cdot n_s(\beta)/ I_{\beta,1} \cdot f(\alpha) \cdot n_s(\alpha)$ on [RNA polymerase II], in which β represents the closed circular form monomer, taken as the reference state, and α represents each of the other DNA forms, in turn. These results demonstrate that RNA polymerase II binds closed circular DNAs with six- to tenfold greater affinity than it does cognate relaxed circular forms.

---

[7] Since each base pair can be the start-point of a new protein binding site, in either orientation, n_s is equal to twice the number of base pairs present in the DNA in question.

Fig. 5A,B. Titration of pBR322 DNA with wheat germ RNA polymerase II. **A** EMSA analysis. Samples *A–G* contained 14.4 nM plasmid (calculated as monomer equivalents), plus 0, 43.4, 86.9, 217, 434, 869, and 0 nM RNA polymerase II, respectively. The binding buffer was 20 mM Tris (pH 7.9 at 21 °C), 5 mM $MgCl_2$, 0.2 mM EDTA, 2 mM dithiothreitol. Samples were incubated at 21 °C for 40 min. prior to gel loading. Electrophoresis was carried out in a 1 % agarose gel equilibrated with binding buffer lacking dithiothreitol. Band designations: *1(c)*, closed circular monomer; *1(r)*, relaxed form monomer; *2(c)*, closed circular tandem dimer; *2(r)*, relaxed form tandem dimer. **B** Normalized binding ratios for the association of wheat germ RNA polymerase II to pBR322 DNA, as functions of the input RNA polymerase:DNA ratio. The normalized binding ratio is equal to $I_{\alpha,1} \bullet f(\beta) \cdot n_s(\beta)/I_{\beta,1} \cdot f(\alpha) \cdot n_s(\alpha)$, as defined in Eq. (10), and gives the relative affinity of polymerase for a test DNA with respect to its affinity for the closed circular (superhelical) monomer form. The input RNA polymerase/DNA ratio is given in units of equivalents of protein (assuming $M_r \sim 575{,}000$; Guilfoyle et al. 1984) and monomer equivalents of pBR322 DNA. *Filled squares* Relaxed circular tandem dimeric DNA; *open circles* closed circular (superhelical) tandem dimeric DNA. *Inset: filled circles* relaxed circular monomeric DNA. (Data from Fried and Daugherty 1997, with permission)

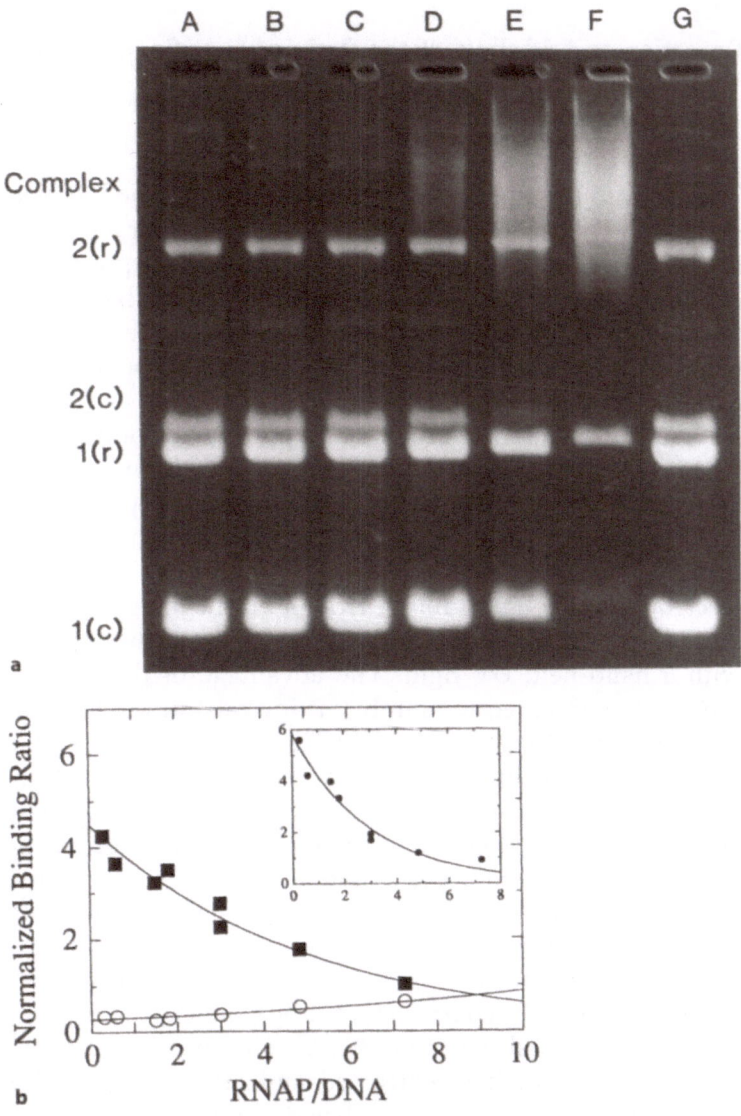

10.1
Detection and Analysis of the Ratio of Specific to Nonspecific Binding

▉ Procedure

Example

To illustrate the principles and methods described above, here we will detail an experiment performed to measure the ratio of specific to nonspecific binding (K_S/K_{NS}) of the galactose (*gal*) repressor protein from *E. coli* . The experiment is performed with two DNA fragments, both of which contain one specific *gal* operator site (the repressor protein has slightly different affinities for the two different operators).

DNA Fragment Preparation

Detailed protocols for all these steps can be found in Maniatis et al. (1983). The 284 bp *Eco*RI-*Pst*I fragment of plasmid pSA509 (Majumdar and Adhya 1983), which contains the two wild-type *gal* operator sites, O_E and O_I, was separated from the plasmid vector using a SeaKem LE agarose (1× TBE, 0.5 µg/ml ethidium bromide) gel. DNA was recovered by electrophoretic elution into a trough cut in the gel in front of the fragment band, after it had clearly separated from the plasmid, as observed with a hand-held UV light. The advantage of the trough elution techniques, as compared to simply cutting a band of DNA out of a gel and eluting by the "freeze and squeeze" method, is that trough elution results in much less contamination of the DNA by co-eluting polysulfate contaminants from the gel. After elution the resulting material was concentrated using a Microcon centrifugal concentrator, phenol/chloroform extracted and ethanol precipitated. Following resuspension, the fragment preparation was further purified by column chromatography using NACS (Life Technologies, Inc.) resin.

The two-operator fragment was further digested with *Sfa*NI to generate fragments with separate O_E (181 bp) and O_I (103 bp) sites, and treated with calf intestinal phosphatase. After each of these steps, the solution was phenol/chloroform extracted and ethanol precipitated twice. The fragments were then 5'-end labeled using the forward reaction of T4 polynucleotide kinase and $[\gamma^{32}P]ATP$. After end-labeling, the DNA was further purified from unincorporated nucleotides, the kinase enzyme, and any contaminating salts using Sephadex spin columns.

Binding Reactions

Galactose (*gal*) repressor protein was the kind gift of Sankar Adhya, and was isolated and stored as described (Majumdar et al. 1987). Poly(dI-dC)·(dI-dC) was purchased from Pharmacia. Repressor binding reactions (to measure the ratio of binding constants between operator and nonspecific DNA, and between repressor binding to the O_E and O_I operators) were carried out in a buffer containing 140 mM KCl, 25 mM Tris-Cl (pH 7.5), 5 mM $MgCl_2$, 1 mM DTT, 0.1 % NP-40, 50 µg/ml acetylated BSA. Total DNA fragment concentration was 0.4 nM, and each reaction volume was 30 µl. It is most convenient to make a concentrated buffer stock (3–10×) containing all components except the macromolecules, and to each reaction tube add (in order) H_2O, concentrated buffer stock, BSA, cold competing DNA, radioactive target DNA, and lastly, protein, prediluted if necessary. The order of addition of the most components is not crucial, although it is convenient to add the radioactive DNA just prior to adding the binding protein or extract (thus minimizing the generation of radioactive pipette tips, etc.). However, it is important that the protein be added last, to ensure that the protein never encounters conditions that deviate significantly from those of the experiment. Reaction mixtures were equilibrated for 45 min 20 °C, prior to loading on the gel.

Polyacrylamide gels (5 %T, 3 %C), cast in 40 mM Tris acetate (pH 7.0), 1 mM EDTA buffer, were used to analyze the reactions. Electrophoresis was carried out in a Hoefer SE-600 gel apparatus held at 20 °C by circulating coolant from an external circulating water bath. The gel was pre-electrophoresed for 30 min to remove impurities and to allow it to temperature equilibrate. Electrophoresis of samples was for ~2 h at 6 V/cm. Under these conditions, the gel temperature does not noticeably increase due to Joule heating. After electrophoresis, gels were dried under vacuum and exposed to X-OMAT AR film at –70 °C, with one LightningPlus Intensifying screen.

Electrophoresis and Visualization

Results

The results of the experiment described above are shown in Fig. 6. The gel clearly shows the degree to which varying the concentration of poly d(I-C)•d(I-C) changes the extent of specific-site binding: An approximately tenfold change in the competitor DNA concentration changes the amount of DNA in each operator complex drastically, even in this case in which one is dealing with a highly purified protein whose specific binding is much tighter than nonspecific (for both operator sites $K_S > 10^4 \times K_{NS}$). This illustrates the importance of very carefully determining the optimal level of concentration of all components in a binding reaction. This becomes especially important when dealing with impure

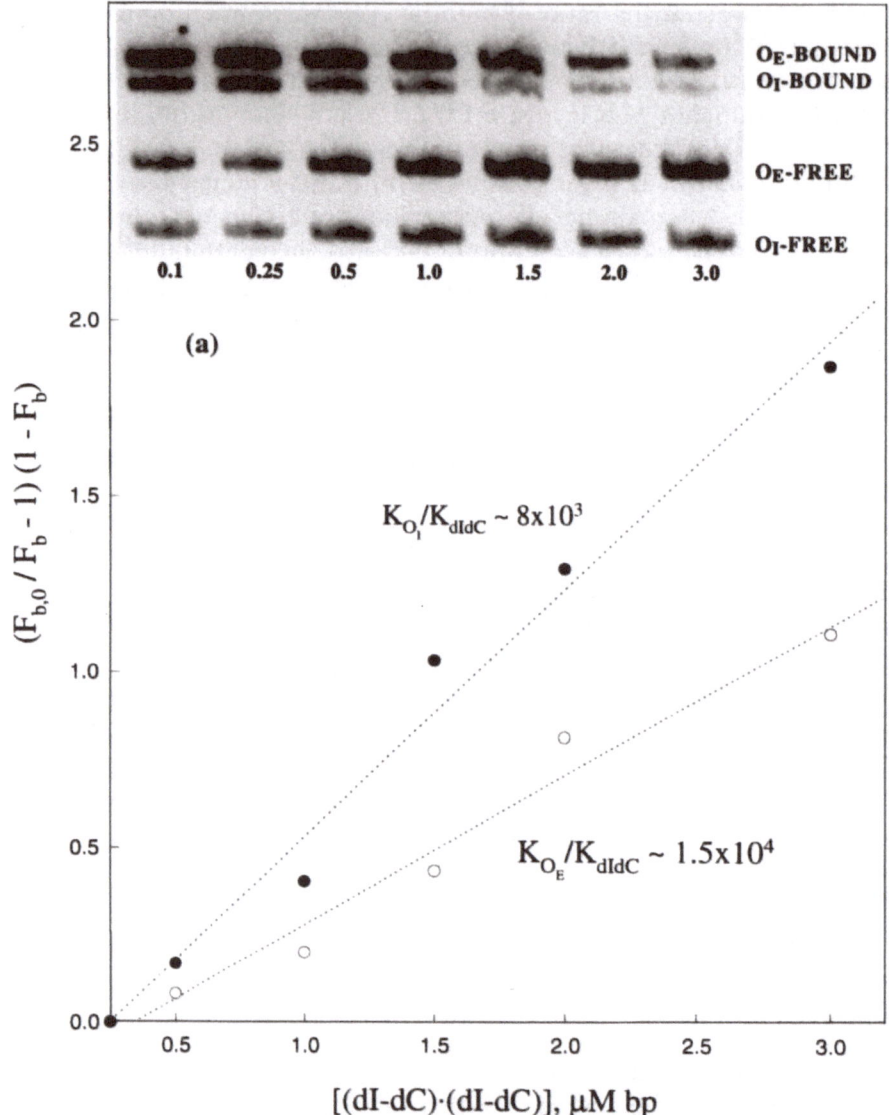

(a)

$K_{O_I}/K_{dIdC} \sim 8 \times 10^3$

$K_{O_E}/K_{dIdC} \sim 1.5 \times 10^4$

$(F_{b,0}/F_b - 1)(1 - F_b)$

$[(dI\text{-}dC)\cdot(dI\text{-}dC)]$, µM bp

Fig. 6. Results of a titration of *gal* operator-repressor complexes with poly d(I-C)•d(I-C). While keeping the concentrations of both repressor and radioactively labeled specific DNA sites constant, increasing amounts of nonradioactive nonspecific DNA are added. *Inset* shows an autoradiograph of the gel The specific complexes are virtually abolished at the highest concentrations of poly d(I-C)•d(I-C). The graph shows the decrease in the amount of each specific complex (F_b) relative to the amount measured at the lowest [poly d(I-C)•d(I-C)], $F_{b,0}$. Such an analysis allows one to determine the ratios of specific to nonspecific binding constant for each specific DNA site independently. (Figure from Garner and Rau 1995, with permission)

binding factors present in nuclear extracts at very low relative abundance. Furthermore, careful quantitative analysis of the data, as illustrated in the graph in Fig. 6, shows an approximately twofold difference in the affinity of the *gal* repressor for these two operator sites.

For quantitative determination of both relative and absolute amounts of free and complex DNA, various imaging systems have been developed to avoid the use of either X-ray or photographic film, with all its inherent limitations (For an extensive discussion of quantitative imaging of electrophoretic gels, see Sutherland 1993). Radioactive bands can be localized by exposing the gel to X-ray film, and then carefully overlaying the X-ray film on the gel (marking the gel or the substrate on which it rests with radioactive or fluorescent ink is very helpful in this superposition). Gel bands located in this manner can be excised with a scalpel and quantitated by scintillation counting. Alternatively, the use of a phosphorescence imager allows one to directly determine the amounts of radioactivity in each band and is linear over almost seven orders of magnitude.

10.2
Detection of Protein-Induced DNA Conformational Changes

One of the most useful features of EMSA is the ability to monitor for DNA conformational changes due to protein binding. It was first observed by Wu and Crothers (1984) that changing the position of the DNA binding site for the CAP protein of *E. coli* on a DNA fragment caused the relative mobility of the CAP-DNA complexes (but not the unbound DNA) to change in a systematic way: complexes with the binding site in the center of the fragment migrate more slowly than those in which the binding site is located close to one end. Kinetoplast DNA, in which static curvature of the DNA causes apparently "aberrant" mobility on a gel (i.e., relative to "normal" DNA standards, a curved DNA fragment migrates more slowly than expected for its length; Marini et al. 1982), behaves in a similar manner, with the location of the apparent center of curvature determining the mobility. Since the X-ray crystallographic structure of the CAP-*lac* promoter complex showed that the protein significantly bent the DNA around itself (Schultz et al. 1991), it was assumed that the difference in mobilities for the CAP-DNA complexes arose from this distortion the helix axis of the DNA.

Although it seems intuitive that the gel mobility of a protein-DNA complex in which the DNA is bent would depend in a sensitive way on the exact location of the bend, it is difficult to exactly model or predict the relationship between angle of bend and relative mobility (discussed by Zimm and Levene 1992). Nevertheless, detecting and measuring DNA bending by bound protein has now become routine by use of the "circular permutation" assay, which provides a way of conveniently moving a DNA binding site to various locations along a linear DNA fragment (an extensive discussion is found in Crothers et al. 1991). DNA bending has been shown to be an important part of, and in some cases the only, biological function of transcription activator or enhancer proteins (Goodman et al. 1992), and its detection and characterization can be an important part of the characterization of a DNA binding protein.

Materials

Agarose vs Polyacrylamide

Although it is conventional to perform EMSA assays on polyacrylamide gels, agarose gels of various types have proven useful for analysis of high molecular weight complexes. In some cases it has even proven possible to separate complexes containing a single, relatively small protein from an unbound DNA fragment on agarose gels (Sidorva and Rau 1996). However, the degree of resolution between complexes and free DNA and the stability of protein-DNA complexes are generally less in agarose than they are in polyacrylamide gels (Chelm and Geidushek 1979; Fried 1989). The reasons for these differences have yet to be discovered. We note however, that the functional pore size of an agarose gel, measured by electrophoretic or chromatographic analysis or by electron microscopic techniques (Serwer and Hayes 1986), is significantly larger than similarly measured pore sizes for polyacrylamide gels (Holmes and Stellwagen 1991; Morris and Morris 1971). In some cases, the larger pore size clearly allows the migration and separation of complexes too large to enter polyacrylamide gels. However, pore size – whether geometric or functional – may not be the only characteristic of a gel responsible for its ability to separate different species (for discussion, see Radko and Chrambach 1997). Furthermore, it has proven possible to separate relatively small protein-DNA complexes in uncross-linked polymer solutions, whose pore size is difficult to define exactly (Xian et al. 1997). Although further understanding of the mechanisms of separation awaits a more complete picture of the gel electrophoretic separation of DNA itself, the shorter running time which often obtains in agarose gels, coupled with the possibility to go to higher molecular weight multiprotein complexes, can be a real practical advantage. For example, reconstituted teleosome complexes (i.e., non-nucleosomal chromatin structures found at the ends of telomeres) from *S. cerevisiae* have been characterized in agarose gels (Wright and Zakian 1995). Given the large size of these assemblies, this analysis would likely not have been possible using conventional polyacrylamide gels.

Procedure

Use of Cellular or Nuclear Extracts

The first step in the identification or cellular localization of a DNA binding protein is to assay for specific binding activity using a cellular or a nuclear extract. Typical protocols for whole cell (Manley), cytoplasmic fraction (S100) or nuclear extract fraction all involve lysis of the cells or nuclei in a high salt buffer, followed by centrifugation to pellet the chromosomal DNA/chromatin, and any cellular or nuclear membranes and organelles (extensive discussions of these procedures are found in Manley et al. 1980; Dignam 1990). The basis of all of these methods is to use a salt concentration high enough to cause chromosome-associated non-histone DNA binding proteins to dissociate without causing the histone proteins to be released: since the histone proteins are in vast excess over all other gene regulatory proteins combined, their extraction would dilute the relative abundance of even a high copy number gene regulatory protein, rendering its detection impossible.

Regardless of the extraction method used or the starting tissue or cell source, the end result is a solution of fairly high salt concentration (typically 0.3–0.5 M K^+) containing a large number of DNA binding proteins. This imposes two special restrictions on studying the specific binding of a single protein: First, the presence of a large complement of DNA-binding proteins that are not of interest means that a large molar excess of nonspecific DNA will be required to act as a sink for unwanted DNA binding activities. Without such a sink, the binding interactions of interest may be obscured or prevented by other competing DNA interactions. Second, since the protein extract is dissolved in high [salt] buffer, it is important to carefully control the amount of extract added or, if a titration is done with extract, to add the same amount of salt to each sample. If the salt concentration is allowed to vary, reproducible binding will not be observed.

Typically, the proteins from cell extracts are assayed using radioactive labeled oligonucleotides containing their specific sequence as a target and nonradioactive poly d(I-C)·d(I-C) as the competitor. A competitive titration assay with poly d(I-C)·d(I-C) can then carried out as above, with the following modifications:

- The poly d(I-C)·d(I-C) should be in large molar excess over the specific target DNA. Although it is difficult a priori to pick relative concentrations, a good initial starting titration range would be 0.01–1 µg poly d(I-C)·d(I-C) per fmole of radioactive DNA (1 ng of a 40 bp long oligo=40 fmoles). If no binding to the target DNA is seen, decrease the poly d(I-C)·d(I-C) by a factor of ten. If all of the target DNA is retained near the top of the gel, increase [poly d(I-C)·d(I-C)] by a factor of ten.

- Because many DNA binding proteins are fairly susceptible to proteases, even in the presence of protease inhibitors, reactions are typically incubated at 4 °C.

- Without knowing the relative abundance, activity, competing activities, and K_S/K_{NS} for the specific protein, it is impossible to specify an amount of cell or nuclear extract to add. For a typical 30 µl reaction containing ca. 10 fmole of oligonucleotide, 1–5 µl of a whole cell extract is a good starting range. This, too, needs to be titrated to determine the optimal amount of protein to add, remembering to control the salt concentration in each reaction carefully. In order to compare different extracts, or different preparations of the same extract, the total protein per unit volume should be determined by standard techniques, i.e., Lowry (1951) or Bradford (1976) assays.

The gel is run, dried, and autoradiographed as above. One often sees a large number of bands in the lower competitor concentration lanes, which progressively diminish as the amount of competitor DNA is increased. These bands are due to proteins binding with low specificity to the labeled DNA. Increasing amount of poly d(I-C)·d(I-C) displace these proteins, causing these bands to disappear. The bands left at the higher concentrations of poly d(I-C)·d(I-C) are specific complexes. This can be confirmed by adding a ten- to 20-fold excess of nonradioactive oligo to a binding reaction in which these bands are seen. At this excess, the cold, specific oligo will compete off only those bands which are due to specific binding.

Comments

Here we have attempted to describe factors that we have found to be important in performing EMSA for both qualitative and quantitative analysis of specific DNA-protein interactions. As stated at the outset, the experimental details of the assay should be optimized for each individual protein-DNA system. No one "cookbook" recipe suffices for all DNA binding proteins. To allow the reader to develop his or her own assay system, we have delineated those factors which need to be carefully controlled to use EMSA successfully, and given an example of one particular protocol. However, this is indeed an example protocol, and should not blindly applied to any DNA binding protein.

In our experience, difficulties with the assay are almost always due to a failure to optimize the binding reaction (concentrations of salt, cofactors, competitor DNA, etc.) rather than problems with the EMSA technique per se. Far from being a limitation, this points up one of the great strengths of EMSA and the main reason for its wide-spread popularity: the assay is robust enough that it is adaptable to a wide variety of reaction conditions with any number of components, provided that the user thoughtfully designs the experiment to optimize, or at least control, the important parameters. More than any particular detail, that is the central conclusion of this chapter.

References

Ackers G, Johnson A, Shea M (1982) Quantitative model for gene regulation by λ phage repressor. Proc Natl Acad Sci USA 79:1129–1133

Boffini A, Prentki P (1991) Identification of protein binding sites in genomic DNA by two-dimensional gel electrophoresis. Nucleic Acids Res 19:1369–1374

Bradford M (1976) A rapid and sensitive method for the quantitiation of microgram quantities of proteins utilizing the principle of protein-dye binding. Anal Biochem 72:248–254

Brenowitz M, Pickar A and Jamison E (1991) Stability of a lac repressor mediated „looped complex". Biochem 30:5986–5998

Brown A, Crothers DM (1989) Modulation of the stability of a gene-regulatory protein dimer by DNA and cAMP. Proc Natl Acad Sci USA 86:7387–7391

Cann JR (1989) Phenomenological theory of gel electrophoresis of protein-nucleic acid complexes. J Biol Chem 264:17032–17040

Cann, JR (1990) Analysis of the gel electrophoresis of looped protein-DNA complexes by computer simulation. J Mol Biol 216:1067–1075

Cann JR (1993) Theoretical studies on the mobility shift behavior of binary protein-DNA complexes. Electrophoresis 14:669–679

Carey J (1991) Gel Retardation. Methods Enzymol 208:103–117

Chelm BK, Geidushek EP (1979) Gel electrophoretic separation of transcriptional complexes: an assay for RNA polymerase selectivity and a method for promoter mapping. Nucleic Acids Res 7:1851–1867

Chodosh LA (1988) Mobility shift DNA-binding assay using gel electrophoresis. In: Ausubel FM, Brent R, Kingston RE, Moore DD, Seidman JG, Smith JA, Struhl K (eds) Current protocols in molecular biology. 12.2.1–12.2.10, John Wiley and Sons, New York

Colclasure GC, Parker JC (1992) Cytosolic protein concetration is the primary volume signal for swelling induced [K-Cl] cotransport in dog red cells. J Gen Physiol 100:1–10

Crothers DM, Gartenberg MR, Schrader TE (1991) DNA bending in protein-DNA complexes. Methods Enzymol 208:118–146

Demczuk S, Harbers M, Vennström B (1993) Identification and analysis of all components of a gel retardation assay by combination with immunoblotting. Proc Natl Acad Sci USA 90:2574–2578

Dignam JD (1990) Preparation of extracts from higher eukaryotes. Meth Enzymol. 182:194–203

Eisenberg D, Crothers DM (1979) Physical chemistry with applications to the life sciences. Benjamin/Cummings, Menlo Park, New Jersey

Fickert T, Muller-Hill B (1992) How *lac* repressor finds lac operator in vitro. J Mol Biol 226:59–68

Fried MG (1989) Measurement of protein-DNA interaction parameters by electrophoresis mobility shift assay. Electrophoresis 10:366–376

Fried MG, Bromberg JL (1997) Factors that affect the stability of protein-DNA complexes during gel electrophoresis. Electrophoresis 18:6–11

Fried MG, Crothers DM (1981) Equilibria and kinetics of lac repressor-operator interactions by polyacrylamide gel electrophoresis. Nucleic Acids Res 9:6505–6525

Fried MG, Crothers DM (1983) CAP and RNA polymerase interactions with the *lac* promoter: binding stoichiometry and long range effects. Nucleic Acids Res 11:141–158

Fried MG, Crothers DM (1984) Equilibrium studies of the cyclic AMP receptor protein-DNA interaction. J Mol Biol 172:241–262

Fried MG, Crothers DM (1984) Kinetics and mechanism in the reactions of gene regulatory proteins with DNA. J Mol Biol 172:263–282

Fried MG, Daugherty MA (1998) Electrophoretic analysis of multiple protein-DNA interactions. Applied and Theoretical Electrophoresis (in press)

Fried MG, Liu G (1994) Molecular sequestration stabilizes CAP-DNA complexes during polyacrylamide gel electrophoresis. Nucleic Acids Res 22:5054–5059

Fried MG, Stickle DF (1993) Ion exchange reactions of proteins during DNA binding. Eur J Biochem 218:469–475

Garner MM (1997) Macromolecular crowding and metabolic channeling. In: Agius L, Sherratt HSA (eds) Channeling and intermediary metabolism. Portland Press, London, pp 41–52

Garner MM, Burg MB (1994) Macromolecular crowding and confinement in cells exposed to hypertonicity. Am J Physiol Cell Physiol:266 C877-C892

Garner MM, Revzin A (1981) A gel electrophoresis method for quantifying the binding of proteins to specific DNA regions: application to components of the *Escherichia coli* lactose operon system. Nucleic Acids Res 9:3047–3060

Garner MM, Revzin A (1982) Stoichiometry of catabolite activator protein/adenosine cyclic 3',5'-monophosphate interactions at the lac promoter of Escherichia coli. Biochemistry 21:6032–6036

Garner MM, Revzin A (1986) The use of gel electrophoresis to detect and study nucleic acid-protein interactions. Trends Biol Sci 11:395–396

Garner MM, Rau DC (1995) Water release associated with specific binding of *gal* repressor. EMBO J 14:1257–1263

Gerstle JT, Fried MG (1993) Measurement of binding kinetics using the gel electrophoresis mobility shift assay. Electrophoresis 14:725–731

Goodman SD, Nicholson SC, Nash HA (1992) Deformation of DNA during site-specific recombination of bacteriophage λ: replacement of IHF protein by HU protein or sequence-directed bends. Proc Natl Acad Sci USA 8:11910–11914

Ha J-H, Capp MW, Hohenwalter MD, Baskerville M, Record MT Jr (1992) Thermodynamic stoichiometries of participation of water, cations and anions in specific and non-specific binding of lac repressor to DNA. J Mol Biol 228:252–264

Ha J-H, Spolar RS, Record MT Jr (1989) Role of the hydrophobic effect in stability of site-specific protein-DNA complexes. J Mol Biol 209:801–816

Hill TL (1985) Cooperativity theory in biochemistry. Springer, Berlin Heidelberg New York

Holmes DL, Stellwagen NC (1991) Estimation of polyacrylamide gel pore size from Ferguson plots of normal and anomalously migrating DNA fragments I. Gels containing 3 % N, N'-methylenebisacrylamide. Electrophoresis 12:253–263

Hudson JM, Crowe LG, Fried MG (1990) A new DNA binding mode for CAP. J Biol Chem 265:3219–3225

Hudson JM, Crowe M, Fried MG (1993) Effects of anions on the binding of the cyclic AMP receptor protein to the lactose promoter. Eur J Biochem 212:539–548

Hudson JM and Fried MG (1990) Co-operative interactions between the catabolite gene activator protein and the lac repressor at the lactose promoter. J Mol Biol 214:381–396

Kristie TM, Roizman B (1986) α4, the major regulatory protein of herpes simplex virus type I, is stably and specifically associated with promotor-regulatory domains of α genes and of selected other viral genes. Proc Natl Acad Sci USA 83:3218–3222

Lane D, Prentki P, Chandler M (1992) Use of gel retardation to analyze protein-nucleic acid interactions. Microbiol Rev 56:509–528

Leirmo S, Harrison C, Cayley DS, Burgess RR, Record MT, Jr. (1987) Replacement of potassium chloride by potassium glutamate dramatically enhances protein-DNA interactions in vitro. Biochemistry 26:2095–2101

Lesser DR, Kurpiewski MR, Jen-Jacobson L (1990) The energetic basis of specificity in the EcoRI endonuclease-DNA interaction. Science 250:776–86

Lowry OH, Rosebrough NJ, Farr AL, Randall RJ (1951) Protein measurement with the Folin phenol reagent. J Biol Chem 193:265–275

Lundbäck T, Härd T (1996) Sequence-specific DNA-binding dominated by dehydration. Proc Natl Acad Sci USA 93:4754–4759

Majumdar A, Rudikoff S, Adhya S (1987) Purification and properties of gal repressor:pL-galR fusion in pKC31 plasmid vector. J Biol Chem 262:2326–22331

Manley JL, Fire A, Cano A, Sharp PA, Gefter ML (1980) DNA-dependent transcription of adenovirus genes in a soluble whole-cell extract. Proc Natl Acad Sci USA 77:3855–3859

Marini JC, Levene SD, Crothers DM, Englund PT (1982) Bent helical structure in kinetoplast DNA. Proc Natl Acad Sci USA 79:7664–7668

Minton A P (1983) The effect of volume occupancy upon the thermodynamic activity of proteins:some biochemical consequences. Mol Cell Biochem 55:119–140

Minton AP, Colclasure CG, Parker JC (1992) Model for the role of macromolecular crowding in regulation of cellular volume. Proc Natl Acad Sci USA 89:10504–10506

Morris CJOR, Morris P. (1971) Molecular sieve chromatography and electrophoresis in polyacrylamide gels. Biochem J 124:517–528

Overman LB, Lohman TM (1994) Linkage of pH, anion and cation effects in protein-nucleic acid equilibria. J Mol Biol 236:165–178

Parsegian VA, Rand RP, Rau DC (1995) Macromolecules and water: probing with osmotic stress. Meth Enzymol 259:43–94

Pierrou S, Enerbäck S, Carlsson P (1995) Selection of high-affinity binding sites for sequence-specific, DNA biding proteins from random sequence oligonucleotides. Anal Biochem 229:99–105

Record MT Jr, Anderson CF, Lohman TM (1978) Thermodynamic analysis of ion effects on the binding and conformational equilibria of proteins and nucleic acids: the roles of ion association or release, screening, and ion effects on water activity. Q Rev Biophys 11:103–178

Record MT Jr, Ha J-H, Fisher MA (1991) Analysis of equilibrium and kinetic measurements to determine thermodynamic origins of stability and specificity and mechanism of formation of site-specific complexes between proteins and helical DNA. Meth Enzymol 208:291–343

Record MT Jr, Lohman TM, deHaseth PL (1976) Ion effects on ligand-nucleic acid interactions. J Mol Biol 107:145–158

Record MT Jr, Mossing MC (1987) Physical-chemical origins of stability, specificity, and control of protein-DNA interactions. In: Reznikoff WS (ed) RNA polymerase and the regulation of transcription. Elsevier, New York, pp 61–83

Record MT Jr, Spolar RS (1990) Some thermodynamic principles of nonspecific and site-specific protein-DNA interactions. In: Revzin A (ed) The biology of nonspecific DNA-protein interactions. CRC Press, Boca Raton, pp 33–69

Saxe SA, Revzin A (1979) Cooperative binding to DNA of catabolite activator protein of *Escherichia coli*. Biochemistry 18:255–263

Schultz S, Shields G, Steitz T (1991) Crystal structure of a CAP-DNA complex: the DNA is bent by 90. Science 253:1001–1007

Serwer P, Hayes SJ (1986) Exclusion of spheres by agarose gels during agarose gel electrophoresis: dependence on the sphere's radius and the gel's concentration. Anal Biochem 158:72–28

Sidorova NY, Rau DC (1996) Differences in water release for the binding of EcoRI to specific and nonspecific DNA sequences. Proc. Natl Acad Sci USA 93:112272–12277

Singh H, Sen R, Baltimore D, Sharp PA (1986) A nuclear factor that binds to a conserved sequence motif in transcriptional control elements of immunoglobin genes. Nature 319:154–156

Spolar RS, Record MT, Jr. (1994) Coupling of local folding to site-specific binding of proteins to DNA. Science 263:777–784

Stickle DF, Vossen KM, Riley DA, Fried MG (1994) Free DNA concentration in *E. coli* estimated by an analysis of competition for DNA-binding proteins. J Theor Biol 168:1–12

Strahs D, Brenowitz M (1994) DNA conformational changes associated with the cooperative binding of cI-repressor of bacteriophage λ to O_R. J Mol Biol 244:494–510

Straney DC, Crothers DM (1985) Intermediates in transcription initiation from the *E. coli lac* UV5 promoter. Cell 43:449–459

Sutherland JC (1993) Electronic imaging of electrophoretic gels and blots. Adv Electrophoresis 6:3–38

Varshavsky A (1987) Electrophoretic assay for DNA-binding proteins. Meth Enzymol 151:551–565

von Hippel PH, Revzin A, Gross CA, Wang AC (1974) Nonspecific DNA binding of genome regulating proteins as a biological control mechanism. 1. The *lac* operon: equilibrium aspects. Proc Natl Acad Sci USA 71:4808–4812

Vossen KM, Fried MG (1997) Sequestration stabilizes lac repressor-DNA complexes during gel electrophoresis. Anal Biochem 245:85–92

Vossen KM, Stickle DF, Fried MG (1996) The mechanism of CAP-lac repressor binding cooperativity at the *E. coli* lactose promoter. J Mol Biol 255:44–54

Vossen KM, Wolz R, Daugherty MA, Fried MG (1997) Role of macromolecular hydration in the binding of the Escherichia coli cyclic AMP receptor to DNA. Biochemistry 36:11640–11647

Wright JH, Zakian VA (1995) Protein-DNA interactions in soluble telosomes from *Saccharomyces cerevisiae* . Nucleic Acids Res 23:1454–1460

Wu HM, Crothers DM (1984) The locus of sequence-directed and protein-induced DNA bending. Nature 308:509–513

Wyman J, Gill SJ (1990) Binding and linkage. University Science Books, Mill Valley, California

Xian J, Harrington MG, Davidson EH (1996) DNA-protein binding assays from a single sea urchin egg: a high sensitivity capillary electrophoresis method. Proc Natl Acad Sci USA 93:86–90

Zimm BH, Levene SD (1992) Problems and prospects in the theory of gel electrophoresis of DNA. Q Rev Biophys 25:171–204

Zimmerman SB, Minton A (1993) Macromolecular crowding: biochemical, biophysical, and physiological consequences. Annu Rev Biophys Biomol Struct 22:27–65

Zimmerman S, Trach SJ (1991) Estimation of macromolecular concentrations and excluded volume effects for the cytoplasm of *Escherichia coli*. J Mol Biol 222:599–620

Chapter 11

Mobility Shift Analysis of Protein-DNA Complexes by Capillary Electrophoresis

Jun Xian and Michael G. Harrington

Introduction

The fundamental process at the onset of embryonic development is the establishment of spatial patterns of differential gene expression. Whereas all cells in the organism contain the same genetic "blueprint," this information is utilized in a highly selective and specific manner. Different types of cells activate different subsets of genes. This specialization of cellular activity lies at the heart of cell biology. Much current research is aimed at understanding the initial processes of development, and the regulatory "programs" for the different parts of the organism.

Transcription factors (TFs) are the molecular switches that, singly or in combination, interact with specific sequence of DNA to activate, repress, or modulate transcription of genes. Highly specific in nature, TFs operate at the most fundamental level within the body to regulate any given cell's biochemical functions.

Measurement of sequence-specific TF-DNA interaction is, therefore, a central experimental procedure in studying the molecular biology of gene regulation. The most commonly used method is the electrophoretic gel mobility shift assay (EMSA), in which a radioactively labeled DNA probe is mixed with a solution containing the TF of interest, and after a brief reaction period, the resultant solutions loaded on an electrophoretic gel. The TF-DNA complex migrates more slowly than does the free probe and, despite the limited stability of such complexes, their recovery

Michael G. Harrington, Huntington Medical Research Institutes, 99 North El Molino Avenue, Pasadena

Jun Xian, Cereon Genomics, L. L. C., One Kendall Square, Building 200, Cambridge, MA 02139, USA

Correspondence to Michael G. Harrington, CA 91101-1830, USA

(*phone* +1-6267954343; *fax* +1-6267955774; *e-mail* mghworks@hmri.org)

is greatly facilitated by the "caging" effect of the gel, which essentially retains the protein in the vicinity of the probe during the rather slow process of electrophoresis (Fried and Crothers 1981). While used extensively (Hudson and Fried 1990; Calzone et al. 1988, 1991), a limitation of EMSA is the amount of material utilized per assay, at least in standard practice. A typical protocol requires $>10^{10}$ molecules of kinase-labeled probe, and at least 1 % of this number of active DNA-binding protein molecules. Since TFs are often present in the range of only 10^3–10^4 molecules per nucleus, the application of EMSA for extracts of only a few hundred, or thousand, cells is precluded. For certain studies this is a severe limitation: for example, in studies of embryonic development, it is possible to dissect out by hand a biologically important element of an embryo containing a few hundred cells, or to collect certain cells from a hundred embryos or so. Invaluable information regarding the spatial activity of key regulatory factors could be obtained from such preparations were it possible to measure interactions with specific DNA probes on this scale.

There are several prior studies in which capillary electrophoresis (CE) was applied to the study of protein-DNA complexes by using a mobility shift assay (CEMSA) (Maschke et al. 1993; Heegaard and Robey 1993; Xian 1994) (Fig. 1). Here, using a detection system based on laser-induced fluorescence, we describe rapid and quantitative procedures that can be carried out using a commercially available CE instrument, and that permit accurate assessment of specific protein-DNA interactions on a scale more than 100-fold below the minimum usually necessary for

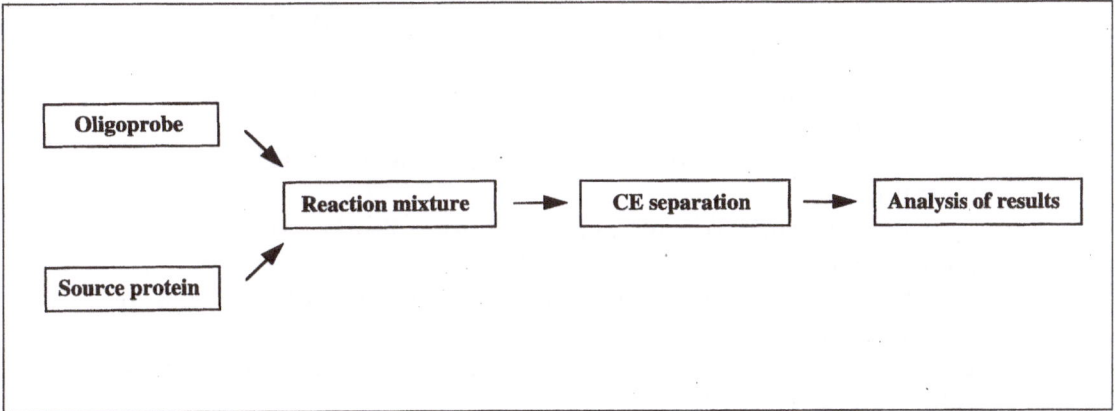

Fig. 1. Flow-sheet demonstrating the capillary electrophoresis mobility shift assay (CEMSA)

EMSA (Xian et al. 1996). This procedure can be used for a variety of purposes, quantitative or qualitative, including studies involving the effects of antibodies on protein-DNA complexes. This method can also be employed as a tool for combinatorial library screening.

The methods presented here are derived from experiments with a sea urchin embryo TF, SpP3A2, that had been cloned and extensively characterized in earlier studies (Calzone et al. 1988, 1991; Hough-Evans et al. 1990). It should be noted that these procedures must be optimized for any alternative protein or oligonucleotide. These variations and options are discussed in the "Troubleshooting" section below.

■ Materials

Equipment
– P/ACE 2100 system (Beckman) upgraded with Gold (version 8.1) software and LIF detector (488 nm excitation and 516 nm emission filter). The separation was performed in a 50 μm×37 cm long (30 cm effective length) neutral coated capillary (eCAP neutral, Beckman).

Note: If the sample can be applied in a larger amount, other CE instrumentation, such as the BioFocus 3000 from BioRad, can be used for this purpose.

Buffers
Note: All buffers used here were filtered through 0.2 μm nylon filter and the CE buffer needs to be degassed before use.

– Buffer A: 10 mM Hepes, pH 7.9; 1 mM EDTA; 1 mM EGTA; 1 mM spermidine-Tris-HCl; 1 mM DTT
– Buffer B: 10 mM Tris, pH 7.4; 1 mM EDTA; 1 mM EGTA; 1 mM spermidine-Tris-HCl; 1 mM DTT; 0.36 M sucrose
– Buffer C: 20 mM Hepes, pH 7.9; 40 mM KCl; 0.1 mM EDTA; 1 mM DTT; 20 % glycerol
– Urea lysis buffer: 50 mM Tris, pH 8.0; 8 M urea; 1 M NaCl; 7.5 mM β-mercaptoethanol
– Lysis buffer: 50 mM Tris, pH 8.0; 5 mM EDTA; 50 mM NaCl; 7.5 mM β-mercaptoethanol
– Buffer 1: 50 mM Tris, pH 8.0; 3 M urea; 100 mM KCl; 7.5 mM β-mercaptoethanol
– Buffer 2: 100 mM potassium phosphate, pH 6.3; 10 mM Tris; 3 M urea; 100 mM KCl; 7.5 mM β-mercaptoethanol
– Buffer 3: 50 mM Tris, pH 7.4; 3 M urea; 100 mM KCl; 10 % glycerol; 7.5 mM β-mercaptoethanol

- Buffer 4: 10 mM Tris, pH 6.3; 300 mM imidazole; 3 M urea; 100 mM KCl; 10 % glycerol; 7.5 mM β-mercaptoethanol
- Binding buffer (5×): 100 mM Hepes, pH 7.9; 375 mM KCl; 25 mM MgCl$_2$; 2.5 mM DTT
- TBE buffer (1×): 89 mM Tris, pH 8.3; 89 mM boric acid; 2 mM EDTA
- CE buffer: 89 mM Tris, pH 8.3; 89 mM boric acid; 2 mM EDTA; 0.05–5 % linear polyacrylamide (MW 750,000–1,000,000 Da)

Procedure

Probe Labeling

The probe we used in this study contains two adjacent SpP3A2 sites of differing affinity for SpP3A2 protein. In addition to the wild-type probes we also constructed a probe on which the strong site was destroyed by mutating its sequence. Wild type:

5'-GATCTTTTCGG*CTTCTGCGCAC*ACCCCACGCGCATGGGGC-3' (sense)

and mutated:

5'-GATCTTTTCGG*CTTCTGCGCAC*ACCCCACATATATGGGC-3' (sense)

There are two methods for probe labeling, covalent and intercalation.

1. Oligonucleotides were synthesized and labeled with 6-FAM fluorescent dye at the 5'-end following the protocol provided by Applied Biosystems   **Covalent Labeling**

(**Note:** Reverse phase or anion exchange HPLC could be used for purifying the oligonucleotides; for details see Warren and Vella 1994).

2. The labeled, single-stranded oligonucleotides were vacuum dried and stored in the dark at –20 °C.

3. Equimolar sense and anti-sense DNAs were annealed in 0.1 M NaCl at 93 °C for 5 min, and allowed to cool at room temperature overnight.

4. Double-stranded DNA was purified from single-stranded DNA on a nondenaturing 8 % polyacrylamide gel, from which the fluorescent bands were excised, electroeluted, purified on Sep-Pak C$_{18}$ columns, vacuum dried and stored at –20 °C. (If the oligonucleotides have been HPLC purified, you can skip over this step.)

Note: The purity of the probe can be checked by capillary gel electrophoresis (CGE) because CGE can easily separate single-stranded from double-stranded DNA.

5. Prior to use, the probes were dissolved in 10 mM Tris-HCl (pH 7.9), 25 mM KCl, and their concentration was determined on a photon counting spectrofluorometer, SLM 8000TMC (SLM Instruments, Inc., Urbana, IL). Stock solutions of 60 nm of wild-type FAM-labeled DNA and 50 nm of mutated FAM-labeled DNA were used.

Intercalation Labeling

1. Double-stranded oligonucleotide were incubated with the desired fluorescent dye at specific molar ratio, in a dark room for 30 min before use.

2. The ratios used in our experiments were 10 bp/1 dye, 20 bp/1 dye, and 30 bp/1 dye with TOTO-1 fluorescent dye. The first ratio yielded a better probe.

Note: These probes do not need further purification because only the double-stranded oligonucleotide has been labeled. We have used these probes to do CEMSA for combinatorial library screening and achieved the same results as the covalent labeled probes. Because the DNA probe is always in excess amount in reaction, the intercalation did not interfere with the protein binding. These probes are not very stable, therefore they need to be prepared freshly. Because quantitation is difficult to achieve by using these probes, we did not use them in the remaining experiments described here, but this method of labeling is more versatile and simpler than covalent labeling in situations in which qualitative results are satisfactory.

Recombinant Proteins

Specific SpP3A2-DNA binding was assayed with purified recombinant protein (rSpP3A2) (Zeller et al. 1995b) from a 300 nm stock solution. The protocol for preparing the recombinant protein as follows:

1. The complete coding sequence of the SpP3A2 protein was subcloned into pRSET expression vector (InVitrogen) in the appropriate reading frame. Then the construct was transformed into BL21 bacterial host cells.

2. The culture of bacteria expressing the construct was grown at 37 °C until OD_{600} ~0.6, which was then induced with isopropyl-β-D-thio-

galactopyranoside (IPTG) at a final concentration of 1 mM and grown for an additional 4 h at 37 °C.

3. The bacteria were harvested by centrifugation and the bacteria pellet was resuspended in 8 ml lysis buffer, and 800 μl of 10 mg/ml lysozyme was added and thoroughly mixed.

4. After incubating at room temperature for 10 min, the protein solution was frozen at −70 °C.

5. On thawing at room temperature, 0.7 times the protein solution volume of urea lysis buffer was added and mixed by vortexing.

6. The mixture was centrifuged for 1 h at 35,000 rpm to remove insoluble matter.

7. A 50 % solution of nickel-agarose resin (Qiagen) was equilibrated with buffer 1 and the crude protein extract was batch-loaded with 8 ml of the equilibrated resin solution for 1 h at room temperature.

8. The loaded resin was then poured into a Bio-Rad Econo-prep column, allowed to settle, and washed with 250 ml buffer 1, then 250 ml of buffer 2, and 50 ml of buffer 3.

9. The purified protein was eluted with buffer 4 and frozen at −70 °C.

Nuclear Extract (Calzone et al. 1988)

1. *S. purpuratus* embryos were cultured by standard methods (Calzone et al. 1991) to blastula-stage, which were collected by filtration with 51-μm Nitex filters and washed one or two times by low-speed centrifugation with ice-cold Ca^{2+}- and Mg^{2+}-free sea water containing 10 mM Tris (pH 7.4) and 1 mM EDTA.

2. The embryo pellet was resuspended in 10–20 times the pellet volume of buffer B, frozen in liquid N_2, and stored at −70 °C.

3. Cells were lysed by vigorous shaking during thawing and nuclei were washed two or three times in buffer B, and two times in buffer B to which 0.1 % Triton X-100 was added.

4. After centrifugation at 3000 g, the nuclei were resuspended in five to ten times the pellet volume of buffer A.

5. While mixing the nuclear suspension, about one-tenth volume of 4 M ammonium sulfate (pH 7.9) was added dropwise (final concentration 0.36 M).

6. After incubation for 30–60 min on ice, chromatin was removed by centrifugation at 35,000 rpm for 1–1.5 h at 4 °C.

7. Protein was precipitated from the supernatant by the addition of 0.3 % g/ml ammonium sulfate.

8. After incubation overnight, the precipitated protein was collected by centrifugation at 10,000 g, and then dissolved in 0.5 times the nuclei pellet volume of buffer C. The proteins were dialyzed against buffer C overnight at 4 °C.

9. Insoluble proteins were removed by centrifugation, and the extracts were stored at –70 °C.

Sea Urchin Egg

1. A known number of fresh *S. purpuratus* eggs suspended in sea water were pipetted into a 1.7-ml microcentrifuge tube.

2. After centrifuging at 13,000 rpm for 5 min, excess sea water was discarded.

3. 1 μl of buffer A was added and the sample stored at –70 °C.

Antibodies

The Caltech Monoclonal Antibody Facility generated a series of monoclonal antibodies to the rSpP3A2. The antigen was full-length rSpP3A2, purified as described above ("Recombinant Proteins"). Two monoclonal antibodies (1 μM stock) were diluted to pmol level and used in this experiment; one, 7B12/1H7, which inhibits the formation of SpP3A2-DNA complexes, and another, 6F1/2E9, which supershifts this protein-DNA complex.

Reaction Conditions

With Recombinant Protein

1. In a 1.7-ml microcentrifuge tube add 3 μl buffer C, 1 μl FAM-labeled DNA (about 50–60 fmol), 1 μl rSpP3A2 (about 300 fmol), 1 μl 50 ng poly(dA:dT), 2 μl 5× binding buffer and 2 μl H_2O.

2. Incubate the mixture on ice for 10–30 min before injection.

Note: The total mass of rSpP3A2 overestimates the amount of the recombinant protein active in DNA binding, as only a minor fraction renatures successfully after extraction from bacteria (Zeller et al. 1995). A certain amount poly (dI:dC) can be used to replace the poly(dA:dT).

With Recombinant Protein and Its Antibodies

1. In a 1.7-ml micro-centrifuge tube add 3 µl buffer C, 1 µl FAM-labeled DNA (about 50–60 fmol), 1 µl rSpP3A2 (about 300 fmol), 1 µl 50 ng poly (dA:dT), 2 µl 5× binding buffer, 1 µl H_2O and 1 µl antibody (about 1 pmol).

2. Incubate the mixture on ice for 10–30 min before injection.

With Nuclear Extract

1. In a 1.7-ml micro-centrifuge tube add 3 µl buffer C, 1 µl FAM-labeled DNA (about 60 fmol), 1 µl sea urchin nuclear extract (about 5 µg), 1 µl 5 µg poly(dA:dT), 2 µl 5× binding buffer, and 2 µl H_2O.

2. Incubate the mixture on ice for 10–30 min before injection.

Extract from Multiple Eggs

1. Immediately before assay, the eggs (about 2 µl in a 1.7-ml microcentrifuge tube) were thawed at room temperature for 5 min to break the cell membrane.

2. Add 3 µl buffer C, 1 µl FAM-labeled DNA (about 60 fmol), 1 µl 5 µg poly(dA:dT), 2 µl 5× binding buffer and 1 µl H_2O to the tube.

3. Incubate the mixture on ice for 30 min before injection.

Reaction with a Single Egg

1. Place a single sea urchin egg in a drop of a 3:1 mixture of filtered sea water and buffer C.

2. Remove the cartridge from the CE machine and insert the inlet of the capillary into the drop.

3. Under the microscope, introduced the egg into the capillary by manually closing/opening the outlet of the capillary. Since the diameter of the egg, about 80 µm, is larger than the internal diameter of the capillary, 50 µm, the egg will be held at the inlet end of the capillary.

4. After replacing the cartridge back into the machine, the egg contents were taken up into the tube by a 10 s high pressure injection of the reaction mixture including 3 µl buffer C, 4 µl FAM-labeled DNA (about 0.24 pmol), 1 µl 5 µg poly(dA:dT) and 2 µl 5× binding buffer.

5. Let the mixture react at the inlet end for 10 min at room temperature.

6. After a second injection of CE buffer to ensure that no sample was lost during the reaction, electrophoresis was commenced at 18 kV.

Electrophoretic Mobility Shift Assay

For the conventional EMSA experiment, electrophoresis was carried out in an 8 % polyacrylamide gel in 1× TBE buffer for 2 h at 200 V which was prerun for 2 h at the same voltage. Probes were labeled with ^{32}P by kinase reaction. The gel was dried, placed under Kodak XAR5 film, and exposed for 4 h.

Capillary Electrophoresis

The capillary was filled with CE buffer. Electrophoresis was run at reversed polarity, i.e., the anode at the detector end, at 18 kV and 18 °C. Between each run, the capillary was rinsed with 1× TBE buffer for 2 min and then in CE buffer for 5 min. Therefore, a complete run lasts 17 min. The sample was introduced by high pressure injection (10 s injection corresponds to ∼10 nl sample) followed by a second injection of CE buffer for 5 s.

CE buffers with different polymer concentrations (range from 0 to 1 %) were investigated to determine the best experimental conditions for CE separation.

■ Results

EMSA and CEMSA with rSpP3A2 Protein

Wild-Type Probe   Figure 2 shows the experiments carried out with wild-type probe that had been reacted with renatured rSpP3A2 protein preparations, in the presence of a large excess of poly(dA:dT) (see reaction conditions and Fig. 2 legend for details). Migration of the probe alone is shown in Fig. 2A. The bimolecular complex formed at one of the two sites (peak 1)

Fig. 2A–H. Resolution of rSpP3A2-DNA complexes by CEMSA and EMSA. In CEMSA, the capillary was filled with CE buffer (1× TBE with 0.2 % linear polyacrylamide). Electrophoresis was run at reversed polarity, i.e., the anode at the detector end, at 18 kV and 18 °C. The sample was introduced by high pressure injection (10 s injection corresponds to ∼10 nl sample) followed by a second injection of CE buffer for 5 s. For the EMSA experiment, electrophoresis was carried out in an 8 % polyacrylamide gel in 1× TBE buffer for 2 h at 200 V which was prerun for 2 h at the same voltage. Probes were labeled with ^{32}P by kinase reaction. A CE electropherogram of

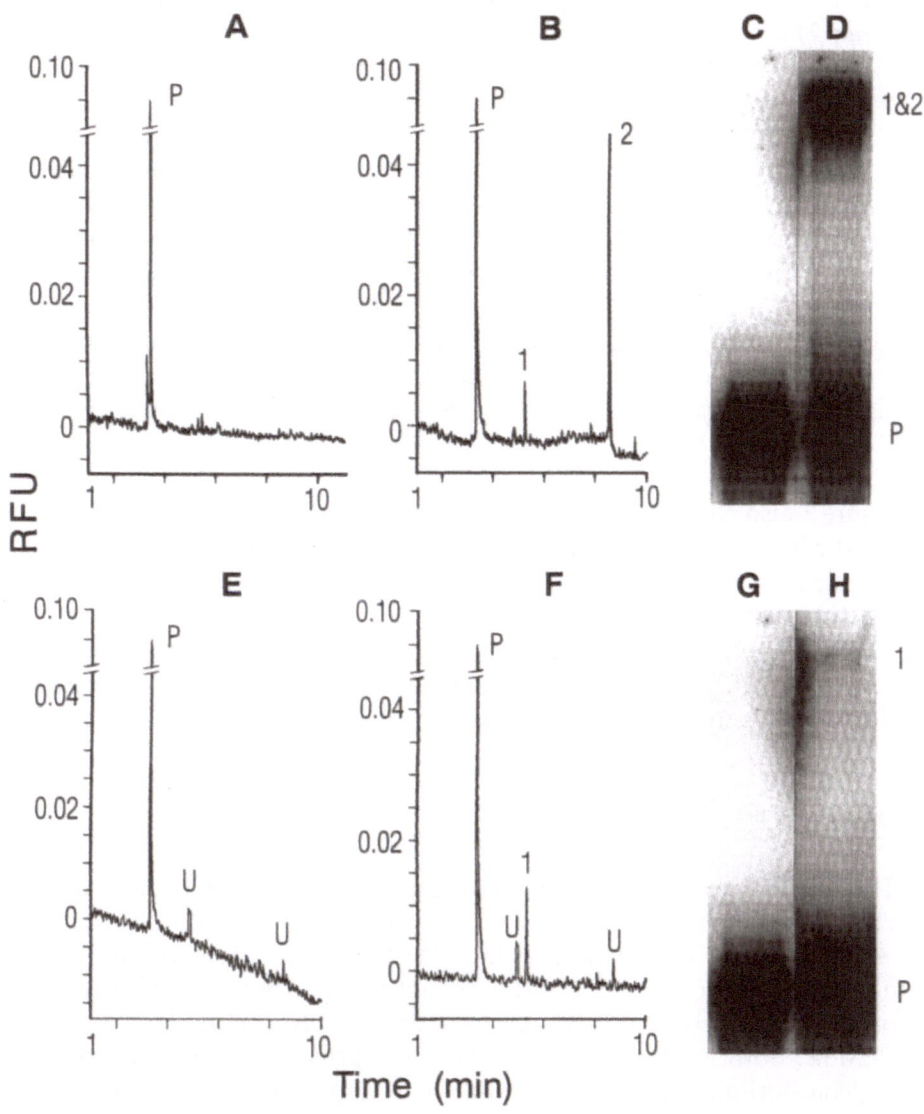

60 fmol wild-type probe; **B** CEMSA of 60 fmol wild-type probe with 300 fmol rSpP3A2; **C** gel electrophoresis of 0.5 ng wild-type probe labeled with ^{32}P; **D** EMSA of 0.5 ng wild-type probe with 15 ng rSpP3A2; **E** CE electropherogram of 50 fmol mutated probe; **F** CEMSA of 50 fmol mutated probe with 300 fmol rSpP3A2; **G** gel electrophoresis of 0.5 ng wild-type probe labeled with ^{32}P; **H** EMSA of 0.5 ng mutated probe with 15 ng rSpP3A2. *RFU* Relative fluorescence unit; *P* free probe; *U* background impurity from the probe; *1* and *2*, *1:1* and *2:1* protein-DNA complexes

and the trimolecular complex formed when both sites are occcupied (peak 2) are shown in Fig. 2B. Because the mobility of the protein-DNA complexes depends directly on their charge/mass ratio, with the field reversed (anode at detection end), the free DNA migrates faster than the protein-DNA complex, which is also shown in conventional EMSA (Fig. 2C, D). However, the bimolecular and trimolecular complexes are difficult to separate in EMSA as shown in Fig. 2D. In addition (not shown) we carried out competition experiments, in which addition of excess unlabeled wild-type probe in the presence of poly(dA:dT) quantitatively abolished the fluorescent peaks (1) and (2). The separation in CEMSA was completed in this case within 10 min (see abscissa), while the EMSA experiment was finished in about 6 h. The samples were loaded in 5–10 nl, and in Fig. 2A the peak shown represents only about 45 attomoles of probe.

Mutant Probe   Figure 2 also shows mutant probes that had been reacted with the renatured rSpP3A2 protein. Figure 2E shows the mutated probe alone, which contains some minor contaminates labeled (U), and Fig. 2F displays the reaction of this probe with rSpP3A2 protein. As expected, only the bimolecular complex is formed (peak 1), Fig. 2G,H show the same reaction in conventional EMSA, respectively. The bimolecular complex in EMSA shows a very light band (Fig. 2H).

EMSA and CEMSA with Nuclear Extract

The superior resolution available in the CEMSA system is illustrated in Fig. 3. These experiments were carried out with unfractionated nuclear extract. SpP3A2 is present in 24 h embryo nuclei at about 10^4 molecules/nucleus (Zeller et al. 1995b). The extract was reacted with the wild-type probe and the complexes were analyzed by conventional EMSA and by CEMSA. Both methods revealed the same protein-DNA complexes, but the conventional method took over 6 h, as opposed to 12 min, and in this case consumed about 1000 times more sample than did the CEMSA. Conventional EMSA using the nuclear extract does not resolve the SpP3A2 complexes clearly because of the presence in 24 h embryo extract of another DNA-binding factor, SpGCF1, which is relatively prevalent, and which interacts weakly at a CCCC site on the wild-type probe that we used (Zeller et al. 1995a). SpGCF1 complexes account for the broad set of bands that extend below the SpP3A2 complexes in the EMSA shown in lane I of Fig. 3A (compare the complexes formed with rSpP3A2

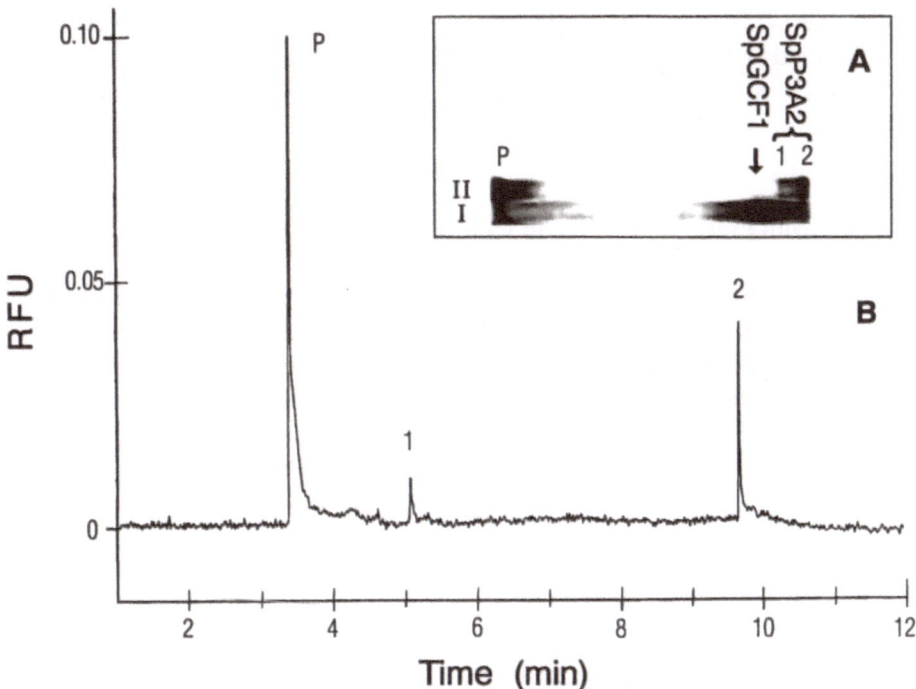

Fig. 3A, B. Identification of SpP3A2:DNA complexes in nuclear extract using conventional and capillary electrophoretic separation procedures. Separation conditions for the CEMSA experiment are as in Fig. 2. **A** Conventional EMSA; *lane I* 10 μg nuclear extract of 24 h sea urchin embryos reacted with 40 fmol wild-type probe together with 5 μg poly(dA:dT); *lane II* 300 fmol rSpP3A2 protein reacted with 40 fmol wild-type probe with 0.5 μg poly(dA:dT). **B** CEMSA of an aliquot of the 5 ng of 24 h embryo nuclear extract reacted with 60 attomole wild-type probe together with 5 ng poly(dA:dT). *RFU* Relative fluorescence unit; *P* free probe; *1* and *2*, *1:1* and *2:1* protein-DNA complexes of SpP3A2; DNA-protein complex of SpGCF1 is indicated by *arrow*

in lane II). In the absence of the EMSA caging effect these complexes do not survive in the CEMSA experiment with nuclear extract shown in Fig. 3B. The CEMSA clearly reveals the same two peaks seen in Fig. 2B. There is a small difference in the retention times of the complex peaks formed in the nuclear extract reaction, as shown in Fig. 3B, compared to those of the rSpP3A2-DNA complexes shown in Fig. 2B. This is not surprising, given the high protein content of the nuclear extract during the CE separation. Furthermore, in the CEMSA the bimolecular and the trimolecular complexes are widely separated, while in this particular EMSA system they are much more difficult to separate. The CEMSA peaks obtained in the nuclear extract are also efficiently competed by excess unlabeled wild-type probe (not shown), and the inhibiting monoclonal antibody again eliminates both complex peaks when added to the nuclear extract (not shown).

CEMSA with rSpP3A2 Protein and Its Monoclonal Antibody

Figure 4B shows that when the reaction mixture includes a monoclonal antibody against rSpP3A2 protein which specifically prevents protein-DNA complex formation, peaks 1 and 2 are absent as shown in Fig. 4A. Figure 4C shows that a different monoclonal antibody, which does not prevent but rather supershifts rSpP3A2-DNA complexes, generates a broad, more slowly moving peak (S), at the expense of peaks 1 and 2 (compare Fig. 4A).

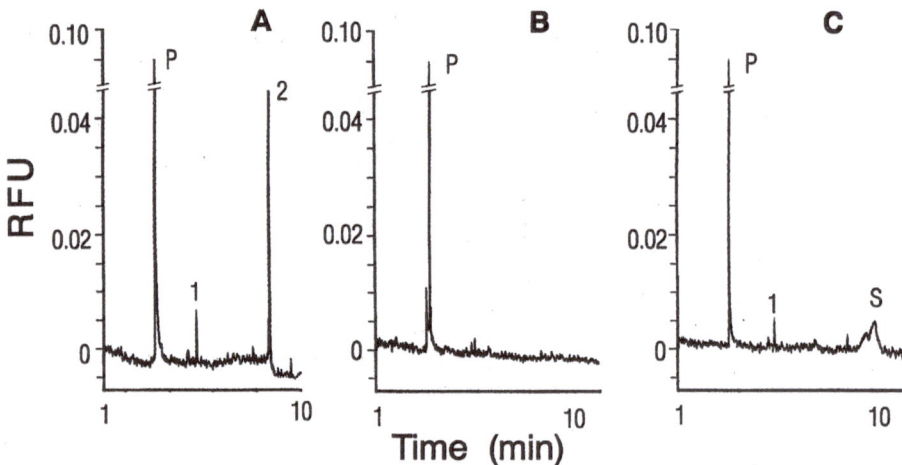

Fig. 4A–C. The function of monoclonal antibodies in CEMSA. The CE separation conditions were as in Fig. 2. **A** CEMSA of wild-type probe with rSpP3A2; **B** CEMSA in the presence of the monoclonal antibody 7B12/1H7, which inhibits the formation of SpP3A2-DNA complexes; **C** CEMSA carried out in the presence of 6F1/2E9, which supershifts the SpP3A2-DNA complex. *RFU* Relative fluorescence unit; *P* free probe; *1* and *2, 1:1* and *2:1* protein-DNA complexes; *S* DNA-protein-antibody complex

CEMSA with Cytoplasm from Multiple Eggs

SpP3A2 is also present in unfertilized egg cytoplasm, at about 2×10^6 molecules per egg (Zeller et al. 1995b). The egg contains in addition a second factor that interacts with the same target site as does SpP3A2, viz. a Zn finger protein, SpZ2-1 (SpP3A1) (this protein is also present in 24 h embryo nuclear extract, but only at very low concentration) (Höög et al. 1991). Figure 5A depicts a CEMSA of the complexes formed with a crude egg cytoplasmic extract. An extract of 1000 *S. purpuratus* eggs in 10 μl of medium was reacted with the wild-type probe, and 10 nl, or the amount equivalent to one egg, were injected into the CE machine. The results were somewhat different from those obtained with nuclear extract or recombinant protein. First, the free probe and the bimolecular protein-

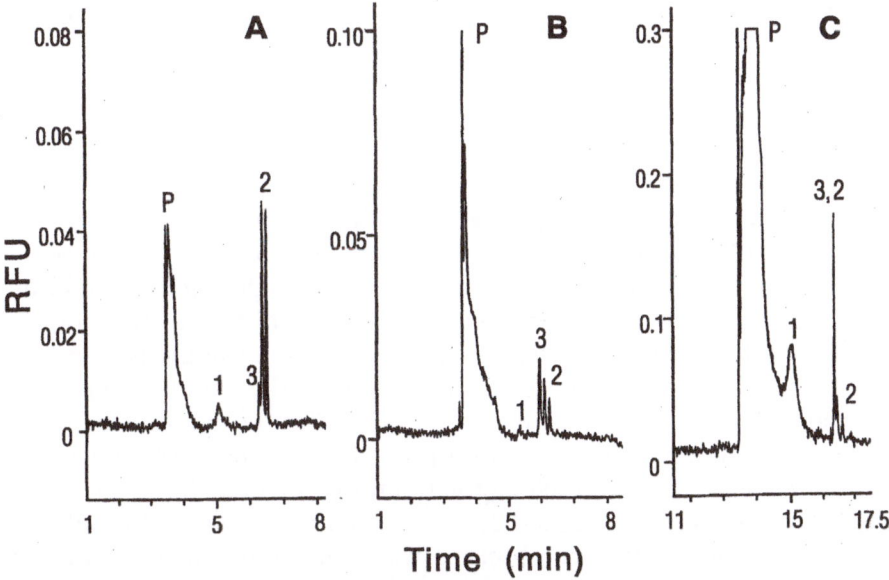

Fig. 5A–C. Complexes formed with sea urchin egg cytoplasm, detected by CEMSA. The CEMSA separation conditions were as in Fig. 2. **A** CE electropherogram of sea urchin egg cytoplasm with wild-type probe; **B** complexes formed in the presence of the monoclonal antibody 7B12/1H7, which inhibits the formation of SpP3A2-DNA complexes; **C** CEMSA carried out on the contents of a single sea urchin egg. The mobility of the complexes are greater in this experiment than with nuclear extract and pure rSpP3A2. We believe this is a consequence of the increase in overall ion content in the sample from the lysed cell contents. *RFU* Relative fluorescence unit; *P* free probe; *1* and *2*, *1:1* and *2:1*, protein-DNA complexes of wild-type probe SpP3A2; *3* protein-DNA complex of wild-type probe SpP3A1

DNA complex peaks became broader, which we think is probably due to the presence of nucleic acids and cell membrane fragments in the crude extract. Second, a triple peak migrated out at 7 min, which is different from the single 2:1 protein-DNA complex peak, as shown in Fig. 2B and Fig. 3B (i.e., peak 2 in these figures). All three peaks of the triplet can be competed by excess wild-type probe DNA in the presence of the non-specific poly(dA:dT) competitor, demonstrating that they are specific protein-DNA complexes. In Fig. 5B the complexes formed in the presence of the inhibitory monoclonal antibody are shown. Addition of the antibody severely depressed the formation of the two peaks labeled 2, while the free probe correspondingly increased. However, peak 3 was not altered (compare Fig. 5A). Peak 3 is probably due to the SpZ2-1, while the twin peaks 2 are due to SpP3A2. However, we do not know why two closely migrating SpP3A2 peaks, rather than one peak, form with the egg cytoplasmic extract. This could of course be of biological interest if it devolves from a covalent difference in a maternal fraction of the SpP3A2 factor.

CEMSA from a Single Cell

Having achieved a detection capability sufficient for assay of SpP3A2 activity from the equivalent of one egg, we undertook to measure complexes formed by the molecules present within a single egg that bind this probe specifically. The egg was lysed at the mouth of the electrophoresis capillary by the external pressure applied when the sample is injected, and the contents of the egg were mixed within the capillary with the probe and other constituents. Figure 5C illustrates the result, which was reproducibly obtained. The retention times for all the peaks are similar to those observed in the 1000 egg extract, except that the SpZ2-1 and SpP3A2 peaks (i.e., peaks 2 and 3) comigrated as one peak. In multiple analyses, the only difference observed was that the quantity of SpP3A2 was slightly different from egg to egg.

Quantitation

Since the binding of protein with the DNA probe does not affect the quantum yield of the fluorescent dye in this assay, it is possible to obtain the DNA and protein quantities in the CEMSA directly from the peak areas, and these can be used for any quantitative measurement that can

Table 1. Sensitivity of capillary electrophoresis mobility shift assay

	Molecules DNA probe used per run	Molecules SpP3A2 detected	Number of S. purpuratus egg equivalents
EMSA[a]	$\sim 10^{10}$	$\sim 5 \times 10^9$	2300
CEMSA rSp3A2[b]	3.6×10^7	1.68×10^7	9
Nuclear extract[c]	3.6×10^7	1.1×10^7	9
Egg cytoplasm extract[d]	3.6×10^7	2.5×10^6	1
Single egg[e]	1.44×10^8	2×10^6	1

[a] From Fig. 3A.
[b] From Fig. 2B.
[c] From Fig. 3B.
[d] From Fig. 5A.
[e] From Fig. 5C.

be carried out with EMSA. Table 1 summarizes the salient quantitative aspects of these experiments. We utilized the data from Fig. 2 to calculate an equilibrium dissociation complex (k_D) for the weak and strong sites on the probe viz 1.9×10^{-7} M and 3.9×10^{-8} M, respectively. These values are within the range of equilibrium constants described by Calzone et al. (1991).

Note: In the experiment of Fig. 2B, the probe (D_0) was present at 6×10^{-9} M; the trimolecular complexes of peak 1 (P_1D) at 1.45×10^{-10} M; the bimolecular complex of peak 2 (P_2D) at 1.35×10^{-9} M, and renatured rSpP3A2 (P) was present at 3×10^{-8} M. In the experiment of Fig. 2F, the bimolecular complex formed at the weak site was present at 2.85×10^{-10} M, and total probe and rSpP3A2 were as in the experiment in Fig. 2B. From the experiment of Fig. 2F, for the weak site $k_{wD} = 1.9 \times 10^{-7}$ M, given that for the nonspecific reaction of rSpP3A2 with the poly(dA:dT) present, $k_D = 1.8 \times 10^{-8}$ M [almost 20 % of the protein is engaged in the complex with the poly(dA:dT)]. From

$$\frac{k_wP + k_sP + k_wk_sP^2}{1 + k_wP + k_sP + k_wk_sP^2} \text{, where } Y = \frac{P_1D + P_2D}{D_0},$$

and $k_w = k_{wD}^{-1}$, $k_s = k_{sD}^{-1}$, $k_{sD} = 3.9 \times 10^{-8}$ M.)

▨ Troubleshooting

● Temperature

Different running temperatures from 18 to 25 °C have been tested. Our results indicated that no significant difference was observed, though the complex is more stable at lower temperature. By adjusting the temperature for protein-DNA formation, we might reduce some nonspecific complexes.

● Buffer Additive

The addition of linear polyacrylamide to the running buffer can prevent the interaction of proteins with the capillary wall. However, its effect on the stability of the complex and any separation between free probe and complex requires further investigation. Our studies show that the addition of linear polyacrylamide can enhance the separation between free probe and the protein-DNA complex due to its sieving effect. Other linear polymers, such as hydroxyethylcellulose, methylcellulose, and polyethylene glycol, can be used in this experiment also. The amount and the length of linear polymer added is determined by the strength of complex formation, the size of protein, and the number of specific binding proteins present in the mixture. Linear polymer might interact with the target protein, and this possibility should be tested before selecting the polymer.

● Complex Formation

The most difficult part of this experiment is to maintain specific protein-DNA complexes and destroy those undesired ones. There are several options to reach this goal. Increasing the temperature or the ionic strength for the binding reaction or increasing CE separation temperature will usually reduce the weak complexes. Use of a smaller polymer or increase in its concentration can exclude the larger size proteins with weak bindings. Different nonspecific DNA, such as poly(dA:dT), poly(dI:dC), poly(dI):poly(dC), and so on, can be used at different concentrations for different complexes.

◾ Applications

CEMSA of TFs in Dissected Population of Cells and Small Samples

The capillary electrophoresis system with laser-induced fluorescence is at least 100 times more sensitive than conventional EMSA utilizing radioactive probes. As Fig. 5C shows dramatically, we can now see protein-DNA interactions from a single egg cell, and no more than about 10^6 molecules of a given TF would be required for quantitative measurement by CEMSA, so long as the relative equilibrium constant for the reaction with its DNA target site is not greatly lower than that of SpP3A2. That is since reactions at the weak SpP3A2 site alone can be detected (Fig. 2F), we should expect that factors displaying a $k_d \leq 4.7 \times 10^{-7}$ M, a relatively modest value, or $k_r s \geq 2 \times 10^4$ M, are detectable by CEMSA. Furthermore, the limits of detection of this method are well below the actual quantities listed in Table 1. This means that such assays can now be performed comfortably on extracts from 10^2 to 10^3 ordinary somatic cells. Perhaps, surprisingly, even very crude extracts such as the whole lysed eggs used in the experiments of Fig. 5 work well. We expect that this method will enable many new experimental explorations of TF activity, in specific cell types or embryonic regions that can be separated out only by hand dissection, or in cell populations separated by fluorescence-activated cell sorting.

Mapping of Transcription Factor Localization in Embryo

Transcription factors are the molecular switches and their concentrations might correlate to the gene expression which they regulate. Their localization in embryos will provide us with a comprehensive understanding of gene expression, cell differentiation, and embryo development at the molecular level. As CEMSA requires a very small amount of sample, we can collect enough material by dissecting a few embryos to do the experiment. The ultimate goal of this method is to map all the TFs in each nuclei of the embryo.

Investigation of Novel DNA Binding Agents

Peptide analogs containing a single imidazole or pyridine ring (ImPImP) which bind in the minor groove of DNA as antiparallel, side-by-side dimers afford a new class of oligopeptides with enhanced sequence-specificity for DNA (Mrksich et al. 1994; Geierstanger et al. 1994). The most striking feature of the ImPImP complex is its sequence specificity. However, the binding of these peptides is different from the interactions between native TFs and DNA which bind mostly in the major groove of DNA. Will these types of compound become potential drugs for gene regulation/expression?

Our experimental results have indicated that CEMSA offers a fast, sensitive tool to investigate which a type of peptide might be useful as a gene regulation reagent.

Combinatorial Library Screening

A recent trend in medicinal chemistry includes the production of vast numbers of compounds of diverse type referred to as combinatorial libraries. The primary objective of producing small-molecule libraries is to provide collections of compounds suitable for both drug-discovery screening and drug-development optimization. The combinatorial technology of synthesis can dramatically accelerate the development of potent compounds for a wide variety of biological targets. It is essential to develop analytical methodology capable of screening compounds from this synthesis in a rapid and sensitive way.

Our results showed that CEMSA can be employed as such a screening tool to determining the potential reagents that regulate gene expression from combinatorial libraries. Each screening process takes only 10 min with very small amounts of sample applied.

Acknowledgement. Both authors performed the experiments for this Chapter while at the California Institute of Technology, and are grateful for working with Eric H. Davidson at that time.

References

Calzone FJ, Thézé N, Thiebaud P, Hill RL, Britten RJ, Davidson EH (1988) Developmental appearance of factors that bind specifically to *cis*-regulatory sequences of a gene expressed in the sea-urchin embryo. Genes and Dev. 2:1074–1088

Calzone FJ, Höög C, Teplow DB, Cutting AE, Zeller RW, Britten RJ, Davidson EH (1991) Gene regulatory factors of the sea-urchin embryo 1. Purification by affinity-chromatography and cloning of P3A2, a novel DNA-binding protein. Development 112:335–350

Fried MG, Crothers M (1981) Equilibria and kinetics of *lac* repressor-operator interactions by polyacrylamide gel electrophoresis. Nucleic Acids Res 9:6505–6525

Geierstanger BH, Mrksich M, Dervan PB, Wemmer DE (1994) Design of a G-C-specific DNA minor groove-binding peptide. Science 266:646–650

Heegaard NHH, Robey FA (1993) Use of capillary zone electrophoresis for the analysis of DNA-binding to a peptide derived from amyloid P component. J Liquid Chromatog 16:1923–1939

Höög C, Calzone FJ, Cutting AE, Britten RJ, Davidson EH (1991) Gene regulatory factors of the sea-urchin embryo 2. Two dissimilar proteins, P3A2 and P3A1, bind to the same target sites that are required for early territorial gene-expression. Development 112:351–364

Hough-Evans BR, Franks RR, Zeller RW, Britten RJ, Davidson EH (1990) Negative spectial regulation of the lineage specific Cy*III*a actin gene in the sea-urchin embryo. Development 110:41–50

Hudson JM, Fried MG (1990) Cooperative interactions between the catabolite gene activator protein and the *lac* repressor at the lactose promoter. J Mol Biol 214:381–396

Maschke HE, Frenz J, Williams M, Hancock WS (1993) Investigation of protein-DAN-interactions by mobility shift assays in capillary electrophoresis. Poster T121 at the Fifth International Symposium on HPCE. Orlando, Florida, Jan 25–28

Mrksich M, Parks ME, Dervan PB (1994) Hairpin peptide motif. A new class of oligopeptides for sequence-specific recognition in the minor groove of double-helical DNA. J Am Chem Soc 116:7983–7988

Warren WJ, Vella G (1994) Analysis and purification of synthetic oligonucleotides by high performance liquid chromatography. In: Agrawal S (ed) Protocols for oligonucleotide conjugates. Humana Press, Totowa, New Jersey

Xian J (1994) Application of capillary electrophoresis for the separation of biomolecules. PhD Thesis, Wichita State University

Xian J, Harrington MG, Davidson EH (1996) DNA-protein binding assays from a single sea urchin egg: A high-sensitivity capillary electrophoresis method. Proc Natl Acad Sci USA 93:86–90

Zeller RW, Coffman JA, Harrington MG, Britten RJ, Davidson EH (1995a) SpGCF1, a sea-urchin embryo DNA-binding protein, exists as 5 nested variants encoded by a single messenger-RNA. Dev Biol 169:713–727

Zeller RW, Britten RJ, Davidson EH (1995b) Developmental utilization of SpP3A1 and SpP3A2-two proteins which recognize the same DNA target site in several sea-urchin gene regulatory regions. Dev Biol 170:75–82

Chapter 12

Polyacrylamide Gel Electrophoresis of DNA/Protein Complexes

ANDREAS KYAS, WINFRIED MÄUELER, JOERG T. EPPLEN

Introduction

Gel mobility-shift assays represent a versatile methodology to study details of the interaction of DNA and proteins. Various experimental procedures allow the study of topics such as binding conditions, thermal stability of observed complexes and the affinity or specificity of the DNA/protein interaction for a given DNA sequence. In recent years we have concentrated our efforts on nuclear protein binding to simple repetitive DNA sequences (Mäueler et al. 1994; Epplen et al. 1996). These impressively regular elements are ubiquitously interspersed in eukaryotic (and some prokaryotic) genomes. The present chapter does not review in detail the broad range of techniques used in gel mobility-shift assays. Rather we present a reliable protocol, which has been proven to generate absolutely reproducible results. The technique described is designed for crude nuclear extracts, but it can also easily be modified for examination of purified proteins.

12.1
Preparation of DNA

Preparation of Target DNA

The basic concept of the gel mobility-shift assay is the differential electrophoretic behaviour of free DNA and DNA complexed with protein(s). Therefore it is necessary to ensure the preparation of high quality target

Andreas Kyas, Winfried Mäueler, Joerg T. Epplen, Molecular Human Genetics, Ruhr-University, 44780 Bochum, Germany
Correspondence to Joerg T. Epplen: *phone* +49-234-700-3839; *fax* +49-234-709-4196; *e-mail* joerg.t.epplen@ruhr-uni-bochum.de

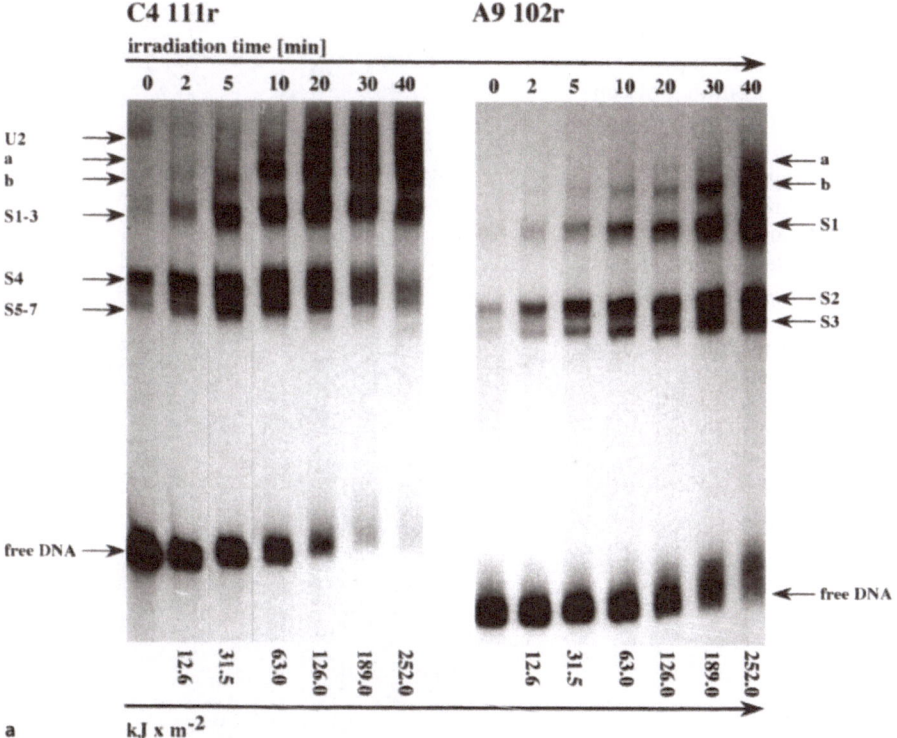

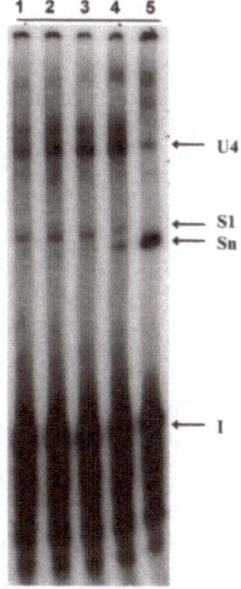

Fig. 1A, B. Effect of UV light and ethidium bromide on the binding of protein to DNA. The target sequences were incubated with 5 mg of HeLa nuclear extract according to the standard procedure (Mäueler et al. 1992a) after exposure of the DNA to increasing amounts UV light (312 nm) (**A**) and ethidium bromide (**B**). Novel retarded bands appeared within 2 min of UV irradiation (12.6 kJ/m^2). At a level of 50 ng/ml of ethidium bromide a novel retarded band "Sn" is visible whereas the intensity of the major retarded band S1 is decreased. Approximately 2000-fold higher ethidium bromide concentrations are recommended in most DNA preparation protocols rendering any subsequent protein binding experiments prone to artifacts

Fig. 2. The effect of strand breaks in DNA as revealed by gel mobility-shift experiments. *Right* Protein binding of control DNAs; these preparations do not harbor nicks. *Left* Protein binding to nicked target DNA exhibiting additional retarded bands (Sni). *Arrowheads* Ladder (lni) of shorter, single-stranded and longer aggregated, double-stranded target DNA fragments in *lanes 1, 2, 3*

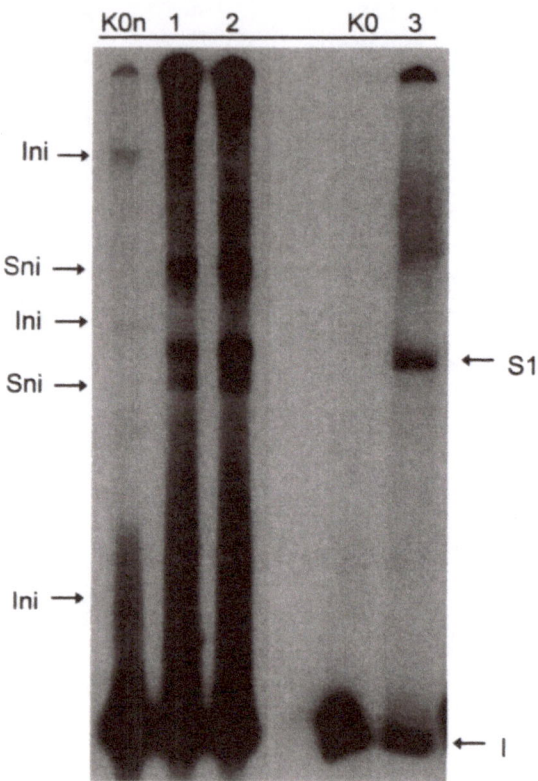

DNA. Many protocols for DNA preparation involve steps seriously affecting the electrophoretic behaviour as well as the protein binding capabilities of the target DNA. This is especially a consequence of the use of ultraviolet (UV) light and ethidium bromide during the isolation and preparation procedures. UV light causes damage to DNA by generating photodimers of neighbouring pyrimidines, especially neighbouring thymines (see Matsunaga et al. 1991 and references cited therein). These photodimers are targets for DNA repair proteins and alter the structure of the DNA molecule (Galloway et al. 1994). Ethidium bromide intercalates into the DNA molecule and thus inflicts minor alterations into the helical structure, which can also become a target for unspecific DNA/protein interaction. Figure 1A depicts alterations of the electrophoretic behaviour of DNA after exposure to UV light. Figure 1B depicts additional DNA/protein complexes produced by residual ethidium bromide intercalated into the double strand of the target DNA. Another potential hazard, which is not directly obvious when performing gel mobility-shift assays, is the damaging of DNA during plasmid preparation. This damage is partially due to residual nucleases which have to be eliminated as early as possible during the preparation. Without proper removal of such nucleases in due time the plasmid DNA is nicked to an unforeseeable extent. Several pitfalls in the subsequent preparation of target DNA can emerge from this ill effect. In particular, radiolabelling of nicked DNA using the Klenow polymerase can cause problems, since at these nicked sites additional labelled nucleotides are incorporated into the target molecules. Nicked DNA leads to the formation of populations of DNA molecules of different length and reduced/altered protein-binding capabilities (Fig. 2). In a similar way, polymerase chain reaction (PCR)-generated target DNA is not suited for gel mobility-shift assays (Mäueler et al. 1992a). Therefore we established protocols for both DNA preparation and recovery from preparative gels that circumvent the use of ethidium bromide and UV light and minimize the risk of nuclease damage.

Preparation of Plasmid DNA and Restriction Fragments

This protocol is designed for large scale preparations of plasmid DNA from *Escherichia coli*. In principle it represents a modification of the original Birnboim-Doley method (Sambrook et al. 1989) including additional precautions.

First the cells are harvested and residual growth medium is removed by washing. After resuspension in the presence of RNase the cells are subjected to alkaline lysis. Most of the cellular proteins and genomic DNA are removed after neutralization by centrifugation and filtration. The clear filtrate, containing plasmid DNA and soluble proteins, is subjected to isopropanol precipitation. After centrifugation the pelleted plasmid DNA is redissolved and subjected to phenol/chloroform extraction. This step has shown to be crucial in the final removal of contaminating nucleases. After a final precipitation followed by two successive ethanol washes this procedure yields 2–6 mg/l overnight culture of plasmid DNA depending on the vector type. Plasmid DNA prepared by this technique has been shown to be essentially free of nicks as evidenced by denaturing PAGE and by electron microscopic monitoring (H.-G. Keyl, pers. comm.).

Materials

– STE buffer: 0.1 M NaCl; 10 mM Tris/HCl, pH 8.0; 1 mM EDTA, pH 8.0
– P1 buffer: 50 mM Tris/HCl, pH 8.0; 50 mM EDTA, pH 8.0; 400 µg/ml RNase A. Always prepare fresh.
– P2 buffer: 0.2 N NaOH; 1 % SDS. Always prepare fresh.
– 7.5 M NH$_4$Ac, pH 7.6: Calibrate with glacial acid and store at –20 °C.
– 2 M H$_4$Ac, pH 7.4: Calibrate with acetic acid.

Procedure

Preparation of Plasmid DNA from *E. coli*

1. Grow cells in 500 ml Luria broth overnight at 37 °C.

2. Harvest cells by centrifugation at 4000 rpm for 15 min at 4 °C (e.g. 500 ml centrifugation bottles, Sorvall GS3); discard supernatant.

3. Redissolve the pelleted cells in 100 ml STE buffer, centrifuge at 4000 rpm for 15 min at 4 °C; discard supernatant.

4. Resuspend cells in 30 ml P1 buffer. Incubate for 5 min at room temperature.

5. Add 40 ml buffer P2, mix carefully by inverting four or five times; incubate on ice for 5 min.

6. Add 30 ml cold 7.5 M NH$_4$Ac (–20 °C). Mix carefully by inverting for four to five times and incubate on ice for 5 min.

7. Centrifuge for 15 min at 4000 rpm and 4 °C. Remove precipitate by filtering through sterilized gauze filter (4 °C).

8. Transfer filtrate to new centrifugation tubes (Sorvall GS3 500 ml tubes or equivalent) and add 3/5 volume isopropanol. Carefully invert tubes two or three times and incubate for 10 min at room temperature. Centrifuge for 15 min at 5000 rpm; discard supernatant.

9. Redissolve pellet in 10 ml 2 M NH$_4$Ac, incubate on ice for 5 min and transfer to 50 ml screw-cap tubes (Falcon).

10. Extract twice with 5 ml phenol/chloroform, once with 5 ml chloroform and transfer to new centrifugation tubes (30 ml Sorvall SS 34 or equivalent).

11. Add 10 ml isopropanol, mix and incubate for 10 min at room temperature.

12. Centrifuge for 10 min at 14,000 g at room temperature; discard supernatant.

13. Wash pellets twice with 5 ml 80 % ethanol (–20 °C); air dry pellets and dissolve in adequate volume of TE buffer (1–2 ml).

Restriction fragments of plasmid DNA are produced according to standard procedures. Double digests yield less optimal results. Incubation buffers should be used according to the manufacturer's recommendations. Enzyme amounts should be calibrated properly according to the amount of target DNA in order to avoid partial and "over" digestions (contaminating nuclease effects in practically all commercial restriction enzyme preparations). A critical step can be the recovery of restriction fragments from preparative gels. As mentioned above, direct visualization of the DNA bands during the preparation by UV irradiation must be prevented. Therefore restriction fragments are separated on agarose or polyacrylamide (PAA) gels appropriate for the required range of separation. Standard protocols for the recovery of intermediate-sized fragments from agarose gels can usually be used without further problems, assuming that the use of UV light/ethidium bromide is avoided. Good recovery rates were achieved using ion exchange chromatography after disrupting the agarose gel in the presence of a chaotrope (Mäueler et al., unpublished data).

When separating restriction fragments on PAA gels, it is necessary to use PAA lots of the utmost purity (three times recrystallized). Otherwise recovery rates may drop dramatically. PAA gels are stained in methylene blue solution at a concentration of 0.04 % (w/vol) or less. After destaining in electrophoresis buffer, the restriction fragments become visible as faint bands, which can easily be excised from the gel. The gel slices are ruptured by centrifugation through a punctured 1.5 ml Eppendorf vessel into a 2 ml reaction tube. Best recovery yields from PAA gels has been achieved by three consecutive elutions of the homogenized gel slices with TE buffer over a total period of 1.5 days. For each elution step the homogenized gel slices are covered with TE buffer to a total volume of 2 ml. This elution takes 12 h. After changing the buffer, the following steps take 6 h each. Elutions are performed by steady gentle rocking movements in the cold room (4 °C). The recovered buffer is lyophilized until a final volume of about 200 µl is reached. This highly concentrated solution is extracted once with 0.5 vol phenol/chloroform and once with

0.5 vol chloroform. Here remaining pieces of PAA as well as the methylene blue stain are removed. The extracted solution is precipitated with ethanol, centrifuged and the resulting pellet is washed with cold 80 % ethanol. It is necessary to submit the restriction fragments to a final purification using ion exchange chromatography (e.g. Qiaex II, Qiagen, Hilden, Germany). Omission of this step results in reduced protein binding capacities of the DNA fragments prepared.

This procedure has been thoroughly tested for preparing restriction fragments in the range of 100–440 base pairs (bp). The protocol avoids the use of ethidium bromide and UV light. Fragments prepared according this method have been used successfully in gel mobility-shift assays as well as in DNase I-footprinting assays and for determining alternative non-B-DNA conformations by chemical modification (Mäueler et al., in press).

Radiolabelling of Target DNA

Target DNA is most conveniently end-labelled by the Klenow fragment of DNA-polymerase I following standard procedures (Sambrook et al. 1989). After phenol/chloroform extraction and ethanol precipitation most of the nonincorporated nucleotides are removed in the precipitation step. In order to remove residual nucleotides an additional gel filtration is performed. To this effect small columns of Sephadex G-100 are prepared.

1. Suspend Sephadex G-100 in TE buffer. In order to prevent bacterial growth NaN_3 is added to a final concentration of 0.04 % (w/vol).

2. Plug a small piece of glass wool into a short 250 mm Pasteur pipette. Pack the pipette with the Sephadex G-100 suspension. Formation of air bubbles should be avoided. The column should be prerun using 2–3 ml TE buffer in order to remove NaN_3.

3. Dissolve the pelleted labelled target DNA in 15 µl TE buffer. Vortex for 30 s to ensure complete solution. Spin shortly and transfer the solution to the column. Let the DNA solution enter the column completely.

4. Then add carefully 200 µl of TE without disturbing the column bed. After 200 µl of buffer have entered the column, a reservoir consisting of a 5 ml disposal syringe connected to the column by a small piece of tubing is put on top of the column. Carefully fill the reservoir with 2 ml TE buffer without swirling up the column material.

5. Monitor the progress of the radioactive front using a hand monitor. When the front nearly reaches the glass wool plug, start collecting fractions, each containing two drops of eluate. Collect 14 fractions.

6. Determine the amount of radioactivity within each fraction by sampling 2 µl of each fraction using a scintillation counter. The first fractions contain labelled target DNA. The latter fractions contain not incorporated nucleotides. Collect fractions containing target DNA.

Results

It is necessary to calculate the volume activity and the specific activity achieved in a labelling reaction. The determination of the volume activity is important for the calculation of the amount of probe to be employed. The specific activity of a probe allows determination of the target DNA concentration relative to an absolute amount of radioactivity.

Calculation

- Determine the total volume of the combined fractions. Calculate the average fraction volume by dividing the total volume by the number of fractions collected.

- Use Eq. (I) to determine volume activity:

$$\text{(I)} \quad \text{volume activity} = \frac{\sum\limits_{\text{fractions}} \text{cpm}}{No.\ (fractions) \cdot 2\ \mu l}$$

- Use Eq. (II) to determine the specific activity:

$$\text{(II)} \quad \text{specific activity} = \frac{\sum\limits_{\text{peak}} \text{cpm} \cdot \overline{V}_f}{\text{pmoles}_{\text{used}} \cdot 2}$$

Note that in Eq. (I) only the cpm of the collected fractions is entered into the calculation whereas in Eq. (II) the cpm of the complete target DNA peak is used. The specific activity can be defined as the amount of radioactivity (measured in cpm) incorporated into 1 pmol target DNA.

12.2
Gel Mobility-Shift Assay

As mentioned above the gel mobility-shift assay is a useful and versatile technique to study different aspects of DNA/protein interactions. The necessary equipment is present in every standard molecular biology laboratory. The basic experiment can be carried out in a single day. This facile method is attractive because of the comparatively little effort in time and expense. An especially interesting feature is the high sensitivity, which extends to a range of atomolar concentrations.

Rational of Gel Mobility-Shift Assay

Gel mobility-shift assays make use of the differential electrophoretic behaviour of free and protein-complexed DNA in nondenaturing PAA gels. The starting point of a series of gel mobility-shift assays is usually a DNA sequence which is to be examined with respect to its protein binding capabilities. Do proteins bind to a given sequence, and if so, do they bind with a certain affinity and specificity or does binding occur at random?

The first experiment to answer these questions includes the quantification of the amount of crude nuclear protein extract necessary to generate retarded bands. Simultaneously the optimal level of unspecific competitor DNA (e.g. pdIdC) relative to the amount of protein is determined in order to avoid unspecific protein binding. This step usually gives an indication whether a DNA fragment binds nuclear protein(s) or not. If retarded bands are observed, it is moreover possible to estimate from the results the minimal amounts of nuclear extract and unspecific competitor DNA needed to generate distinct retarded bands.

The following experiments allow definition of optimal binding conditions. Protein binding to DNA is strongly influenced by the ionic milieu. Some proteins require the mediation of specific ionic ligands. In the last decade it has been shown that certain DNA elements undergo a number of conformational changes in the presence of bivalent ions such as Mg^{2+} or Zn^{2+}. This effect can give rise to new protein binding targets (for reviews see Wells et al. 1988; Palecek et al. 1992). The influence of NaCl and/or KCl as well as the influence of the most potent structure-inducing salts, $MgCl_2$ and $ZnCl_2$, should therefore be examined. The intensity of retarded bands can be markedly enhanced under certain ionic conditions. In additional experiments the optimal incubation time and incubation temperature are determined. This is done by incubating the target DNA with protein under optimal binding conditions for times ranging from 2 to 90 min at both room temperature and at 4 °C.

Further very important characteristics are the affinity and sequence specificity of the retarded bands. These data are revealed by competition experiments. Defined amounts of labelled target DNA are incubated with protein and synthetic competitor DNA under previously determined optimal binding conditions in the presence of increasing amounts of unlabelled competitor DNA. To work out an equilibrium constant K_{eq} for the complex-forming DNA/protein interaction, unlabelled target DNA is employed as a specific competitor. With increasing amounts of unlabelled specific competitor, the intensity of the retarded bands will decrease. Figure 3 depicts a typical specific competition experiment. The equilibrium constant of every individual complex band is determined according to Scatchard (1949). The sequence specificity of individual retarded

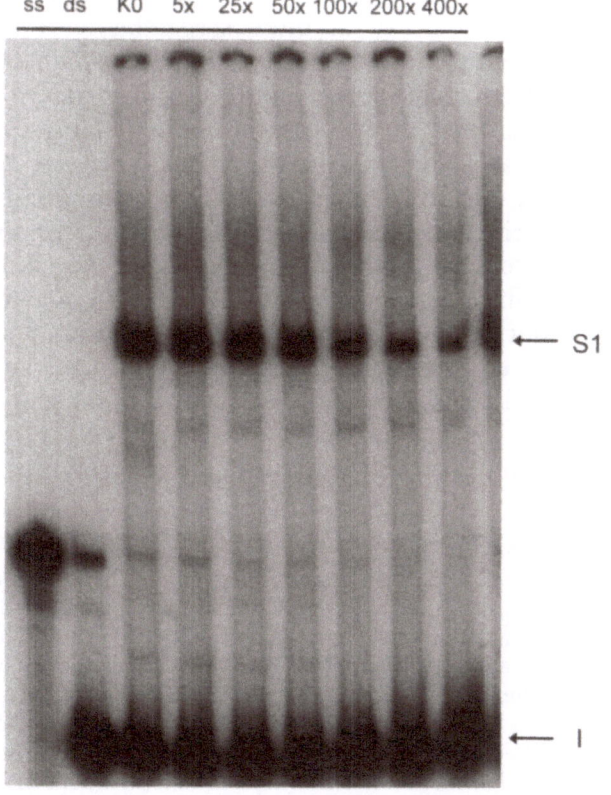

Fig. 3. A representative protein binding competition experiment demonstrating specific DNA/protein interaction. The target DNA (A9 138r), isolated from a human *HLA-DRB4*0101* gene harboring a $(gt)_{22}(ga)_{15}$ simple repeat, was incubated with increasing amounts of unlabelled A9 138r DNA (in a range from 5- to 400-fold molar excess). *Arrow* The main retarded band on the right-hand side. *ss* single-stranded control; *ds* double-stranded control target; *KO* control without unlabelled competitor; *I* free insert band; *S1* major retarded band

bands is determined by competition experiments with increasing amounts of unlabelled competitor DNA consisting of unrelated sequences. Note the difference between the synthetic, unspecific competitor pdIdC and the unspecific competitor with a sequence unrelated to the target DNA. DNA binding proteins exhibit a general DNA binding capacity apart from their binding to a distinct DNA sequence or structural element. This unspecific DNA protein interaction is competed and thus reduced in the experiments by pdIdC. The competitor DNA with an unrelated sequence may exhibit specific protein binding itself. It is used to examine whether target DNA and unspecific competitor DNA bind to the same nuclear protein(s). If the intensity of the retarded band in question is in fact diminished, this means that the complex-forming protein binds any DNA fragment with the affinity determined in the preceding specific competition experiment, yet without sequence specificity.

For the interpretation of data generated by gel mobility shifting it is important to take into consideration that only the complete set of experiments described above allows identification of the major retarded band(s) that represent complexes of affinity and sequence specificity. Probably in all of these cases the patterns of retarded bands vary immensely depending on the experimental conditions employed. As indicated above, the aim of the optimization of binding conditions concerns the sharpness and intensity of the high affinity and specificity bands. Sharp and intense bands, however, do not mean by themselves that they are indeed bands representing complexes of high affinity or specificity. By varying the incubation conditions the unspecific DNA/protein interaction can be enhanced as well. In Fig. 4 a striking example of this effect is depicted. C4 111r DNA, a genomic 111 bp fragment harbouring a $(gaa)_{24}$ simple repeat, was examined for protein binding. In the course of establishing optimal binding conditions, a certain retarded band, Sn, showed distinct maxima of intensity for KCl, $ZnCl_2$ and $MgCl_2$. As a consequence it was attempted to even further enhance the intensity of Sn using combinations of the electrolytes. As shown in Fig. 4 A, and B

Fig. 4A–C. The effects are shown of $ZnCl_2$ and KCl in different concentrations on DNA/protein interaction of the genomic DNA fragment C4 111r. **A** Autoradiogram: *K0* control under standard conditions without further addition of salt; *K1* control with an additional concentration of 250 mM KCl; *K2* control with an additional concentration of 2.5 mM $ZnCl_2$. All other combinations of salts are indicated on the top of the lanes. *I* Free insert DNA; *Sn, U2* the main DNA/protein complexes of the fragment C4 111r. **B** The intensity of the retarded band Sn relative to the control K0 is plotted vs the concentrations of KCl and $ZnCl_2$. **C** The intensity of the retarded band U2 relative to K0 is plotted vs the concentrations of KCl and $ZnCl_2$

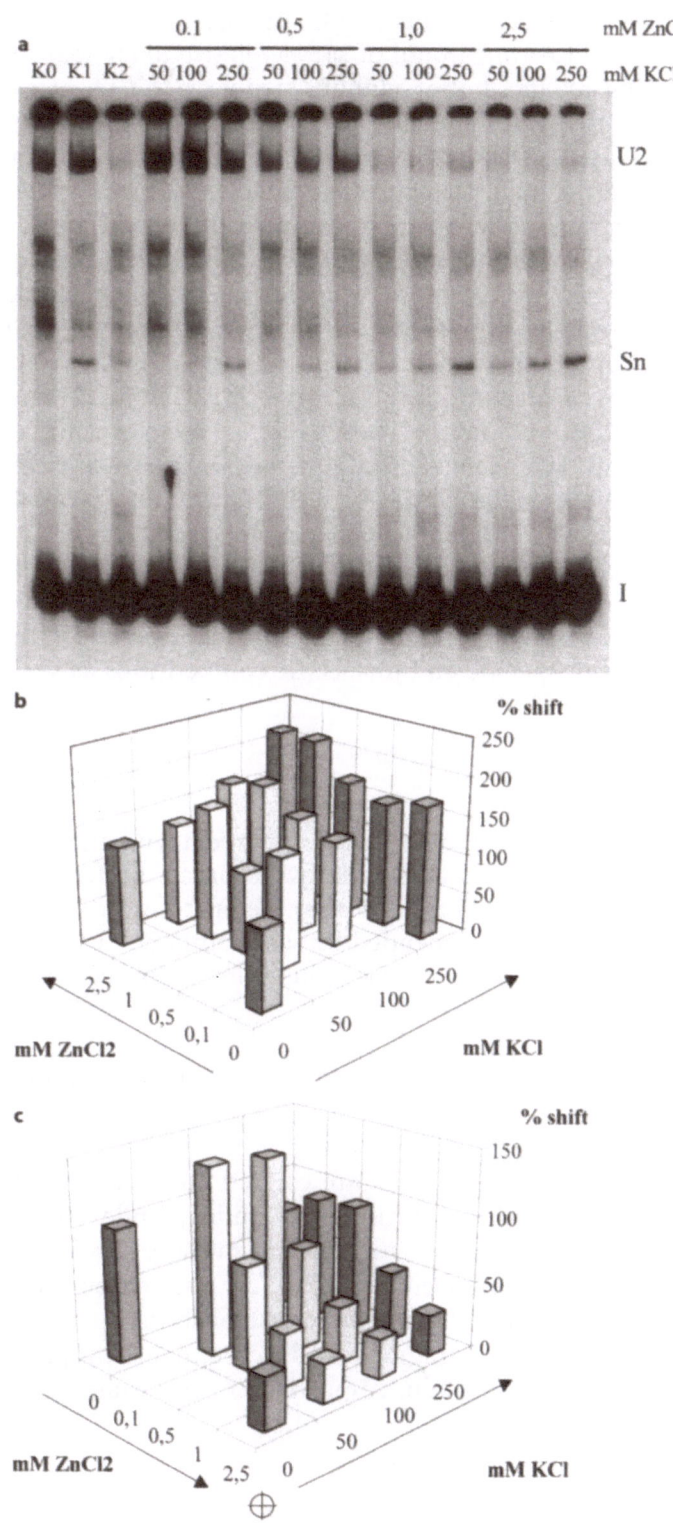

it was possible to enhance the intensity of Sn by combination of $ZnCl_2$ and KCl to a maximum of 215 % relative to standard binding conditions (using 5× inc. buffer without further addition of electrolytes). Simultaneously the intensity of a second retarded band, U2, gradually decreased, as shown in Fig. 4 C. It was assumed that the band Sn represented the main DNA/protein complex formed by the fragment C4 111r. Surprisingly the specific competition experiments conducted under binding conditions optimal for band Sn revealed an equilibrium constant of $K_{eq}=1.289\times10^{-7}$ mol/l, whereas the equilibrium constant of the U2 band was $K_{eq}=3.544\times10^{-8}$ mol/l. When conducted under standard binding conditions (5× inc. buffer without additional electrolytes) the equilibrium constant for Sn was $K_{eq}=1.105\times10^{-7}$ mol/l and for U2 $K_{eq}=1.024\times10^{-8}$ mol/l. Thus band Sn was revealed as a complex band without significant affinity generated by unspecific DNA/protein interaction, whereas the slightly blurred band U2 represented the main complex formed with affinity. The sequence specificity of U2 was confirmed by unspecific competition experiments (Mäueler et al., in press). Thus increased intensity of a certain band can be misleading. The identification of the main retarded band(s) is only possible with regard to both the results of specific and unspecific competitions as verification.

Materials

Equipment
- Power supply providing direct current with an output of at least 350 V
- Electrophoresis tank with two separate buffer tanks for vertical slab gels (e.g. 20×25 cm×1.1 mm). Temperature controlled electrophoresis tanks are recommended by some authors, but usually gels can be run at 4 °C in the cold room. Therefore additional expenses are not mandatory.
- Slab gel dryer
- For quantification of the data a scintillation counter should be available.

Buffers and Solutions
Gel mobility-shift assays are carried out in a special incubation buffer which supplies a certain ionic milieu, under which the equilibrium between specific and nonspecific protein/ DNA interaction is shifted mainly towards specific binding. As pointed out above, several experiments have to be carried out to establish the specific binding conditions for a certain DNA fragment and a certain nuclear protein preparation (or for a purified protein, respectively). As a starting point for the deter-

mination of these conditions a universal incubation buffer is employed which provides an ionic milieu of intermediate concentration. In this protocol an incubation buffer is used in five times the required concentration. Apart from the ionic milieu unspecific DNA/protein interaction is reduced by unspecific competitor DNA, e.g. sheared calf thymus or herring sperm DNA. Since our protocol was devised for the examination of the protein binding capacities of simple repetitive elements, no genomic DNA containing simple repetitive sequences should be used. Instead, synthetic DNA pdIdC·pdIdC was mainly employed. Protein extract is diluted with DP buffer in a concentration of 2.5–5 µg protein/µl.

- 5× incubation buffer (5x inc. buffer): 25 % glycerol; 100 mM HEPES/KOH, pH 7; 100 mM KCl; 25 mM $MgCl_2$
- polydIdC (pdIdC): 1 µg/µl
- DP buffer: 25 mM HEPES, pH 7.6; 40 mM KCl; 0.1 mM EDTA, pH 8.0; 10 % glycerol; 1 mM dithiothreitol
- PAA (40 % stock solution): 39.5 % polyacrylamide (w/vol); 0.5 % Bis-acrylamide (w/vol)
- Tris/glycine electrophoresis buffer (10×): 250 mM Tris; 1.92 M glycine
- Gel loading buffer (10×): 0.4 % bromophenol blue; 0.4 % xylene cyanol FF; 25 % Ficoll
- Gel fixing solution: 12 % methanol; 10 % acetic acid

Procedure

The final procedure has to be adapted to the special purposes of the experimenter. The usual procedure should comprise the following steps:

- The employed amount of protein is incubated with the unspecific competitor (e.g. pdIdC) in the presence of 1× incubation buffer for a constant period of time.

- The preincubated protein solution is thoroughly mixed with the target DNA. In the following incubation the DNA/protein complexes are formed.

- Then, 1/10 volume of 10× gel loading buffer and 1/10 volume of 10× electrophoresis buffer are added to the reaction mixture and the whole mixture is transferred to the gel.

When examining the influence of additional electrolytes or competitor DNAs on the DNA/protein interaction, the target DNA should be preincubated with the respective concentrations of the additional components. Make sure that the total reaction volume remains constant between individual samples.

1. Set up a 5 % PAA gel. For a 20 cm×25 ×1.1 mm gel, the following recipe can be used yielding a gel with a PAA:bis-acrylamide ratio of 79:1.

7.5 ml PAA from 40 % stock solution
6.0 ml 10× Tris/glycine electrophoresis buffer
46 ml double-distilled water
200 µl TEMED
200 µl ammonium persulfate (1 mg/ml)

Fill up to a 60 ml solution.

2. Prerun gel at 4 °C and 150 V for at least 30 min.

3. Preincubate desired amount of protein with unspecific competitor pdIdC and 1× incubation buffer for 20 min on ice.

4. Add desired amount of target DNA. For an overnight exposure of the gels at least 50,000 cpm/lane of labelled DNA should be employed, so that also the fainter bands become visible on the autoradiogram. Mix thoroughly using a pipette tip and incubate on ice for 20 min.

5. Add 1/10 vol 10× gel loading buffer and 1/10 vol 10× electrophoresis buffer. The final reaction volume should not exceed 25 µl, otherwise the samples cannot be loaded on the gels properly. Load samples onto the gel under current of ∼150 V.

6. When all samples are loaded onto the gel, increase current to 350 V. Run gels at this current until the bromophenol blue front has reached about the last quarter of the gel (this should take about 3.5 h).

7. Remove front plate and put a piece of 3MM Whatman paper (or equivalent) onto the gel. Fix the gel for 15 min in fixing solution. Dry the gel and expose to X-ray film over night.

8. For quantification of radiosignals mark the positions of the retarded bands according to the autoradiogram. Cut out the individual bands from the gel, put the gel pieces into scintillation vials and add 3 ml scintillation counter solution (e.g. LSC cocktail, Baker). Count each band for 1 min.

As pointed out above, some details of the general procedure have to be modified during the course of successive experiments. This refers mainly to the amounts of protein and unspecific competitor pdIdC, but also to the amounts of additional electrolytes. In some cases the use of additional components such as polyamines (e.g. spermidine, spermine), glycerol or nonionic detergents (dithiothreitol) are necessary to ensure stabile protein binding and to prevent proteins from aggregating in large complexes forming smears in the gels, or that do not enter the gel at all.

Results

Determination of Equilibrium Constants According to Scatchard

Assuming a simplified complex formation reaction of the type shown in Eq. (1), it is possible to formally divide the reaction in Eq. (1) into a bimolecular association reaction and a monomolecular dissociation reaction.

$$\text{DNA} + \text{protein} = \text{DNA/protein complex} \tag{1}$$

The reaction velocities are given by Eqs. (2) and (3):

$$\text{DNA} + \text{protein} \rightarrow \text{complex} \qquad v_{ass} = k_{ass} \times [\text{DNA}] \times [\text{prot}] \tag{2}$$

$$\text{Complex} \rightarrow \text{DNA} + \text{protein} \qquad v_{diss} = k_{diss} \times [\text{PD}] \tag{3}$$

At equilibrium, both reaction rates are equal as expressed in Eq. (4) in which only equilibrium concentrations of the reaction partners appear:

$$v_{ass} = v_{diss}$$

$$K_{eq} = \frac{k_{ass}}{k_{diss}} = \frac{[PD]_{eq}}{[DNA]_{eq}[P]_{eq}} \tag{4}$$

Equilibrium concentrations of both DNA and protein can be expressed as the difference between the initial starting reactions $[\text{DNA}]_0$ and $[\text{P}]_0$ minus the equilibrium concentrations of the complex formed, $[\text{PD}]_{eq}$. This yields Eqn. (5):

$$K_{eq} = \frac{[PD]_{eq}}{([DNA]_0 - [PD]_{eq})([P]_0 - [PD]_{eq})} \tag{5}$$

Equation (5) can be reformulated to Eq. (6):

$$K_{eq} = \frac{[PD]_{eq}}{([DNA]_0 - [PD]_{eq})([P]_0 - [PD]_{eq})}$$

$$\rightarrow K_{eq} \cdot ([P]_0 - [PD]_{eq}) = \frac{[PD]_{eq})}{([DNA]_0 - [PD]_{eq})} \quad \quad (6)$$

$$\rightarrow K_{eq} \cdot [P]_0 - K_{eq} \cdot [PD]_{eq} = \frac{[PD]_{eq})}{([DNA]_0 - [PD]_{eq})}$$

The expression $[DNA]_0$-$[PD]_{eq}$ represents the concentration of the remaining "free," that is not complexed DNA molecules, whereas $[PD]_{eq}$ is the concentration of the DNA bound in a complex. Therefore Eq. (6) can be rearranged to render Eq. (7):

$$\frac{[DNA]_{bound}}{[DNA]_{free}} = K_{eq} \cdot [P]_0 - K_{eq} \cdot [PD]_{eq} \quad \quad (7)$$

According to Scatchard (1949), the plot of the ratio $[DNA]_{bound}/[DNA]_{free}$ vs the concentration of the complex $[PD]_{eq}$ yields a straight line with a slope of K_{eq}.

With respect to the data collected by scintillation counting, the ratio of radioactivity within a certain retarded band to the radioactivity within the free target DNA band is plotted versus the concentration of the retarded band. In order to determine the concentration of a retarded band, one has to perform the following calculations.

Calculation of the Actual Amount of DNA for Each Lane (fmol$_{tot}$)

In a competition experiment, constant amounts of labelled target DNA are incubated with increasing amounts of competitor DNA (specific or unspecific) (**Note:** competitor DNA to reduce unspecific protein binding is not entered into this calculation). Competitor DNA is added in fixed amounts of molar excess. According to the specific activity achieved in the labelling reaction, a given amount of radioactivity equals a certain target DNA concentration. For instance with a specific activity of 4×10^6 cpm/pmol, each reaction with 80,000 cpm/lane contains 20 fmol of target DNA. When adding competitor DNA to a molar excess of 500× target DNA, the reaction contains a total amount of fmol$_{tot}$ equal to 10.02 pmol of DNA.

Calculation of the Ratio of DNA/Protein Complex to Total Radioactivity Complex/Total

To calculate the fraction of DNA contained within a complex band, all individual bands of a lane have to be cut out and scintillation counted. Calculate the sum of all bands yielding the total counts per lane.

Calculation of the Molarity of the Complex Band in Question

The amount of DNA contained in a complex band is calculated by the amount of total DNA multiplied by the fraction of DNA in the complex band complex/total. A correction for the total reaction volume has to be included. This yields Eq. (8):

$$pM \text{ (complex)} = fmol_{tot} \cdot \frac{radioactivity \text{ (complex)}}{radioactivity \text{ (Total)}} \cdot \frac{1}{ml \text{ reaction volume}} \quad (8)$$

Troubleshooting

- Avoid damage to target DNA. Avoid nuclease damage during the plasmid DNA preparation. Avoid UV light and intercalating dyes.

- Take care that target DNA is of utmost purity. Target DNA extracted from preparative gels has to be purified by ion exchange chromatography. Phenol/chloroform extraction including subsequent ethanol precipitation is insufficient.

- All steps of the mobility-shift assays are carried out on ice. Even if in subsequent binding condition refinements the complexes of interest are stable at room temperature, it is advantageous to work on ice as all complexes formed will be even more stable at the lower temperature. Keep to the incubation times.

- Even if certain retarded bands gain in intensity and sharpness during the optimization of the assay conditions, the equilibrium constants should be determined first using standard conditions without additional components.

- When examining the effect of an additional component on a system it can make a difference in which order the components of the mobility-shift assay are pipetted together. Add salt in appropriate concentration to the target DNA solution so that all components except protein extract and pdIdC are present in the reaction mixture. Once a pipetting scheme for a certain experiment is chosen, stick to it when pipetting the individual samples.

- Mark the sides of the gel before fixation or load asymmetrically. Loading double and single strand controls avoids misinterpretations of the lanes. Every batch of labelled target DNA is controlled by these means.

- You are working with radioactivity. Check for contamination regularly, collect all radioactive reaction vessels and pipette tips and discard into the appropriate radioactive waste container. Do not mix radioactive waste with other laboratory waste!

References

Epplen JT, Kyas A, Mäueler W (1996) Genomic simple repetitive DNAs are targets for differential binding of nuclear proteins. FEBS Lett 389:92-95

Galloway AM, Liuzzi M, Paterson MC (1994) Metabolic processing of cyclobutyl pyrimidine dimers and (6–4) photoproducts in UV-treated human cells. J Biol Chem 269:974–980

Matsunaga T, Hieda K, Nikaido O (1991) Wavelength dependent formation of thymine dimers and (6–4) photoproducts in DNA by monochromatic ultraviolet light ranging from 150 to 365 nm. Photochem Photobiol 54:403–410

Mäueler W, Frank G, Siedlaczck I, Epplen JT, Melmer G (1992a) PCR amplification products are of limited use for the study of DNA/protein interaction. Electrophoresis 13:641–643

Mäueler W, Muller M, Köhne AC, Epplen JT (1992b) A gel retardation assay system for studying protein binding to simple repetitive DNA sequences. Electrophoresis 13:7–10

Mäueler W, Frank G, Muller M, Epplen JT (1994) A complex composed of at least two HeLa nuclear proteins protects preferentially one strand of the simple $(gt)_n(ga)_m$ containing region of intron 2 in the *HLA-DRB* genes. J Cell Biochem 56:74–85

Mäueler W, Kyas A, Keyl H-G, Epplen JT Genome-derived $(gaattc)_{24}$ trinucleotides form triple helices and bind HeLa nuclear protein(s) specifically. Gene (in press)

Palecek E (1992) Probing of DNA structure in cells with osmium tetroxide-2,2'-bipyridine. Methods Enzymol 212:305–318

Sambrook J, Fritsch EF, Maniatis T (1989) Molecular cloning – a laboratory manual. Cold Spring Harbor Laboratory Press, Cold Spring Harbor, NY

Scatchard G (1949) The attractions of proteins for small molecules and ions. Ann NY Acad Sci 51:660–672

Wells RD, Collier DA, Hanvey JC, Shimizu M, Wohlrab F (1988) The chemistry and biology of unusual DNA structures adopted by oligopurine/oligopyrimidine sequences. FASEB J 2:2939–2949

Detection of DNA Curvature Using Transverse Pore Gradient Polyacrylamide Gel Electrophoresis

DAVID WHEELER

▓ Introduction

DNA Curvature

The phenomenon of sequence-directed DNA curvature was first detected electrophoretically in the case of a 212 base pair (bp) fragment of DNA cut from the mini-circle kinetoplast DNA of the trypanosome *Crithidia fasciculata*. The DNA migrates as if it were over 1000 bp long in 6 % poly-acrylamide (Marini 1982; Kitchin 1986). This anomalously slow migration is believed to be due to sequence-directed curvature. The sequence of the kinetoplast DNA contains several runs of adenine, producing a planar curvature (Fig. 1), which has been visualized under the scanning electron microscope (Griffith 1986). For a review of sequence directed curvature in DNA, see Hagerman (1990). DNA curvature is of importance in the study of gene expression where curvature per se has been shown to regulate the activity of promoters (Gourse 1986; Plaskon 1987).

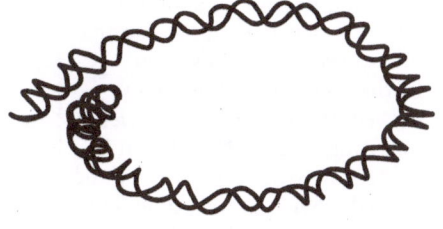

Fig. 1. Predicted structure of the 212 bp kinetoplast DNA fragment showing the planar, circular curvature which gives rise to its slow migration on polyacrylamide gels. The figure was generated by DNACurve, using values for the three dinucleotide wedge angles determined empirically. (From Kabsch 1982; Bolshoy 1991)

David Wheeler, PhD, National Center for Biotechnology Information, Bldg 38A, 8N800, The National Institutes of Health, Bethesda, MD 20892, USA (e-mail: wheeler@ncbi.nlm.nih.gov

Detection of Curved DNA on Transverse Pore Gradient Polyacrylamide Gels

The reduced mobility of curved DNA during polyacrylamide gel electrophoresis is more pronounced at higher gel concentrations (%T); hence on a transverse pore gradient gel, DNA curvature is easily detected as a trace intersecting those of non-curved size standards (Fig. 2).

Transverse pore gradient polyacrylamide gels have been used to distinguish between conformationally distinct lariat and linear RNAs (Copertino 1991). An anomalously migrating DNA molecule containing the curved upstream activating region of the *E. coli* rrnB P1 promoter was detected using transverse pore gradient polyacrylamide gel electrophoresis (Zacharias 1990). The subject of transverse pore gradient gel electrophoresis of DNA is reviewed in Chrambach and Wheeler (1994).

Advantages and Disadvantages of the Method

The transverse gel gradient method has some advantages over single-concentration gel retardation methods. First of all, it is possible, through an unfortunate choice of gel concentration, to find that a curved DNA fragment of interest comigrates with another, even though they are of different lengths, because the curvature of the larger is greater than that

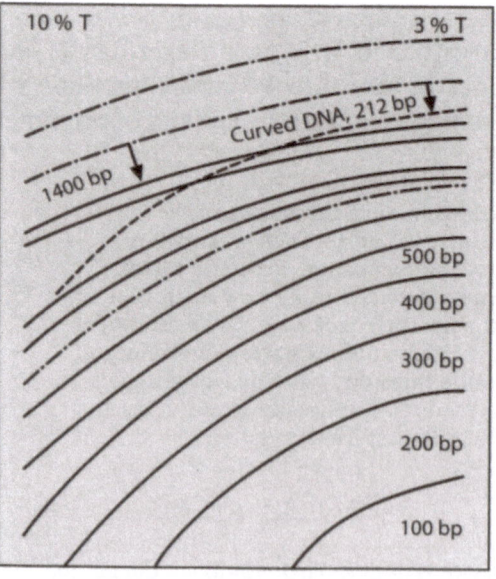

Fig. 2. A typical transverse pore gradient gel. The gel gradient runs from 10 %T on the left to 3 %T (%T=concentration of acrylamide monomer) on the right and is approximately linear between the two extremes. The Bethesda Research Laboratories 100 base pair ladder, consisting o non-curved DNA, has been loaded along with the highly curved 212 bp *Crithidia fasciculata* mini-circle DNA, shown in Fig. 1. The trace representing the curved DNA is *dotted*; note that i crosses the other DNA traces. Some closely spaced traces from the 100 bp ladder between 800 and 1400 bp have been omitted for clarity

of the smaller. Although this difficulty can be overcome by choosing another gel concentration (if the error is realized!) such ambiguity cannot exist using a transverse gradient gel since any two DNA fragments are seen as continuous traces over a range of gel concentrations, as is evident Fig. 2. Although the traces may cross at a single gel concentration (the trace of the 1400 bp size standard crosses that of the 212 bp kinetoplast DNA at about 6 %T), they are clearly separable at others.

Another advantage of the transverse gradient is that it allows derivation of instantaneous retardation coefficients, Krs (Kr is the change in mobility with respect to gel concentration), for a DNA fragment at many gel concentrations. This information is useful since Kr has been related to a number of physical parameters such as particle radius(Tietz and Chrambach 1992).

The main disadvantage of the method is that it is technically more difficult than electrophoresis at a single gel concentration and that there is some uncertainty in determining the actual acrylamide monomer concentration at a particular point in the gel. This uncertainty stems from difficulty in producing gradients which approximate linearity.

13.1
Mini-Gel Protocol

Materials

Precast continuous pore gradient polyacrylamide mini-gels are available from Jule Inc. New Haven, Connecticut 06511 (cat. no. 0310N75IT) and Gradipore, Pyrmont, Australia (cat. no. TRANSVERSE-315, G900623).. Useful gradients run from 3 % acrylamide monomer to 10 % or 15 % monomer (%T) with a constant bis-acrylamide cross-linker concentration (%C)of 3 %.

Precast Transverse Pore Gradient Mini-Gels

Two representative electrophoresis chambers suitable for transverse pore gradient electrophoresis are listed below. Other mini-gel chambers may also be suitable and should be chosen based on the dimensions of the gels to be used.

Mini-Gel Chambers

- Mighty Small electrophoresis chamber: Hoefer Scientific Instruments, San Francisco, California 94102
- Novex X-cell electrophoresis chamber: Novex Experimental Technologies, 4202 Sorrento Valley Blvd., San Diego, California 92121; 1-800-456-6839

Stock Solutions of Running Buffers

The following stock buffers for the electrophoresis of nucleic acids are suitable for transverse pore gradient gels (Maniatis 1982).

- TBE (5×): 54 g Tris base; 27.5 g boric acid; 20 ml 0.5 M EDTA, pH 8.0
- TAE (50×): 242 g Tris base; 57.1 g glacial acetic acid (17.4 M); 100 ml 0.5 M EDTA, pH 8.0)

High-Density DNA Loading Solutions

Two standard high-density DNA loading solutions are given below which are appropriate for the loading of transverse gradient mini-gels (Maniatis 1982).

- Glycerol (6×): 0.25 % bromophenol blue; 0.25 % xylene cyanol; 30 % glycerol in deionized water. Store at 4 °C.
- Ficoll (10×): 0.25 % bromophenol blue; 0.25 % xylene cyanol; 25 % Ficoll (type 400) in deionized water. Store at room temperature.

Note: Avoid loading solutions containing sucrose since these solutions are unstable and can cause electrophoretic artifacts.

Staining Solutions

The colloidal Coomassie blue stain is useful in the verification of gel gradients as described below.

- Colloidal Coomassie blue staining solution (Novex Instruction Manual, IM 6025): 10 ml colloidal Coomassie blue (Novex cat. no. LC6025); 40 ml stainer A solution; 40 ml methanol; 110 ml deionized water

Ethidium bromide is a fluorescent stain for nucleic acids and is the preferred stain for the staining of DNA on transverse gradient mini-gels.

- Ethidium bromide DNA staining stock solution (10 mg/ml): Weigh out 1 g ethidium bromide in the hood and mix with 100 ml deionized water (Maniatis 1982).

Note: The storage container should be protected from light with aluminum foil or other means since ethidium bromide is light sensitive. Ethidium bromide is a powerful mutagen so gloves and dust mask are recommended when making the stock.

Procedure

Adaptation of Conventional Gradient Gels to Transverse Gradient Gel Electrophoresis

It is recommended that precast transverse gradient gels be purchased since the fabrication of these gels with any hope of reproducibility requires a complex apparatus. Methods for the fabrication of polyacrylamide gradient gels are given in several articles (Margolis and Kenrick 1968; Margolis 1969; Altland and Altland 1984) but will not be discussed further since the existence of precast gels renders the use of homemade gels impractical.

Conventionally oriented pore gradient mini-gels may also be purchased and reoriented so that the pore gradient runs orthogonally to the direction of electrophoresis. To adapt a normal pore gradient gel to transverse pore gradient, one must remove the side spacers and install small dams on either side of what will become the top of the gel when it is rotated 90° from its usual orientation. These dams may be made of the same material as the spacers and may be cemented in place with a small amount of grease or agar. Commercial gradient gels to be adapted to transverse orientation should be as close to the dimensions of a square as possible in order to fit correctly into most electrophoresis chambers after a 90° rotation.

The steps involved in the adaptation are as follows and are performed on a conventional gradient mini-gel such as that depicted in Fig. 3.

1. Remove the sample comb from the top of the gel and break off all the plastic teeth except the outer pair. These plastic teeth can serve as the dams for the new sample well.

2. If the sides of the gel are secured with tape, use a razor blade to remove the tape and carefully extract the side spacers. Take care not to separate the gel from the glass plates.

3. Chose one of the original "sides" of the gel for the new sample well and insert two of the teeth from the comb between the glass plates on this side (from now on, the top) to form the sample dams.

4. Between the glass plates on the side of the gel which contains the old sample wells, insert the original comb so that the two remaining teeth fit into their original wells.

5. On the side of the gel opposite the old sample wells, insert several comb teeth at strategic locations such as the gel corners and midpoint

Fig. 3. Conventional
gradient mini-gel

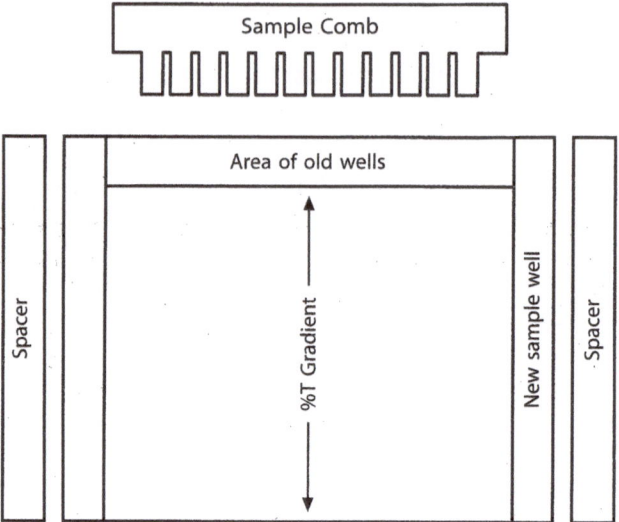

to provide mechanical stability. It will be necessary to penetrate the gel so that this area of the gel will not be useful. Reorient the sample dams accordingly so that the sample will travel over the unblemished sector of the gel.

The gel is now ready for mounting into the electrophoresis chamber.

Empirical Determination of the Gel Gradient

It is important to be sure that the conventional gradient gel purchased was made with a continuous pore gradient, as many are made using only a few discrete gel concentrations covering the range indicated for the gradient. The quality of the gradient may be ascertained by soaking the gel in colloidal Coomassie blue and destaining lightly in deionized water. Colloidal Coomassie blue is a high molecular weight colloid of the conventional dye which penetrates polyacrylamide gels to varying degrees depending on gel concentration. A quantitative measure of the existing gradient may be obtained by performing densitometry on a gel stained with colloidal Coomassie blue to an extent such that the lowest gel concentration is deeply stained while the highest is only lightly stained.

To determine the gel gradient using colloidal Coomassie blue, the gel is soaked overnight in the staining solution and destained in deionized water or buffer 10–20 min until the gel concentration gradient is visible as a blue color gradient running from very dark to light.

Loading the Gel

To load a mini-gel, 100 ng of DNA is mixed with 1× loading solution to a total volume of 20–40 µl and layered across the entire loading surface of the gel taking care to make sure the sample is evenly distributed.

Running the Gel

A voltage gradient of 5–10 V/cm is sufficient for transverse gradient gels run in TBE or TAE buffers. It is important that the gel be maintained at room temperature or below since the electrophoretic retardation of curved DNA decreases markedly above room temperature and disappears by 37 °C.

Staining the Gel

The staining solution should be 0.5 µg/ml in ethidium bromide and is made by mixing 50 µl of the stock with 100 ml of deionized water or buffer. Remember that ethidium bromide is a powerful mutagen and that gloves should be worn when handling solutions containing even small amounts of ethidium. The gel is soaked in this solution for 10–20 min and the image is acquired during irradiation at 305 nm through an orange filter. If destaining is required to remove background, this may be accomplished by soaking for 30 min in deionized water or buffer. Typically, a 2 s exposure at F-stop 4.5 is sufficient for Polaroid type 55 film. Several seconds may be required for direct image acquisition with some CCD (charge coupled device) cameras.

13.2
PhastGel Protocol

Materials

The following protocol is adapted from Buzas 1994. The materials are available from Pharmacia (Piscataway, New Jersey).

- PhastSystem apparatus (cat. no. 18-1657-01)
- Sample applicator (TC cat. no. 18-1657-01)
- Polyacrylamide gradient gel: 4–15 % (cat. no. 17-0678-01)

- Gel buffer: 0.112 M Tris acetate, pH 6.4
- Native buffer strips (cat. no. 17-0517-01): 0.88 M alanine; 0.25 M Tris-acetate, pH 8.8
- PhastGel silver kit (cat. no. 17-0617-01)

▊ Procedure

Mounting, Pre-electrophoresis, Loading and Running

The PhastGel is positioned such that the pore gradient runs orthogonally to the direction of electrophoresis. This positioning corresponds to a rotation of 90° from the usual orientation for a gradient gel. The gel is pre-electrophoresed for 80 V*h after which a sample of 200 ng of DNA in a volume of 2 ml is loaded across the top of the gel with the sample applicator and electroinjected into the gel for at 2 V*h. The resolving electrophoresis is conducted at 400 V (100 V/cm) for about 7.5 min at a temperature of 15 °C.

Silver Staining the PhastGel

1. Fix the gel for 5 min at 20 °C in 20 % TCA followed by 5 min at 50 °C in 5 % glutaraldehyde.

2. Wash gel for 2 min in deionized water.

3. Repeat water wash of step 2.

4. Soak gel for 8 min at 40 °C in 0.4 % silver nitrate.

5. Wash gel for 30 s at 30 °C in deionized water.

6. Repeat wash of step 5.

7. Develop gel in solution of 55 ml 36 % formaldehyde in 150 ml 2.5 % sodium carbonate.

8. Repeat step 7 for 4 min.

9. Soak gel for 2 min at 50 °C in 5 % acetic acid.

10. Soak gel in 10 % acetic acid/10 % glycerol for 3 min at 50 °C.

Results

Acquisition of Gel Image

For the acquisition of gel images, a computer coupled to either a CCD camera or a scanner is required. Two sample configurations are given:

- Pentium based computer running Windows95 and coupled to a CCD camera 8 bit frame grabber using ImageTool(see below for WWW address) software for image acquisition.
- Macintosh IIC running NIH Image (see below for WWW address) to acquire gel image or photograph of gel image from a CCD camera via a frame grabber.

Hardware Requirements and Configurations

From Gel Image to Cartesian Coordinates

The raw data collected from a transverse pore gradient gel are in the form of a continuous trace of migration distances over a range of gel concentrations. Once the image of the stained gel is acquired, a set of coordinates must be obtained for each trace of interest on the gel. These coordinates will ultimately be of the form (%T, migration distance). A number of image analysis programs are available which will allow the user to read Cartesian coordinates from a TIFF file, and three free packages appropriate for the prevalent computer platforms and operating systems are given in Table 1.

Data Analysis

Table 1. Software for image analysis and the extraction of Cartesian coordinates of DNA migration traces recorded on TIFF images of transverse pore gradient gels[a]

Software title	Operating system	Features
ImageTool 1.23	Windows95, Windows NT	Image acquisition; image processing; supports many image formats
DNA Simdex 2.0F	Windows 3.1, Windows 95	Reads gel images stored as bitmap (.BMP) files; automatic generation of coordinates for bands or traces on; must convert TIFF to BMP before using program
NIH Image 160	MacIntosh System 7	Image acquisition and processing

[a] Alternatively, coordinates may be determined by placing a transparency of a grid on the image. For linear gradients, the conversion of x-coordinates to %T values is straight forward. If the gradient is determined to be nonlinear by a method such as the colloidal Coomassie method outlined in the text, then the non-linear function empirically determined must be used.

Curve Fitting

It is not necessary to accumulate large numbers of points to define a trace since the traces obtained tend to be monotonic and smooth. Experience has shown that a second or third order polynomial fits most data well so that only 8–20 points need be collected to approximate the data. A procedure for polynomial curve fitting is presented below in a form which can be implemented using a computer math package such as MathCad (MathSoft Inc., Cambridge, Massachusetts)or a calculator capable of matrix operations.

Calculation of Instantaneous Krs from Fitted Curves

From the best fit polynomial, values of instantaneous Kr may be computed as shown above by evaluating the first derivative(Kr)of the mobility function (μ). Kr, the retardation coefficient, represents the change in mobility with respect to gel concentration of a migrating molecule. In the case of curved DNA, the change in Kr with respect to gel concentration is more pronounced than that of non-curved DNA. Retardation coefficients can be related to physical properties of both the migrating particle and the gel matrix using the electrophoretic analysis program Elphofit (Orban 1993; Tietz and Chrambach 1992; Wheeler 1993).

References

Altland K, Altland A (1984) Forming reproducible density and solute gradients by computer-controlled cooperation of stepmotor-driven burettes. Electrophoresis 5:143–147

Bolshoy A, McNamara P, Harrington RE, Trifonov EN (1991) Curved DNA without A-A: experimental estimation of all 16 DNA wedge angles. Proc Natl Acad Sci USA 88:312–2316

Buzás Z, Wheeler D, Garner M M, Tietz D, Chrambach A (1994)Transverse pore gradient electrophoresis, using the PhastSystem. Electrophoresis 15:1028–1031

Chrambach A, Wheeler DL (1994) Capabilities and potentialities of transverse pore gradient gel electrophoresis. Electrophoresis 15:1021–1027

Copertino DW, Favreau MR, Hallick RB (1991) Transverse gradient gel electrophoresis: a gel system for resolving complex mixtures of lariat and linear RNA molecules. Nucleic Acids Res 18:7451–7452

Gourse RL, de Boer HA, Nomura M (1986) DNA determinants of rRNA synthesis in E. coli: growth rate dependent regulation, feedback inhibition, upstream activation, antitermination. Cell 44:197–205

Griffith J, Bleyman M, Rauch CA, Kitchin PA (1986) Visualization of the bent helix in kinetoplast DNA by electron microscopy. Cell 46:717–724

Hagerman PJ, (1990) Sequence-directed curvature of DNA. Annu Rev Biochem 59:755–781

Kabsch W, Sander C, Trifonov EN (1982) The ten helical twist angles of B-DNA. Nucleic Acids Res 10:1097–1104

Kitchin PA, Klein VA, Ryan KA, Gann KL, Rauch CA, Kang DS, Wells RD, Englund PT (1986) A highly bent fragment of *Crithidia fasciculata* kinetoplast DNA. J Biol Chem 261:11302–11309

Koo HS, Drak J, Rice JA, Crothers DM (1990) Determination of the extent of DNA bending by an adenine-thymine. Biochemistry 29:4227–4234

Koo HS, Wu HM, Crothers DM (1986) DNA bending at adenine-thymine tracts. Nature 320:501–506

Maniatis T, Fritch EF, Sambrook J (1982) Molecular cloning: a laboratory manual. Cold Spring Harbor Press, Cold Spring Harbor, New York

Margolis J (1969) A versatile gradient-generating device. Anal Biochem 27:319–22

Margolis J, Kenrick KG (1968) Polyacrylamide gel electrophoresis in a continuous molecular sieve gradient. Anal Biochem 25:347–362

Marini JC, Levene SD, Crothers DM, Englund PT (1983) A bent helix in kinetoplast DNA. Cold Spring Harb Symp Quant Biol No 47, 1:279–283

Novex Instruction Manual IM-6025

Orbán L, Garner MM, Wheeler D, Tietz D, Chrambach A (1993) Characterization of the electrophoretic properties of nucleosome core particles by transverse polyacrylamide pore gradient gel electrophoresis. Electrophoresis 14:720–724

Plaskon RR, Wartell RM (1987) Sequence distributions associated with DNA curvature are found upstream of strong *E. coli* promoters. Nucleic Acids Res 15:85–795

Tietz D, Chrambach A (1992) Concave Ferguson plots of DNA fragments and convex Ferguson plots of bacteriophages: evaluation of molecular and fiber properties, using desktop computers. Electrophoresis 13:286–294

Wheeler D, Tietz D, Chrambach A (1993) Information on DNA conformation derived from transverse pore gradient gel electrophoresis in conjunction with an advanced data analysis applied to capillary electrophoresis in polymer media. Electrophoresis 13:604–608

Zacharias M, Wagner R, Goringer HU (1990) Polyacrylamide gradient gel electrophoresis for the detection of bended DNA fragments. Nucleic Acids Res 18:2827–2827

Software References

DNACurve (1.0) David L. Wheeler, Laboratory of Molecular and Cellular Biology, The National Institutes of Health, LMCB, Bldg. 8, 2A22, Bethesda, Maryland 20892

DNA Simdex (2.0F) Scott Archer (http://www.hku.hk/zoology/SIMDEX2/simdex2/htm)

Elphofit (2.32) Dietmar Tietz (Email: djt@his.com, http://www.his.com/~djt/elphofit.html)

ImageTool (1.23) C. Donald Wilcox, S. Brent Dove, W. Doss McDavid and David B. Greer, Department of Dental Diagnostic Science at The University of Texas Health Science Center, San Antonio,Texas (http://ddsdx.uthscsa.edu/dig/itdesc.html, ftp://maxrad6.uthscsa.edu)

NIH Image (160) Wayne Rasband, Research Services Branch, National Institute of Mental Health, National Institutes of Health (http://rsb.info.nih.gov/nih-image/)

◼ Algorithm for the calculation of polynomial fits to transverse gradient gel data

Specify voltage of electrophoresis, electrode separation (cm) and time of electrophoresis (seconds)

electrode-separation := 4

T := 3600

$$\text{voltage-gradient} := \frac{\text{voltage}}{\text{electrode-separation}}$$

Read in a set of gel concentrations and migration distances.

M := READPRN (a)

N := length $\left(M^{<0>}\right)$ – 1

Set the order of the polynomial to 3 and set a range variable, d, to run from a power of 0 to 3 to allow for computation of the best fit polynomial below.

D := 3

d := 0..D

I := 1..N

Extract gel concentrations (%T) from input matrix M into a vector, x. Extract migration distances from electrophoretic origin and convert into mobilities taking into account voltage gradient and electrophoresis time, T. Set range variable, j, to run first to last gel concentration.

x := $M^{<0>}$

$$y := \frac{M^{<1>} \cdot 10^5}{\text{voltage-gradient} \cdot T}$$

J := 1..length $\left(M^{<0>}\right)$ – 1

Create a matrix X from the input gel concentrations, x, in which each column, d, represents the values of x raised to the power of d. For a 3rd order polynomial, X is a three column matrix.

$$X^{<d>} := \overrightarrow{(x^d)}$$

Create a vector, b, consisting of the coefficients for the 3rd order polynomial of best fit by dividing the dot product of the transpose of X and y by the dot product of the transpose of X and X.

$$b := (X^T \cdot X)^{-1} \cdot (X^T \cdot y)$$

Create a mobility function, mu, to allow interpolation of mobility at any %T. Also, create a function to compute instantaneous Krs.

$$mu\,(x) := \sum_d b_d \cdot x^d$$

$$Kr\,(x) := \frac{d}{dx}\,mu\,(x)$$

$$ymean := mean\,(y)$$

Compute a statistical parameter for goodness of fit by taking the ratio of ssu/sst (sum of squares unexplained/sum of squares total). The closer this ratio is to 1.0 the better the fit.

$$sse := \sum_j \left(y_j - mu\,(x_j)\right)^2$$

$$sst := \sum_j \left(ymean - mu\,(x_j)\right)^2$$

$$ssu := sst - sse$$

The SSU/SST value is close to 1.0; the fit is good.

$$\frac{ssu}{sst} = 0.986$$

Figure 4 shows the mobilities computed from the migration distances plotted against gel concentration. The squares are the data and the line represents the best fit 3rd order polynomial. Figure 5 shows the instantaneous Krs derived from the best fit polynomial plotted in Fig. 4, for the range of gel concentrations.

In Fig. 6 are the values, after calculation, of some of the matrices and vectors referenced above for sample data consisting of eight pairs of gel concentration and migration distance.

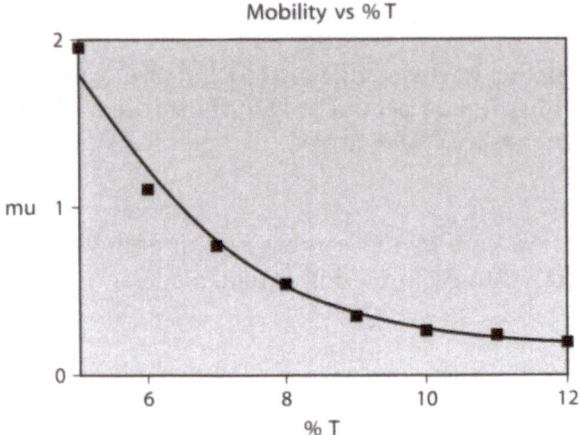

Fig. 4.

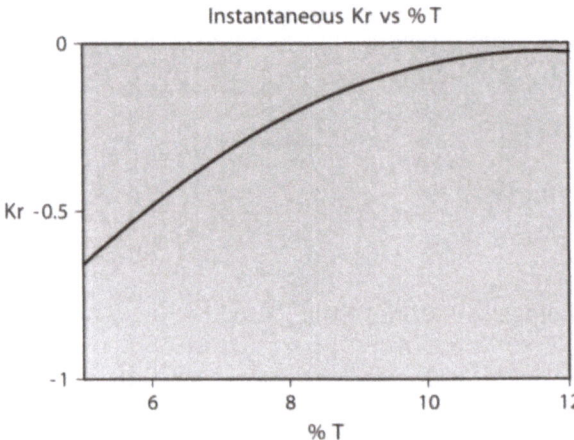

Fig. 5.

$$M = \begin{bmatrix} 4 & 9 \\ 5 & 7 \\ 6 & 4 \\ 7 & 2.8 \\ 8 & 2 \\ 9 & 1.3 \\ 10 & 1 \\ 11 & 0.9 \\ 12 & 0.7 \end{bmatrix} \quad x = \begin{bmatrix} 4 \\ 5 \\ 6 \\ 7 \\ 8 \\ 9 \\ 10 \\ 11 \\ 12 \end{bmatrix} \quad y = \begin{bmatrix} 2.5 \\ 1.944 \\ 1.111 \\ 0.778 \\ 0.556 \\ 0.361 \\ 0.278 \\ 0.250 \\ 0.194 \end{bmatrix} \quad X = \begin{bmatrix} 1 & 4 & 16 & 64 \\ 1 & 5 & 25 & 125 \\ 1 & 6 & 36 & 216 \\ 1 & 7 & 49 & 343 \\ 1 & 8 & 64 & 512 \\ 1 & 9 & 81 & 729 \\ 1 & 10 & 100 & 1000 \\ 1 & 11 & 121 & 1331 \\ 1 & 12 & 144 & 1728 \end{bmatrix} \quad b = \begin{bmatrix} 8.176 \\ -2.01 \\ 0.171 \\ -0.005 \end{bmatrix}$$

Fig. 6.

Subject Index